MiG

**FIFTY YEARS
OF SECRET
AIRCRAFT
DESIGN**

MiG

**R. A. Belyakov
and J. Marmain**

FIFTY YEARS OF SECRET AIRCRAFT DESIGN

Airlife
England

Published in the United States in 1994 by United States Naval Institute,
Annapolis, Maryland

This edition published in the United Kingdom in 1994 by Airlife Publishing Ltd. England.

An earlier edition of this work appears in French, published by Editions Larivière in 1991.

British Library Cataloguing in Publication Data
A catalogue record of this book is available
from the British Library.

ISBN 1 85310 488 4

Printed in the United States of America on acid-free paper.

9 8 7 6 5 4 3 2
First printing

Airlife Publishing Ltd.

101 Longden Road, Shrewsbury SY3 9EB, England.

◀ *Frontispiece:* The MiG-31, a Mach 3 interceptor, was intended to counter the threat of American B-52 bombers carrying long-range cruise missiles.

Contents

Foreword

WHO WOULD HAVE THOUGHT that one of the most secret sanctuaries of world military affairs, the MiG design bureau, would open its doors and allow material from its archives to be published in the West? It took the clear-sightedness of one of the most important figures in the Soviet aerospace industry, R. A. Belyakov, combined with the obstinacy—and patience—of a French journalist and expert on Soviet aviation, Jacques Marmain. It took two years of negotiations to overcome the many difficulties raised by this project, especially the declassification of thousands of documents, figures, and photographs—a process complicated by the very poor state of the Soviet communication network.

Not long ago, in 1989, the Soviet military celebrated the fiftieth anniversary of the creation of the OKB, an experimental construction bureau still bearing the names of its founding fathers, Mikoyan and Guryevich. This was an important milestone in the history of aviation worldwide. For half a century the experimental prototypes, performance tests, and production techniques developed by MiG all centered on a single product: the fighter aircraft. This specialized knowledge, backed up by numerous research institutes and branches of industry, is what earned the organization its present reputation.

During the Cold War, MiG always had to find a way to counter the latest American program—whether it was the MiG-9 versus the F-80 Shooting Star, the MiG-15 versus the F-86 Sabre, the MiG-19 versus the F-100 Super Sabre, or the MiG-25 versus the Lockheed A-11. All of its efforts became landmarks of technical progress on both sides of the Atlantic Ocean. For fifty years MiG built fighters primarily for the Soviet air force (VVS) and its air defense branch, the PVO. Some 180 machines (basic aircraft and their derivatives) were designed, the most famous being the MiG-3, MiG-9, MiG-15, MiG-17, MiG-19, MiG-21, MiG-23, MiG-25, MiG-27, MiG-29, and MiG-31.

MiG aircraft have set fifty-five world records, twenty-one of which are still standing as this book goes to press. The MiG design bureau has always used the most advanced technologies, developing the first afterburning chamber, the first Soviet air-to-air missile, the first sweepback wing, pressurized and air-conditioned cockpits, tricycle landing gears, automatic brakes for gear wheels, ejection seats—and this list is far from exhaustive.

The technical firsts that can be credited to the OKB are countless: servo-controls (MiG-15 and MiG-17), the automatic stabilizer (MiG-19), welded-steel integral tanks (MiG-25), the variable geometry wing (MiG-23), and the lifting-body principle of the MiG-29 as well as its complex navigation and fire control systems, to list a few. Each new project drew on intensive research by specialized institutes, engine makers, and weapon manufacturers. A number of experimental prototypes were then built to test new ideas and to advance the state of aerospace technology in the Soviet Union. The core of this process was the foundation of scientific and technical knowledge on which the design bureau relied.

It is only in recent years that the Soviet military has put its best aircraft on display in international air shows—in Great Britain first, then in France, Canada, and the United States. The OKB welcomed the opportunity to show the MiG-29 and, more recently, the MiG-31 (first seen in Paris in 1991). Specialists from all over the world were able to assess the aircraft and the skill of their Soviet pilots.

The MiG design bureau can rely on a cadre of highly qualified engineers, technicians, scientists, and workers. Test pilots, because of their extensive experience, are involved in the design process from the very first drawing. All of these individuals are heirs to the tradition of excellence established by the team that founded MiG in 1939 (some members of that team are still active).

With the advent of international détente and disarmament, the design bureau engineers see the necessary conversion of the aerospace industry as a good thing, an opportunity to redirect their years of experience into the air transport field. The international market for airliners is huge, and MiG engineers are already working on projects designed in cooperation with foreign manufacturers. Meanwhile, let us examine the work of the world's most secretive design bureau with this book, whose exclusive documentation is enriched at the end by a series of tables detailing power plants, radars, and armaments that have been used aboard MiG aircraft.

DOCAVIA

Acknowledgments

THIS BOOK WOULD NOT HAVE BEEN POSSIBLE without the considerable amount of work (research and synthesis) done by a group of MiG OKB engineers: A. M. Savelyev, L. I. Egenburg, A. I. Saprikin, and I. G. Sultanov, under the leadership of V. A. Arkhipov, chief constructor.

All illustrations, with very few exceptions (and these are duly noted), come from the MiG OKB archives.

You will not find in this book any of those conditional sentences, unwarranted suppositions, and pure inventions usually found in books—including the newest—dealing with Soviet aviation. Each fact, each figure, and each picture have been checked and cross-checked by the authors and the MiG engineers mentioned above.

NATO designations for aircraft and U.S./NATO designations for missiles are listed in the index.

MiG

FIFTY YEARS OF SECRET AIRCRAFT DESIGN

Artyom Ivanovich Mikoyan (1905–1970), doctor of technical sciences (1959), member of the USSR Academy of Science (1968), two-time Hero of Socialist Labor, winner of the Lenin Prize, and six-time winner of the USSR State Prize.

Interview: Birth of a Design Bureau

THIS INTERVIEW WITH R. A. BELYAKOV was conducted and recorded by Jacques Marmain in Moscow on 6 March 1991.

The MiG OKB celebrated its fiftieth anniversary some months ago. Today you hold the collective memory of this big enterprise, founded at the very moment that World War II broke out. This event having overshadowed the circumstances surrounding the birth of the OKB, can you tell us how all this started?

In early 1939 the threat of war loomed large over Europe and the USSR knew that it would not be spared. Having to face the growing power of Nazi Germany, the Party and the government implemented measures necessary to reinforce the Soviet Army, of which the VVS (air force) was a major component. At this time the fighters in service with our squadrons, the Polikarpov I-153s and I-16s, were greatly outclassed by the Bf 109Es built by Messerschmitt. There was no time to waste. To create a competitive spirit, several design bureaus were asked to develop and build prototypes. The one with the best flying qualities and combat capabilities was to be mass-produced. Simultaneously, the NKAP (Commissariat of the People for the Aviation Industry) urged the creation of an experimental construction bureau able to attract young, talented specialists.

Which manufacturers were involved in the new program?

There were nine of them: Polikarpov, Yakovlev, and Lavochkin, assisted by Gorbunov and Gudkov, Sukhoi, Borovkov and Florov, Shevchenko, Kozlov, Yatsenko, and Pashinin. All these engineers had to report on their projects personally to Stalin, who was following this program closely. This is why N. N. Polikarpov, chief constructor, P. A. Voronin, manager of the Polikarpov production plant, and P. V. Dementyev, chief engineer, were summoned to the Kremlin in the summer of 1939. The talks were heated. Voronin and Dementyev, supported by Stalin, wanted to give priority to the monoplane, while Polikarpov preferred the biplane for its handling qualities. A work crew was formed within the Polikarpov OKB to create a preliminary design (coded Kh). The crew was composed of a structure specialist, Ya. I. Selyetskiy, an airframe constructor, N. I. Andrianov, and an aerodynamicist, N. Z. Matyuk. These three men are still alive: Selyetskiy is today an assistant to G. Ye. Lozino-Lozinskiy in the space shuttle program, Andrianov is retired, and Matyouk is one of my faithful assistants as chief constructor. The three went to the Kremlin, where Stalin approved the project in the absence of Polikarpov, who was

Mikhail Iosifovich Guryevich (1892–1976), Hero of Socialist Labor (1945), doctor of technical sciences (1964), and winner of the Lenin Prize and the USSR State Prize on several occasions.

then in Germany. You will notice that the name of Mikoyan has not come up yet. There is a very good reason for this: he was not aware of any of these developments. At that time Mikoyan, who was the VVS representative at Polikarpov, was in charge of producing and upgrading the I-153.

Voronin and Dementyev were the first to believe that Mikoyan was the right man to manage the new team. They confided their thoughts to Stalin, who at first snarled and said, "What! Anastas's brother!" and then finally agreed—"OK. It's up to you." The two men did not choose Mikoyan at random. To them, he offered many advantages. They had noticed his spirit, his mind for organization, and his popularity in the VVS and the NKAP, and he was the brother of Anastas Ivanovich Mikoyan, member of the Politburo since 1935 and people's commissar in various economic commissariats since 1926. But they still had to persuade the surprised Artyom Ivanovich Mikoyan, who finally agreed to this unexpected proposal on one condition: that he could take on his old friend Guryevich as his assistant.

In the meantime, several other engineers had joined the team in November 1939. They all came from the Polikarpov OKB: V. A. Romodin, A. G. Brunov, D. N. Kurguzov, and one or two others whose names I have forgotten.

How did the new team succeed in winning its autonomy?

Simply by government decree. But it was quite an event because, with the blessing of the authorities, Voronin and Dementyev exploded the Polikarpov system. The OKB, factory no. 1, and the new team were assembled in an OKO, or design and experiment section. A. I. Mikoyan was put in charge of this OKO, assisted by M. I. Guryevich and V. A. Romodin. In November–December 1939 the new OKO attained full strength and submitted the final design of the Kh project renamed the I-200, which was planned initially with a AM-37 engine. The I-200 preliminary design was quickly approved by the central committee of the Communist party, the people's commissariat of the aircraft industry, and the air force. The existence of the OKO was formalized on 8 December 1939.

Could you tell us more about Mikoyan, a man whose name is famous all over the world? We do not really know much about him.

Artyom Ivanovich Mikoyan was born on 5 August 1905 in Sanain, a small Armenian village. His father was a carpenter. He learned to read and write at the village school and then was sent to Tbilisi high school in Georgia.

In 1923 he was admitted to the training school of the Krasniy Aksai factory in Rostov, and the following year he was hired as a mechanic by the railway workshop in the same town. In 1925, still a mechanic, he was employed at the Dynamo factory in Moscow. In December 1928

Rostislav Apollosovich Belyakov (1919–   ), winner of the Lenin Prize (1972), doctor of technical sciences (1973), member of the USSR Academy of Science (1981), recipient of the A. N. Tupolev Gold Medal (1988), two-time Hero of Socialist Labor, and two-time winner of the USSR State Prize.

he left the factory to do his national service. Discharged two years later, he returned to Moscow and found work at the Kompressor factory. He entered the air force academy named after N. Ye. Zhukoskiy in 1931. There things became much more difficult, but young Artyom Ivanovich stuck with it, became keen on parachuting, and learned to fly. In 1935 he and two fellow students used a 22-horsepower American engine to build a light sport aircraft dubbed Oktyabryonok, which apparently was quite a good flier. In 1937 he passed his exams at the academy with flying colors and, his first-grade diploma in the bag, found himself representing the military client (in this case, the VVS) at the no. 1 Aviakhim factory. The Polikarpov design bureau was located inside this factory, which produced the I-153 Chaika (seagull) fighters. Mikoyan was in charge of the acceptance checks on behalf of the VVS and then was appointed the permanent representative of the customer within the design bureau. He was therefore in regular contact with Polikarpov, head of the design bureau.

Less than two years later, in March 1939, Polikarpov asked Mikoyan to help him organize and update the Chaika production line. That is where Voronin and Dementyev noticed him. You know what followed.

How was the OKB organized during the first months of its existence?

In the beginning, everybody worked like hell at producing all of the drawings for the I-200 and preparing the assembly of three prototypes. In December 1940 the aircraft received its production designation: MiG-1. The actual assembly of the prototypes began in January 1940, and fewer than a hundred days later, on 30 March, the first machine was moved to the airfield. The first MiG took off on 4 April.

In the meantime, in March, Mikoyan was appointed chief constructor at the no. 1 Aviakhim factory, and Guryevich was named deputy chief constructor. The careers of both men developed along the same lines. On 16 March 1942, by decree of the State Commissariat of Defense, the OKO was reorganized within the no. 155 factory, in Moscow—that is, in the very place where we are talking today. Mikoyan became manager and then chief constructor of the new OKB. On 20 December 1956 Mikoyan was appointed general designer, a position that he held until 27 May 1969, when he was struck down by a myocardial infarction. For his part, Guryevich was appointed chief constructor, a position that he retained until he retired in 1964.

Guryevich had by then become no. 2 in the OKB hierarchy. Could you tell us more about his career?

Mikhail Iosifovich Guryevich was born in 1892 in Kharkov, a big Ukrainian city. His father worked in a distillery. After finishing high school he entered the advanced mathematics program at Kharkov University but was expelled for taking part in a student revolutionary movement. In 1913 he left for France to study math at Montpellier Uni-

This photograph of the Mikoyan family has never been published before. *Right to left:* Artyom Ivanovich Mikoyan; Ovanes, his son; Svetlana, his daughter; Talida Otarovna, his mother; Zoya Ivanovna, his wife; and Natasha, his eldest daughter.

versity. After the October Revolution he returned to the USSR and resumed his studies at the technological institute in Kharkov. He organized a group of students there into an aeronautics circle that soon became a formal faculty of aviation, and he left in 1925 with a diploma. After working here and there for four years without finding what he was after, he decided to devote all his energies to aviation (his first love), either in a factory or in a design bureau. He spent time in the OKB managed by P. E. Richard, a French engineer who had been invited by the Soviet government to work in the USSR, and in the one headed by S. A. Kocherigin.

In 1937 Mikhail Iosifovich was sent to the United States to negotiate a manufacturing license for the Douglas DC-3 airliner. When he returned to the USSR a year later he took an active part in the preparation of the production line for this aircraft (known here as the PS-84 or Li-2) and the development of new manufacturing techniques. Toward the end of 1938 he joined the Polikarpov design bureau to take charge of the study project department. And that is where Mikoyan noticed him. The loop was looped, and the couple who were to marry their initials—MiG—was formed.

If we can trust the morphology of their faces, Mikoyan and Guryevich undoubtedly had very different natures. Could you tell us about these two men as individuals?

You are right. They had very different personalities—but often they complemented one another. Mikoyan was as open, outspoken, and convivial as Guryevich was modest, even unassuming. Mikoyan took good care of himself; for Guryevich, clothes were the least of his worries. Mikoyan had no experience in aircraft construction. He learned on the job. Because of his great knowledge of common manufacturing problems, he was able to assess a project quickly—and better still, to submit one of his own. Guryevich was erudite, the math expert, the mastermind, who because of his experience and the wide range of his technical knowledge had a talent for drawing up preliminary designs. Without Guryevich, Mikoyan probably would not have succeeded—but the reverse is also true. Guryevich was an engaging person, enamored with literature, always shocked by impurity, impoliteness, and coarseness. He wouldn't have hurt a fly and was never angry or cross with anyone. He was a married man with no children. Everyone at the design bureau was fond of him, and he proved to be a good social negotiator. Little was heard of him toward the end of his professional life because he was handed responsibility for the "set of themes B," an innocent name for one of the most secret OKB activities: the development of winged missiles. He retired in 1964 and died of old age in 1976 in Leningrad.

Mikoyan was married and had three children, one son and two daughters. His wife Zoya Ivanovna is still alive. His son, Ovanes Artyomovich, worked at the OKB as an engineer for a long time; he has a passion for sport aircraft and now belongs to the team of Lozino-Lozinskiy.

Some examples of Mikoyan's other traits are now coming back to me—I offer them to you in a jumble. For instance, he waited on his pilots hand and foot. He loved to take care of people. When the woman in charge of the workers' council drew his attention to the poor health of the design bureau's engineers during the war—at this time we seldom got enough to eat—he decided to send us to the country. That is how circumstances led me to make hay, when others organized fishing competitions. We were all able to regain some strength. Mikoyan also liked to fall in with his Armenian friends such as Marshal Bagramian and Tumanyan, Anastas's brother-in-law. He also struck up friendships with other general designers, including Ilyushin, Tupolev, Yakovlev, and Tumanskiy, an engine specialist who was also a cultured and interesting man.

Mikoyan had a cardiac defect, an abnormal thickness of the pericardium, that caused a lot of problems toward the end of his life. He was strongly affected by the accidental death in 1968 of Gagarin, who was thirty-six at the time, and by the death in April 1969 of Kadomtsev, the PVO commander-in-chief, a man whom he knew well, in a MiG-25 crash. One month later, on 27 May 1969, Mikoyan suffered a stroke that forced him to retire. On 9 December 1970 he had to be sent to the hospital and died on the operating table.

Mikoyan was a friend of Tupolev, another famous aircraft manufacturer. They are together here during an air show in July 1949. *Right,* Zoya Ivanovna, Mikoyan's wife.

That very day, you were assigned the heavy responsibility of taking over under difficult conditions. I'm sorry to ask, but you must tell us a little about yourself.

It is always much easier to talk about others, but as you wish to know every detail, I would have you know that Rostislav Apollosovich Belyakov was born on 4 March 1919 in Murom, in the Vladimir region. My father was a civil servant. After graduating in 1936 I entered the Moscow Aviation Institute (MAI) named after S. Orjonikidze, where I was taught by such prestigious professors as Yuryev and Zhuravchenko. I left with a diploma in 1941 and became an engineer at the MiG OKB. My first task was to work on updating the armament of the MiG-3 under the leadership of Volkov, a weapons specialist of great repute. Soon Mikoyan asked me to work with him and put me in

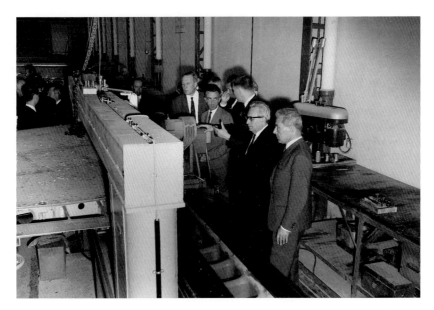

In June 1965 Mikoyan quietly visited the Dassault factory in Talence, France. In the company of Paul Deplante, technical manager of the Mérignac Dassault factory, Mikoyan examines the spot-facing of a Mystère 20 wing socket.

charge of the high-lift devices department. In 1955 I was made director of the research department, in 1957 deputy to the chief constructor, and in 1962 first deputy to the general designer. In 1971, a few weeks after Mikoyan's death, I was named general designer.

As an engineer and a scientist, what are your favorite activities?

I have devoted much of my time and energies to fluid mechanics, aeroelasticity, problems of strength, flight control modes and consequently to flying controls, design of aircraft-plus-missiles systems, power plants, airborne systems, aircraft design, and all advanced aeronautical technologies. I also pay special attention to materials and survival equipment. As a general designer I coordinate the work of the several design bureaus and research centers involved in the creation of a new prototype.

Could we return briefly to 1971? Tell us how you took over for Mikoyan and how you managed to preserve the spirit of initiative that prevailed in the OKB.

I succeeded in taking over without much trouble. After all, I had grown up within the OKB and had worked in practically every department. Do not forget that since 1957 I had been working with Mikoyan, in charge of the gears, flying controls, and various other systems and devices, and that since 1962 I had been his first deputy. One day Mikoyan told me, "You will have to decide on everything for yourself.

1

3

2

On 16 March 1942, by government decree, the OKB managed by Mikoyan was reorganized within the no. 155 factory area in Moscow. The area was in fact a wasteland near the Leningrad causeway that held a few crumbling huts, several barracks, an antiquated boiler room, and a small one-story building. It was in the latter that the design bureau set up shop on a primitive level in April 1942 after its return from Kuybyshev, to which it had been evacuated the year before. Quite rapidly, critical restoration and construction projects were completed. (*1*) The site after initial repairs in 1942. (*2*) The first assembly workshop. (*3*) Construction of the building planned to accommodate the design bureau. (*4*) The site in 1943.

4

This photograph was taken in March 1969, a few weeks before Mikoyan suffered the stroke that ended his career. He hands R. A. Belyakov a present for his fiftieth birthday: a sculpture symbolizing Belyakov's two passions, aviation and skiing.

So keep calm." If I had doubts about something, I could always knock at his door. I took part in all important meetings—Mikoyan, who was prone to headaches, abhorred business meetings and frequently sent me in his place, many times at the last minute. Because of my different activities, I was well known in all the departments. As you can see, I was in the best position to take over.

Well then, is aviation the one and only passion in the life of Rostislav Apollosovich?

Of course not. When I was a student I loved skiing, especially Alpine skiing and ski jumping. I was a five-time champion in the USSR, once in ski jumping and four times in downhill racing, and I am quite proud of that. With one of my professors, Zhuravchenko, I even "tested" a man in a wind tunnel to calculate the best descent attitudes. In 1940 I skied down the eastern side of the Elbrus, which reaches some 5,600 meters (18,600 feet) at its highest point. But I should not forget my wife, Ludmilla Nikolayevna. I met her in Kuybyshev, a port on the Volga River where the OKB withdrew in 1941 at the time of the German breakthrough. I returned to Moscow with our engineering office staff in 1942, but my wife-to-be did not return until later. We got married in 1945. She is now retired from her career as a television engineer specializing in large screens. And while I was at MAI I of course learned to fly—either on a Polikarpov Po-2 or a Yakovlev UT-2. I have also made parachute jumps.

You wish to know my shortcomings? I do not know how to relax, how to keep away from my work and think of something else. I like classical music, but I never have time to go to a concert. Besides, I have always regretted not knowing how to play an instrument. I never learned a foreign language either. In my everyday life, I go out for the newspaper and my dog takes me for a walk twice a day. That's all. There are two great sources of satisfaction in my life: Sergei, my son, who is thirty-nine years old and an engineer at MiG (previously in the automation department, today in the external economic relations department), and Olga Sergeievna, my granddaughter, who is a student at MAI. As you can see, my succession is already assured!

MiG-1 and MiG-3 Series

I-200 / Kh

In 1939 Soviet aircraft manufacturers, research scientists, and air force officers were taught a lesson by the poor performance of their planes in the Spanish civil war and against Japanese fighters in the Khalkin-Gol area of Mongolia. That lesson led to the acknowledgment of a severe failure.

It was just about time to update the fleet of the Soviet air force (VVS). Most of its planes were old or obsolete. The Soviet government knew that Germany was preparing for war. Discreet reconnaissance flights conducted at extremely high altitude by Junkers Ju-86P-2s from the Luftwaffe's Aufklärungsgruppe, a special unit commanded by Lt. Col. Theodor Rowehl, had not passed unnoticed. These flights were terminated after the signing of the Soviet-German "Friendship Pact" on 23 August 1939. It was in this context that the VVS asked its suppliers to develop a single-seat interceptor offering its best performance at altitudes above 6,000 m (18,300 feet). A preliminary design, the Kh, was prepared by the Polikarpov OKB. Then in October 1939 the establishment of OKO-1, an experimental research unit outside the Polikarpov OKB, brought together M. I. Guryevich, V. A. Romodin, and A. I. Mikoyan. The new team took up the preliminary design of the Kh project and added a new and promising engine then being tested, the Mikulin AM-37, which was expected to generate 1,030 kilowatts (kW) or 1,400 chevaux (ch) of power above 5,000 m (16,400 feet). Unfortunately, the development of this engine proved to be much trickier than expected, so the design team had to make do with the only available supercharged in-line engine that offered the necessary power, the AM-35A.

All of the team's efforts were concentrated on the new design, which was designated the I-200. The program was of such urgency that only one hundred days elapsed between the first set of production drawings and the maiden flight of the first prototype on 5 April 1940.

The I-200 was built with the materials that were at hand: pine and birch plywood, fabric, and as little metal as possible, because Soviet production of special steel alloys and duralumin had fallen well short

---

At rollout the I-200 no. 01 prototype had only one oil cooler, on the left side of the engine (see three-view drawing). In these photographs there are two oil coolers, one on each side of the engine. The aircraft was modified in this way before its second flight. *Bottom,* pilots disliked the side-hinged canopy because it could not be jettisoned.

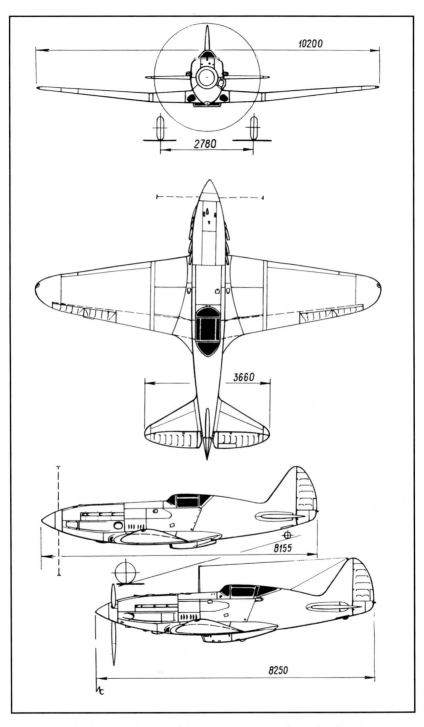

I-200 (Kh); *bottom,* side view of the MiG-3 (MiG OKB three-view drawing)

of demand. A detailed look at this machine is warranted because it is representative of Soviet technology and materials used at that time.

The wing had a single spar and a Clark YH-type airfoil section with a thickness ratio of 14 percent at the wing root and 8 percent at the wing tip. The all-metal structure of the wing center section consisted of an I-shaped spar with flanges in 20-KhGSA heat-treated steel and a web made of two 2-mm-thick stiffened light-alloy sheets; thirteen dural-pressed ribs, two of which were reinforced at the wing roots; two auxiliary spars, one on each side of the main spar; and five formers under the flush-riveted skin of the upper surface. The two 70-l (18.5-US gallon) wing fuel tanks were located between the main spar and the rear auxiliary spar. The 110-l (29-US gallon) fuselage tank was situated between the engine and the cockpit. Fuel gauges were mechanically controlled.

The wing center section was attached to the fuselage by 6-mm bolts placed 60 millimeters apart. The skin panels underneath the fuel tanks were removable. The two single-spar wooden outer wing panels attached to the center section had a 5-degree dihedral. The spar was of the box type with a web made of seven 4-mm-thick plywood sheets (five sheets at the wing tips). The 14- to 15-mm-wide spar flange was made of delta-drevyesina, a densified wood. The spar box tapered from 115 millimeters wide to 75 millimeters at the wing tips. Front and rear auxiliary spars were made of pine plywood, and all ribs were made of duralumin, a light, strong alloy of aluminum, copper, manganese, and magnesium. The skin consisted of five 4- to 2.5-mm-thick sheets of Bakelite plywood applied diagonally (except in the leading edge area) and homogenized with casein glue. Both ailerons were metal-framed and fabric-covered. The all-metal, two-segment split flaps were pneumatically positioned at two settings, 18 degrees and 50 degrees. The wing aspect ratio was 5.97.

The fuselage structure was also of the mixed type. The front section, which extended from the nose to behind the cockpit, was made of welded 30-KhGSA steel tubes with twelve duralumin skin panels held by Dzus-type fasteners. The rear section (including the fin) featured a monocoque wooden structure made of four pine longerons, gusset plates in Bakelite plywood under the attachment points, eight hollow frames with a Bakelite plywood web, and pine stringers. The skin was made of five 0.5-mm-thick plywood sheets coated on the inside with strips of calico impregnated with a nitrocellulosic varnish and on the outside by glued calico. The cockpit was made of molded Plexiglas panels but was not bulletproof. Its hinges opened to the right.

Both legs of the main landing gear retracted inward into the center part of the wing, and there was an electromechanical position indicator

(1) Assembly of the tail section.

(2) Forming the skin with a plywood sheet.

THE BIRTH OF THE FIRST MiG

(5) Gluing it to the fuselage structure.

(7) Checking the engine before it is installed.

(8) Assembly of the engine cowl panels.

(3) Fitting a bell crank to the center section of the wing.

(4) Covering the fuselage skin with glue.

(6) The center section of the wing, gear down.

(9) Positioning the attachment fittings.

(10) Putting the finishing touches on the tail section.

Exploded view of the I-200 prototype (MiG OKB drawing)

on the instrument panel. Shock struts were made of 30-KhGSA steel (130–150 kgf/mm²). The hydraulic fluid was a dialcohol (70 percent glycerin and 30 percent alcohol). The shock absorber stroke was 270–250 mm (10.6–9.8 inches) with an operating pressure of 39^{+1} atmospheres (573 psi). Main gear wheels were fitted with 600 x 180 tires. Wheel wells were closed by doors attached to the gear legs, two of them hinged to take into account the shock absorber stroke. The door covering the lower half-wheel was open 90 degrees when the gear was down and locked. The tail wheel (170 x 90 tire) retracted in a well closed by two small doors.

The cockpit panel was quite soberly equipped with only thirteen instruments. Noteworthy equipment and accessories included a GS-350 generator for the electrical system, a 12A-5 battery, and a KPA-3bis oxygen system. No radio was installed. Rod-operated mechanisms handled roll and pitch control (stick), while cable-operated ones provided yaw control (rudder bar). Trim control was maintained by means of a Bowden cable.

Through a reduction gearbox, the AM-35A engine drove a VISh-22Ye variable-pitch, three-bladed propeller 3 meters in diameter and fitted with a spinner made of *elektron,* a magnesium alloy. It rested on welded 30-KhGSA steel mounts. This engine delivered 993 kW (1,350 ch) at takeoff but was particularly heavy at 830 kg (1,830 pounds). By comparison, the Klimov VK-105 weighed 600 kg (1,323 pounds), the Daimler-Benz DB-601 575 kg (1,268 pounds), and the Rolls-Royce Merlin 605 kg (1,334 pounds). In fact, the AM-35A was meant to power not a fighter but rather the TB-7 heavy bomber. For the same reason, it was fitted with a single-stage, single-speed supercharger that permitted a maximum continuous rating of 883 kW (1,200 ch) at 6,000 m (18,300 feet). The honeycomb radiator contained 40 liters (10.6 US gallons) of glycol. It was located in a ventral bath with a frontal area of 23 dm² (2.5 square feet) and an airflow control flap at its rear end. There was only one oil cooler on the prototype, on the left side of the engine. Supercharger air intakes were located in wing roots, and the aircraft's pneumatic system fed the air starter. The six exhaust pipes—one for every two cylinders—were made of EYa1-TL-1 heat-resistant steel.

The first unarmed I-200 prototype was moved from the factory to the Khodinka airfield on 30 March 1940. After taxiing tests, in which the aircraft briefly rose a few feet into the air, MiG chief pilot A. N. Yekatov made the first flight on 5 April 1940. Later he would be assisted by other test pilots such as M. N. Yakushin, A. I. Zhukov, M. K. Martselyuk, and M. N. Yakushin. The program's chief engineer was A. G. Brunov, assisted by A. T. Karyev. On 1 May 1940 Yekatov was at the controls of I-200 no. 1 during the yearly air parade above the Red Square in Moscow. Two more prototypes were hastily assembled. The second made its first flight on 9 May 1940, and on 6 June 1940 I-200 no. 3 had its turn.

In an exception to the normal practice dictated in large part by the urgency of the program, factory tests and state acceptance trials proceeded concurrently on the three prototypes. Test pilots and military engineers worked in close collaboration with the factory engineers. Some who deserve mention are S. P. Suprun, squadron leader; A. G. Kubishkin, flying officer; A. I. Filin, deputy chief of staff and manager of the air force test center; A. I. Kabanov, colonel; P. M. Stefanovskiy, squadron leader; A. G. Proshakov, captain; and A. G. Kochetkov, colonel and engineer.

The chief engineer in charge of the state acceptance trials was V. I. Nikitenko, first-class military engineer. They were carried out over a very short period of time. On 22 May 1940 Yekatov attained a speed of 648.5 km/h (349.9 kt) in level flight at 6,900 m (22,640 feet). Suprun took the second prototype past that benchmark, hitting 651 km/h (351 kt) at 7,000 m (22,965 feet).

Factory tests ended on 25 August 1940 after 109 flights lasting a total of forty hours and forty-nine minutes. State acceptance trials were resumed on 29 August and ended on 12 September 1940. Here, with all the starkness of its style, is an extract from the certification document signed by Lt. Gen. P. V. Richagov of the VVS: "The I-200 fighter, powered by a AM-35A engine and built by Mikoyan and Guryevich at the NKAF factory no. 1, has a top speed of 628 km/h [339 kt], a better figure than those reached by other aircraft tested in our country, and is second to none when compared with foreign fighters in the same category when flown above 5,000 m [16,400 feet]. The aircraft has passed its acceptance trials successfully."

Though it was certified, the I-200 was not an aircraft for inexperienced pilots. The flight tests had revealed poor longitudinal stability, heavy controls, and a tendency to stall at the slightest provocation and go into a spin, from which it was nearly impossible to recover. Moreover, due to the rearward position of the cockpit forward visibility when taxiing was very poor. Yet the need was urgent, and despite all these flaws the government authorized an initial production run of one hundred aircraft. They received the service designation MiG-1.

Specifications

Span, 10.2 m (33 ft 5.6 in); length, 8.155 m (26 ft 9.1 in); wheel track, 2.78 m (9 ft 1.4 in); wing area, 17.44 m^2 (187.72 sq ft); empty weight, 2,475 kg (5,456 lb); takeoff weight, 2,968 kg (6,543 lb); fuel, 190 kg (418 lb); oil, 28 kg (62 lb); wing loading, 170.2 kg/m^2 (34.9 lb/sq ft).

One of the first eight MiG-1s built in 1940. The side-hinged canopy was removed.

MiG-1

The first production aircraft did not differ greatly from the prototype. Because of the simultaneity of acceptance trials and the commitment to production, only minor modifications could be made:

— the single oil cooler proved inadequate, so a second one was placed on the right side of the engine
— a double outlet flap replaced the single control flap at the shroud forward end of the coolers
— the ventral radiator bath was enlarged and extended forward
— the fuel tanks received a measure of bulletproofing as rubber-based sheathing was installed to act as a self-sealing material
— the hinged lower halves of the main wheel doors were transferred from the gear legs to the fuselage, on the side of each wheel well

None of those modifications, as can be seen, could improve the aircraft's maneuverability or flying qualities.

The first eight MiG-1s were fitted with a sideways-hinged canopy that could not be jettisoned in flight. Starting with MiG-1 no. 9, a new aft-sliding canopy was introduced. The last MiG-1 rolled out of the fac-

This MiG-1 was equipped for spin tests. Note the control surface deflection sensors on the rudder and elevator.

tory in December 1940. Its armament was grouped above the engine and consisted of a 12.7-mm UBS machine gun with 300 rounds and two 7.62-mm ShKAS machine guns with 375 rounds per gun (rpg). Two store stations under the wing could receive either two 50-kg (110-pound) FAB-50 or two 100-kg (220-pound) FAB-100 bombs. The pilot was provided with a PBP-1 gunsight.

The first MiG-1s reached service units in April 1941, less than three months before the German invasion.

Specifications
Span, 10.2 m (33 ft 5.6 in); length, 8.155 m (26 ft 9.1 in); height, 3.3 m (10 ft 9.9 in); wheel track, 2.78 m (9 ft 1.4 in); wing area, 17.44 m^2 (187.72 sq ft); empty weight, 2,602 kg (5,736 lb); takeoff weight, 3,099 kg (6,832 lb); fuel, 190 kg (418 lb); oil, 28 kg (62 lb); wing loading, 177.7 kg/m^2 (36.4 lb/sq ft).

Performance
Max speed, 628 km/h at 7,200 m (339 kt at 23,600 ft); max speed at sea level, 486 km/h (262 kt); landing speed, 141 km/h (76 kt); climb to 5,000 m (16,400 ft) in 5.9 min; service ceiling, 12,000 m (39,400 ft); range at 550 km/h (297 kt) with 70% W and 10% fuel reserve, 580 km (360 mi); takeoff roll, 238 m (780 ft); landing roll, 400 m (1,310 ft).

One of the 3,170 MiG-3s built between December 1940 and December 1941.

MiG-3

The inadequacies of the I-200 (as recorded in various factory and military test reports) led to an intensive research program, including full-scale tests in the new T-101 and T-104 TsAGI large wind tunnels. The modifications introduced as a result of this effort did not remedy all the aircraft's faults—and could not have, given its time constraints. But they certainly counted as improvements in several areas:

— the dihedral of the outer wing panels was increased from 5 to 6 degrees to improve stability
— the engine was moved 100 millimeters (3.9 inches) forward to off-set a tail-heavy trimming
— a new 250-l (66-US gallon) fuel tank was introduced beneath the pilot's seat to increase the aircraft's range
— the fuel system was protected from fire by transferring inert gases (cooled exhaust gases) in the fuel tanks
— a second oil tank was added
— the engine cooling system was improved
— a new VISh-61Sh propeller was installed, enhancing efficiency by increasing the pitch range to 35 degrees
— larger main gear wheels were used (650 x 200 tires), enabing the MiG-3 to operate from grass strips; the gear doors were consequently modified
— the wheel brakes were improved
— an 8-mm-thick armor plate (later made 9-mm thick) was added to the back of the pilot's seat

The oil cooler mounted under the exhaust collector, its exhaust flap open to the full angle.

— the fuselage decking behind the cockpit was modified and glazed to improve aft vision
— the PBP-1 gunsight was replaced by the more advanced PBP-1A model
— an RSI-3 single-channel receiver was installed (later replaced by an RSI-4)
— the external weapon load was increased: four store stations under the wings thus enabled the MiG-3 to carry mixed loads of 8- to 100-kg (18- to 220-pound) bombs up to a maximum of 220 kg (485 pounds), two VAP-6M/ZAP-6 chemical/incendiary spray containers, or eight unguided RS-82 rockets

The built-in armament was similar to that of the MiG-1 but was sometimes complemented by two 12.7-mm BK machine guns in slipper pods beneath the wing. With five weapons, firepower more than doubled, the weight of a salvo per second increasing from 1.38 to 3 kg (49 to 106 ounces). But those supplementary weapons pushed the takeoff weight of the MiG-3 up to 3,510 kg (7,738 pounds), increasing the wing loading to 201.3 kg/m^2 (41 pounds per square foot).

A few MiG-3s were equipped with two 12.7-mm UBS machine guns. Tests were also conducted with two 20-mm ShVAK cannons, but this weapon was not retained. Other MiG-3s flew unarmed as photo-

MiG-3 detail. (*1*) One of the two oil tanks. (*2*) Machine gun ammunition box. (*3*) One of the two fuselage fuel tanks, 110 l (29 US gallons). The oil cooler exhaust flap is open to the minimum angle.

reconnaissance aircraft above the front line. The first MiG-3s rolled out of Aviakhim factory no. 1 in December 1940. At the end of that month eleven aircraft had been built. The production rate rose quickly: no fewer than 140 aircraft left the assembly line in January 1941. In June, on the eve of Operation Barbarossa, the factory operated around the clock and produced twenty-five aircraft per day. The first MiG-3s and MiG-1s reached the IAPs (fighter aviation regiments) in April 1941.

On 4 December 1940 ten MiG-3s were set aside for combat training missions. These flights took place partly in Katcha, Crimea, where there are more sunny days than in the Moscow area. On 13 March 1941 during one of these missions test pilot Yekato lost his life. An investigation revealed that the supercharger compressor wheel had come loose, going right through the fire wall and the front fuel tank before fatally wounding the pilot.

Because of the volatile situation in Europe in early 1941, the first series of modern Soviet fighters (Yak-1, LaGG-3, MiG-1, and MiG-3) were sent first and foremost to the border zones of the western USSR, especially to the air bases of Kaunas, Lvov, Byelstok, Kishinev, Byeltsy (Moldavia), and Eupatoria. As of 1 June 1941, three weeks before the German invasion, thirteen fighter regiments were equipped entirely with MiG-3s, and another six regiments had received partial comple-

A few MiG-3s had their built-in armament supplemented with two 12.7-mm BK machine guns podded beneath the wing and firing outside the propeller disk.

OFFICIAL TELEGRAM

December 1941   Two addressees in Kuybyshev
SHENKMAN, factory no. 18 manager
Copy to: TRETYAKOV, factory no. 1 manager

You are playing a shabby trick on our homeland and on our Red Army
STOP Until now you did not consider it advisable to produce Il-2s STOP Il-
2s are today as essential to our Red Army as are air and bread STOP
Shenkman produces only one Il-2 every day, and Tretyakov one or two
MiG-3s STOP This is an insult to our homeland, to the Red Army STOP We
need Il-2s, not MiGs STOP If factory no. 18 intends to weaken our homeland by delivering only one Il-2 per day, they make a heavy mistake and
they will have to pay for it STOP I ask you not to abuse the patience of
our government and to produce more Il-2s STOP Let this be a last warning
to you both STOP   No. P553 STALIN

---

ments. On 22 June 1941, the first day of Operation Barbarossa, MiG-1s
and MiG-3s represented only 37 percent of the total number of operational fighters; the other 63 percent were I-15s, I-15bis, I-153s, and even
a handful of obsolete I-5s still used in training centers. But 89.9 percent
of the new fighters were MiG-3s distributed in fighter regiments of the
VVS (air force), the PVO (air defense units), and the VMF (navy).

In October 1941 the German breakthrough threatened the OKB
and factory no. 1. They were hastily moved to Kuybyshev, a major port
on the Volga River. Two months later, the fate of the MiG-3 was sealed.
Its AM-35A engine was built in the same factory as the AM-38, which
powered the Ilyushin Il-2 Shturmovik ground attack aircraft. This
model now received priority so that the production rate of this aircraft
could be accelerated. Without an engine, the MiG-3 was dead. Stalin
had intervened personally. In a cable that became famous in the Soviet
aircraft industry but was unknown in the West (until now), Stalin stern-
ly accused two factory managers of slowing down production. On 23
December 1941 the MiG-3 program was terminated; however, all aircraft then on the assembly line were to be completed, the last one
rolling out of the factory at the beginning of 1942. A total of 3,120 MiG-
3s had been delivered in just over one year. After the OKB and factory
no. 1 returned to Khodinka in March 1942, fifty MiG-3s were assembled
from components that had been hidden away at the time of the evacuation. These aircraft were allotted to the Moscow PVO.

Early in the war MiG-3s often operated in a less than optimal flight
envelope. This high-altitude interceptor was used first as a "frontal
fighter" for low- and medium-altitude interception and even for close-
support missions below 5,000 m (16,400 feet), flight levels where German fighters were superior in terms of performance. But at that time
the Soviets had to make the most of what they had: the German
attacks on their airfields at daybreak on 22 June 1941 had had a devas-
tating impact. But once mastered by its pilot the MiG-3 regained the

A rare photograph, taken in 1942, shows a MiG-3 fitted with launch rails for RS-82 rockets. This aircraft belonged to the 12 IAP (12th fighter aviation regiment) of the Moscow PVO (air defense).

Three MiG-3s are delivered to their pilots at a VVS maintenance base in 1942.

upper hand when operating above 5,000 m thanks to its outstanding speed and ceiling—but there the German fighters would not dare to cross swords with it. So MiG-3s were usually assigned to PVO units for high-altitude interception of bombers and reconnaissance aircraft. They even flew as night fighters. It was in a MiG-3 that A. I. Pokryshkin, future air marshal, gained the first of his fifty-nine kills when he shot down a Messerschmitt Bf 109E. It was also in a MiG-3 that A. N. Katrisht, future general-colonel, distinguished himself during night missions.

To conclude, we ought to mention the surprising proposal made just after the German invasion by S. P. Suprun, the test pilot responsible for the I-200 state acceptance trials during the summer of 1940. Since Soviet fighter regiments had already sustained heavy losses on the ground (mostly) and in the air, Suprun suggested that two fighter regiments equipped with MiG-1s and MiG-3s be staffed mostly by factory test pilots, GosNII VVS (state scientific research institute of the air force) test pilots, and military pilots in charge of acceptance flights. Two regiments were formed as a result—Suprun's 401 IAP, assigned to the western front near Smolensk, and the 402 IAP under the command of P. M. Stefanovskii, a test pilot of great skill, on the northwestern front.

Specifications
Span, 10.2 m (33 ft 5.6 in); length, 8.25 m (27 ft 0.8 in); height, 3.3 m (10 ft 9.9 in); wheel track, 2.78 m (9 ft 1.4 in); wing area, 17.44 m² (187.72 sq ft); empty weight, 2,699 kg (5,950 lb); takeoff weight, 3,350 kg (7,385 lb); fuel, 385 kg (849 lb); oil, 55 kg (121 lb); wing loading, 92 kg/m² (39.3 lb/sq ft).

Performance
Max speed, 640 km/h at 7,800 m (346 kt at 25,600 ft); max speed at sea level, 505 km/h (273 kt); climb to 8,000 m (26,250 ft) in 10.28 min; service ceiling, 12,000 m (39,400 ft); range at 550 km/h (297 kt) with 70% W and 10% fuel reserve, 820 km (510 mi); takeoff roll, rated power, no flaps, 305 m (1,000 ft); rated power, 15° flaps, 268 m (880 ft); full power, no flaps, 234 m (770 ft); landing roll, 400 m (1,310 ft).

MiG-3 with AM-37 Engine

During the second half of 1941, to improve the aircraft's maximum speed at high altitude, the OKB engineers equipped a MiG-3 with a Mikulin AM-37, an engine then under development that was rated at

MiG-3 pilots taking the oath to the flag of the Guards when the 27 IAP was reorganized and became the 12 GvIAP (12th fighter aviation regiment of the Guards).

1,030 kW (1,400 ch) at 5,000 m (16,400 feet). Matching the new engine was no problem because both the AM-35A and the AM-37 had the same design dimensions and attachment points. Test pilot A. I. Zhukov made a few flights with this reengined MiG-3. But in the end the whole program was canceled, development of the AM-37 having proved too difficult. In some OKB documents the MiG-3 with the AM-37 engine is referred to as the MiG-7.

MiG-3 with AM-38 Engine

In an attempt to correct the inadequacy of the MiG-3 at low and medium altitudes, the AM-35A was replaced by the AM-38, the only available engine that could give the aircraft the means to stand up to the Messerschmitt Bf 109Fs first encountered in June 1941. Installing the AM-38 in the MiG-3 required few modifications, since the weight and overall dimensions of both engines were almost identical. The AM-38 afforded 1,178 kW (1,600 ch) at takeoff, and its maximum rating at 2,000 m (6,560 feet) was 1,141 kW (1,550 ch).

A production MiG-3 was reengined in this way and somewhat modified by reshaping the exhaust pipes, removing the bomb racks, and suppressing the inert-gases transfer system. The aircraft was ready at the end of July 1941, and it went up for its maiden flight at the beginning of August with an LII pilot, Yu. K. Stankevich. The flight test program was carried out by NII VVS pilots A. G. Kochetkov, A. G. Kubishkin, and A. M. Popyelnushenko as well as LII pilots such as G. M. Shiyanov and A. V. Yumashev. Tests ended on 17 August. They showed that the reengined fighter was a sound machine if flown in ambient temperatures lower than 16–20° C (61–68° F), but the engine cooling system had to be modified for use in higher temperatures. The

An anachronistic scene: supplies were delivered on this MiG-3 air base on sleds pulled by reindeer.

test report concluded, "Below 4,000 m [13,100 feet], the AM-38–engined MiG-3 offers new tactical possibilities and can successfully face the enemy fighters at low and medium altitudes." The time required for a 360-degree turn had dropped to between twenty and twenty-one seconds. Considering this laudatory appraisal, series production of the aircraft was recommended. But this plan too was thwarted by the unavailability of the AM-38 engine, which was still reserved for the Il-2 assembly lines. However, a repair workshop fitted a small number of MiG-3s with overhauled AM-38s because no more AM-35As could be found. Some of these MiG-3s were equipped with two synchronized 20-mm ShVAK cannons (100 rounds each) to fight in PVO units.

Specifications
Span, 10.2 m (33 ft 5.6 in); length, 8.25 m (27 ft 0.8 in); height in level flight position, 3.325 m (10 ft 10.9 in); wheel track, 2.78 m (9 ft 1.4 in); wing area, 17.44 m² (187.72 sq ft); empty weight, 2,582 kg (5,692 lb); takeoff weight, 3,225 kg (7,110 lb); fuel, 463 kg (1,020 lb); oil, 45 kg (99 lb); wing loading, 185 kg/m² (37.9 lb/sq ft).

Performance
Max speed, 587 km/h at 3,000 m (317 kt at 9,800 ft); 592 km/h at 4,000 m (320 kt at 13,000 ft); max speed at sea level, 547 km/h (295 kt); climb to 5,000 m (16,400 ft) in 7.95 min; service ceiling, 9,500 m (31,200 ft); takeoff roll, 380 m (1,250 ft); landing roll, 400 m (1,310 ft).

The OKB tried to prolong the life of the MiG-3 at all costs, but matching a radial engine with an airframe engineered for an in-line engine raised too many difficulties.

MiG-3 with M-82 Engine / I-210 / IKh

When series production of the MiG-3 stopped in December 1941 for lack of engines, the Perm power plant factory was producing the M-82A, a 14-cylinder, double-row, air-cooled radial engine. Its power output was high at takeoff—1,251 kW (1,700 ch)—but it lost power rapidly with altitude. At 6,500 m (21,300 feet) its rating dropped to only 979 kW (1,330 ch).

Despite this shortcoming the MiG OKB redesigned the MiG-3 with an M-82 engine in a risky attempt to prolong the aircraft's life. Chief engineer I. G. Lazarev tried to match the engine with the front fuselage structure with a minimum of modifications and designed an engine cowling to minimize load losses resulting from its internal turbulent flow.

The cockpit canopy was enlarged and deepened to improve rear and lateral visibility. The fin area was increased slightly to improve yaw stability, and the wing was fitted with leading-edge slats tested on a production MiG-3. The three 12.7-mm UBS machine guns (one on top of the engine cowling and one on each side of it) were controlled by a single trigger.

Five I-210s were completed in November–December 1941, and the first flight took place in December with NII VVS pilot V. Ye. Golofastov in the cockpit. From the very beginning of the test flights, strong vibrations occurred in the tail unit, and the aircraft's maneuverability

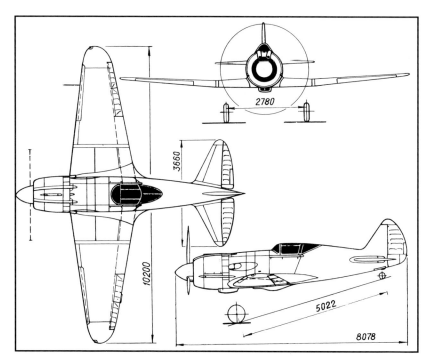

I-210 (IKh) (MiG OKB three-view drawing)

The M-82A was easy to service thanks to the four hinged access panels of the engine cowl.

proved to be rather scanty. The size of the engine master cross section and the poor airtightness of the engine cowling generated vortex flows that increased the drag significantly. Despite the high power of the M-82, the maximum speed of the I-210 was less than that of the MiG-3 with an AM-35A engine.

The aircraft was tested in a TsAGI full-scale wind tunnel and returned to the OKB for modifications. The wind tunnel tests confirmed the decisive effect of the lack of cowling airtightness on the aircraft's aerodynamic drag. They led at last to the design of a new aircraft code-named Ye and powered by a M-82F, which was not completed before the end of 1942. Despite its poor performance, the I-210 was combat-proven on the Kalinin front. Two 7.62-mm ShKAS synchronized machine guns were added to the basic armament.

In some OKB documents the I-210 is referred to as the MiG-9 (the first of many).

Specifications

Span, 10.2 m (33 ft 5.6 in); length, 8.078 m (26 ft 6 in); wheel track, 2.78 m (9 ft 1.4 in); wheel base, 5.022 m (16 ft 5.7 in); wing area, 17.44 m² (187.7 sq ft); empty weight, 2,720 kg (5,995 lb); takeoff weight, 3,382 kg (7,454 lb); fuel + oil, 360 kg (795 lb); wing loading, 193.9 kg/m² (39.71 lb/sq ft).

Performance

Max speed, 565 km/h at 6,150 m (305 kt at 20,170 ft); max speed at sea level, 475 km/h (257 kt); landing speed, 146 km/h (79 kt); climb to 5,000 m (16,400 ft) in 6.7 min; range, 1,070 km (665 mi); takeoff roll, 410 m (1,345 ft); landing roll, 535 m (1,755 ft).

MiG-3U / I-230 / D

The aims of this program were not very different from those of the previous designs—that is, improving the aerodynamics, handling qualities, and production processes of the MiG-3 even though its assembly lines were now closed. The motivation of this new attempt was to give the MiG-3 a successor without departing too much from the original design.

Compared to the MiG-3, the I-230's fuselage was 370 mm (14.57 inches) longer. The main landing gear was modified and equipped with more efficient and reliable shock absorbers, and the pilot's view was greatly improved. Aware of the scarcity of light alloys in 1942, the designers called for the fuselage to be made entirely of wood except at the engine mounting.

The I-230 would have entered mass production, but the engine was not available. Prototype no. 01 had no antenna mast.

At the time of the preliminary design, the utmost was done so that the I-230 might be manufactured on the same assembly lines as the MiG-3, with the same production tooling and methods. The quality of the skin finishing and the aerodynamic cleanness of the airframe were superb. The aircraft was really a beauty.

The ventral radiator bath underneath the wing center section was moved forward and made much smaller. A slightly bigger wing was used on an experimental basis to increase the aircraft's ceiling by 500 m (1,640 feet). But this was not considered enough of an improvement to justify acceptance of the new wing.

The team in charge of this program ran into an old snag. The AM-35A, the only suitable engine for this type of aircraft, was no longer being produced. The AM-38F was certainly not the ideal engine for a fighter—and besides, its entire production run was reserved for the Il-2. Consequently, the I-230 was fitted with a salvaged AM-35A. The flight reports filed by MiG-3 pilots were taken into account: firepower was increased by mounting two synchronized 20-mm ShVAK (SP-20) cannons on top of the engine (150 rpg). The I-230 became the first single-engine MiG equipped with cannons.

The aircraft was first flown and flight-tested in August 1942. Even though the results were positive, there was uncertainty in 1943 over whether series production could be launched. After the prototype roll-out, five preproduction aircraft were completed in the OKB workshop and delivered to the 1 GvIAP (1st Guards fighter aviation regiment) for

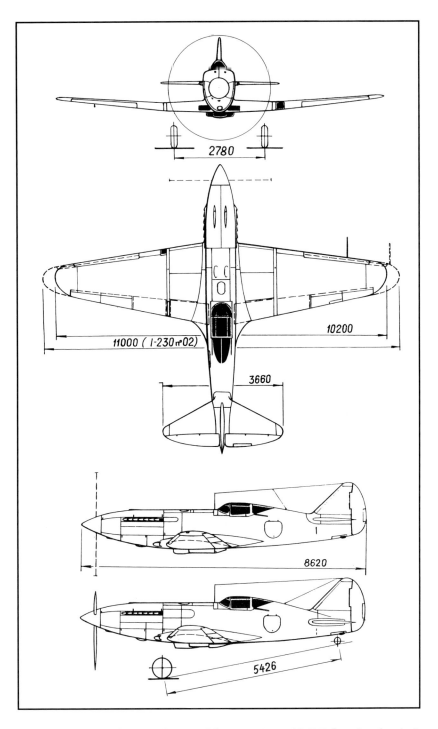

I-230 (D)/I-231 (2D); *bottom,* side view of the I-231 (2D) (MiG OKB four-view drawing)

I-230 prototype no. 02 had a larger wingspan.

combat proving on the Kalinin front, where they performed with distinction. Nonetheless, the chaotic circumstances of most engine manufacturers in this period (evacuation of production factories, establishment of new production facilities in distant places, and difficulties in managerial staff recruitment) made it impossible to resume production of the AM-35A. Despite all its good qualities, the I-230 did not succeed the MiG-3.

The I-230 is sometimes called the MiG-3D (*Dalnostniy,* long range) or MiG-3U (*Ulushchenniy,* improved).

Specifications
Span, 10.2 m (33 ft 5.6 in); I-230 no. 02 span, 11 m (36 ft 1 in); length, 8.62 m (28 ft 3.4 in); wheel track, 2.78 m (9 ft 1.4 in); wheel base, 5.426 m (17 ft 9.6 in); wing area, 17.44 m^2 (187.7 sq ft); I-230 no. 02 wing area, 18 m^2 (193.75 sq ft); empty weight, 2,612 kg (5,757 lb); takeoff weight, 3,285 kg (7,240 lb); fuel, 324 kg (714 lb); oil, 56 kg (123 lb); wing loading, 188.4/182.5 kg/m^2 (38.57/37.36 lb/sq ft); max operating limit load factor, 8.

Performance
Max speed, 660 km/h at 6,000 m (356 kt at 19,680 ft); max speed at sea level, 560 km/h (302 kt); climb to 5,000 m (16,400 ft) in 6.2 min; service ceiling, 11,500 m (37,700 ft); I-230 no. 02 service ceiling, 12,000 m (39,360 ft); range, 1,350 km (840 mi).

The I-231—as beautiful as the Spitfire—was not the MiG-3's successor despite its outstanding performance. In 1943 there was not a single factory available to build it.

I-231 / 2D

OKB engineers were desperately trying to find a successor to the MiG-3. In 1943 they thought they had hit on something when a new engine, the AM-39A, became available. Its takeoff power reached 1,325 kW (1,800 ch), and it was rated at 1,104 kW (1,500 ch) at 5,850 m (19,190 feet). The engine was installed in an airframe almost similar to that of the I-230 with the same armament (two 20-mm ShVAK cannons above the engine with 160 rpg).

The I-231 prototype was completed and test-flown in 1943 with Yu. A. Antipov and later P. M. Stefanovskiy at the controls, but it was destroyed as a result of a mislanding. It could have been a chance mishap, the aircraft having shown great capabilities justifying its mass production. But once more a shortage of engines put an end to what would be the final attempt to extend the life of the MiG-3: series production of the AM-39 had to be stopped a short time after it began. Also, no production unit had the capacity to manufacture the I-231. They were all busy, day and night, turning out Yakovlev and Lavochkin fighters.

Specifications
Span, 10.2 m (33 ft 5.6 in); length, 8.62 m (28 ft 3.4 in); height in level flight position, 3.275 m (10 ft 8.9 in); wheel track, 2.78 m (9 ft 1.4 in); wheel base, 5.426 m (17 ft 9.6 in); wing area, 17.44 m² (187.7 sq ft); empty weight, 2,583 kg (5,693 lb); takeoff weight, 3,287 kg (7,245 lb); fuel, 333 kg (734 lb); oil, 34 kg (75 lb); wing loading, 188.5 kg/m² (38.6 lb/sq ft); max operating limit load factor, 8.

Performance
Max speed, 707 km/h at 7,100 m (382 kt at 23,290 ft); climb to 5,000 m (16,400 ft) in 4.5 min; service ceiling, 11,400 m (37,400 ft).

I-211 / Ye

As mentioned earlier, this program was a direct result of the I-210 tests in one of the full-scale TsAGI wind tunnels. The goal was still to prolong the MiG-3 series with an updated product. But to improve the level and climbing speeds of the aircraft, it was necessary to find a more powerful engine than the M-82A and to reduce its takeoff weight significantly. The only available engine alternative was the M-82F, delivering 1,362 kW (1,850 ch) at takeoff and 979 kW (1,330 ch) at

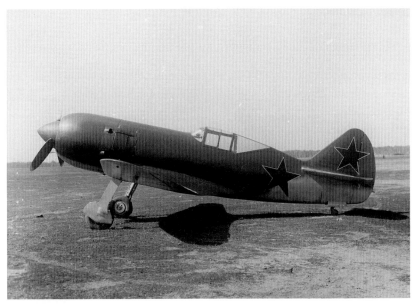

The I-211 marked a new attempt to match a radial engine with the MiG-3 airframe. This time it was a complete success. But, strangely enough, the project was not pursued, and Lavochkin inherited most of the technical innovations tested with this machine.

5,400 m (17,700 feet). The I-211 differed from the I-210 in several other respects as well. The front fuselage cross section was increased to make the junction of the engine cowling and the fuselage smoother, and the adjustable flaps of the engine exhaust outlet were moved to the sides of the fuselage. The cockpit was moved 245 mm (9.64 inches) back and the fin chord was extended forward, increasing its area and improving the aircraft's yaw stability. The oil cooler air inlets were moved into the wing root fairings. The shape of the engine cowling was carefully designed to cope with the problem of airtightness and to achieve the best junction possible with the fuselage. All of these modifications resulted in an outstanding increase in the aircraft's speed. In 1942 most of these technical innovations (engine cowling design and airtightness, proper positioning of the engine itself, and the I-210 wing leading edge slats) were passed on—at the order of the Narkomavprom (state commissariat of the aviation industry)—to the Lavochkin OKB, which adapted them successfully to the La-5, a mass-produced fighter.

The armament was also modified: the I-210's machine guns were replaced by two synchronized 20-mm ShVAK cannons at the bottom of engine cowling. Assembly of the I-211 began in December 1942 and finished in August 1943. Golofastov was the first pilot to fly it. Two prototypes were built, followed by eight preproduction aircraft. The shortcomings of the first I-210 and its engine became nothing more than a bad memory, and the I-211 proved to be the best Russian fighter of the time. Compared with the 1942 version of the La-5, its level speed was 40 to 166 km/h (21 to 90 kt) higher, depending on the altitude. Compared with the 1942 version of the Yak-9, it was 65 to 73 km/h (35 to 39 kt) faster. To climb to 5,000 m (16,400 feet), the La-5 took 1.4 to 2.2 minutes longer, and the Yak-9 0.9 to 1.5 minutes longer. After the prototypes passed the factory flight tests, ten preproduction I-211s were delivered to the VVS to prove themselves in combat. They engaged successfully in air battles over the northwestern front near Kalinin. Air force pilots and NII VVS test pilots spoke out in favor of adding the I-211 to the VVS fleet. But in spite of their recommendations and the aircraft's remarkable flying qualities, the GKO (defense state committee) gave up mass-producing the aircraft because two factories were already building the La-5FN (an La-5 with an ASh-82FN engine).

Specifications
Span, 10.2 m (33 ft 5.6 in); length, 7.954 m (26 ft 1.1 in); height, 3.63 m (11 ft 10.9 in); wheel track, 2.78 m (9 ft 1.4 in); wheel base, 5.015 m (16 ft 5.4 in); wing area, 17.44 m² (187.7 sq ft); empty weight, 2,528 kg (5,572 lb); takeoff weight, 3,100 kg (6,830 lb); fuel + oil, 385 kg (848 lb); wing loading, 177.75 kg/m² (36.38 lb/sq ft).

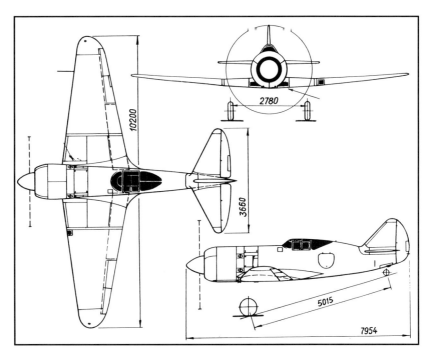

I-211 (Ye) (MiG OKB three-view drawing)

Performance
Max speed, 670 km/h at 7,000 m (362 kt at 22,960 ft); climb to 5,000 m (16,400 ft) in 4 min; service ceiling, 11,300 m (37,065 ft); range, 1,140 km (710 mi).

PBSh-1 and PBSh-2 Series

PBSh-1

Throughout the 1930s the doctrines relating to the use of big armored units were completely reappraised. Because every new threat demanded an immediate answer, a new type of aircraft appeared in the USSR: the *shturmovik,* or assault aircraft. Several of the best-known aircraft manufacturers worked on this new weapon. At the start of the decade Tupolev proposed two heavy shturmoviks, the ANT-17 and ANT-18, but they were never built. The TsKB (central construction bureau) built four aircraft designed by D. P. Grigorovich, the LSh-1, TSh-1, TSh-

The PBSh-1 main gear was directly inspired by that of the I-200.

2, and ShON. In 1933 a design brigade under the leadership of S. A. Kocherigin—and assisted by the forty-one-year-old M. I. Guryevich, Mikoyan's future right-hand man—built the TSh-3, also called the TsKB-4. N. N. Polikarpov, for its part, developed the R-ZSh.

All of these attempts led in 1936 to the Ivanov program ("Ivanov" was Stalin's cable address), an air force initiative that resulted in the construction of such prototypes as the KhAI-5 (R-10) and KhAI-52, designed by Nyeman, winner of the contest; and the ANT-51 and ShB, both designed by Sukhoi. But in the end the true winner was a relatively unknown outsider, S. V. Ilyushin. More than 40,000 of its BSh-2s, renamed the Il-2 in 1940, were built because of the war.

Mikoyan and Guryevich started the preliminary design of their first assault aircraft in 1940. The PBSh-1 (*Pushechniy Bronirovaniy Shturmovik:* armored assault aircraft with cannons) was a single-seater designed to attack frontline ground targets such as troops, strong points, and armored vehicles. It had a cantilever inverted gull wing and was to be powered by a 1,178 kW (1,600 ch) Mikulin AM-38 engine. Sensitive parts, the engine, and the cockpit would be protected by armor plating. But the weight of the armor alone was 1,390 kg (3,065 pounds), or 30 percent of the aircraft's takeoff weight. The design had to be completely reconsidered to integrate the armor into the stressed structure. It was to be equipped with two 23-mm cannons in fairings beneath the wing (96 rpg) and six 7.62-mm ShKAS machine guns on the wing leading edge (750 rpg). All of these weapons would fire outside the propeller disc.

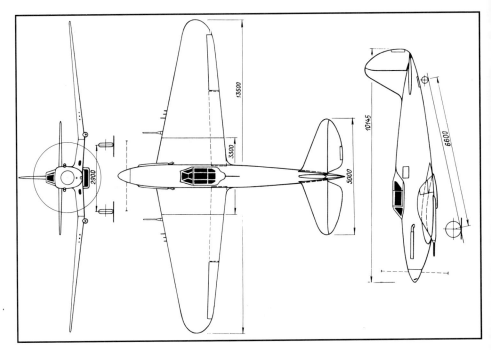

PBSh-1 (MiG OKB three-view drawing)

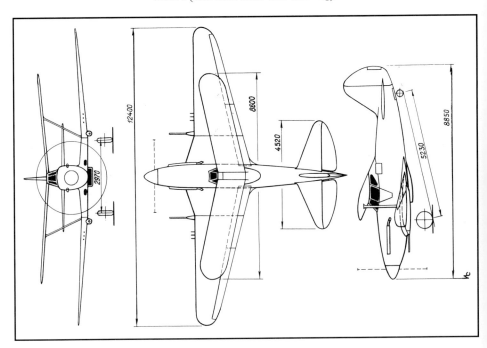

PBSh-2 (MiG OKB three-view drawing)

46

The bomber version of the aircraft was to carry in its fuselage either 24 10-kg (22-pound) FAB-10 or 24 8-kg (17.6-pound) FAB-8 bombs and either 280 2.5-kg (5.5-pound) FAB-2.5 or 120 1-kg (2.2-pound) ZAB-1 bombs. (FAB indicates demolition bombs, while ZAB refers to incendiary bombs. The number after the acronym reflects the bomb's weight.) For dive-bombing missions, two FABs ranging in weight from 25 to 250 kg (55 to 550 pounds) could be added beneath the wings.

The preliminary design by N. Z. Matyuk, chief of the aerodynamic design department, was approved on 24 July 1940 by OKO chief engineer Mikoyan, his assistant Guryevich, and P. V. Dementyev, manager of Aviakhim factory no. 1. The OKO started work on a full-scale model of the aircraft in the fall. But as soon as the Ilyushin Il-2 was approved for series production, OKB engineers halted all work on the PBSh-1 and started in immediately on the preliminary design for the PBSh-2.

In some OKB documents the PBSh-1 is referred to as the MiG-4.

Specifications
Span, 13.5 m (44 ft 3.5 in); length, 10.145 m (33 ft 3.4 in); height, 3.2 m (10 ft 6 in); wheel track, 2.9 m (9 ft 10 in); wheel base, 6.6 m (21 ft 7.9 in); wing area, 30.5 m^2 (328.3 sq ft); takeoff weight, 4,850 kg (10,690 lb); max takeoff weight, 6,024 kg (13,277 lb); wing loading, 159 kg/m^2 (32.6 lb/sq ft).

Design Performance
Max speed, 441 km/h at 5,000 m (238 kt at 16,400 ft); max ground speed, 449 km/h (242 kt); range, 900 km (560 mi); service ceiling, 7,600 m (25,690 ft).

PBSh-2

A note came with the PBSh-2 preliminary design created by the factory no. 1 OKO in July 1940. It said: "Considering that hedge-hopping flying in a highly wing-loaded aircraft is rather demanding and moreover that monoplanes seldom forgive pilots' mistakes at very low altitudes, we have chosen the biplane configuration for the PBSh-2, even though it has been designed to carry out the same missions as the PBSh-1. Biplanes are much easier to fly. They have far better stability and maneuverability."

The PBSh-2 silhouette was quite unusual. Areas of the two wings of this biplane were quite different. The smaller upper wing had a forward sweep angle of 12 degrees and no ailerons. A light dihedral was applied to the lower wing outer panels. The whole trailing edge was

The unorthodox PBSh-2 was no more successful than the PBSh-1. The VVS chose instead the Ilyushin *shturmovik,* the Il-2.

occupied by large ailerons and two-segment flaps. The wings were braced by I-type struts.

Like the PBSh-1, the engine was to be a 1,178 kW (1,600 ch) AM-38 protected (along with the fuel tanks and cockpit) by case-hardening homogenized armor plates whose thickness varied from 7.5 to 15.5 mm. The planned armament was similar to that of the PBSh-1: two 23-mm cannons and six 7.62-mm machine guns.

The center section of the lower wing contained two bays for 1- to 10-kg (2.2- to 22-pound) bombs. As with the PBSh-1, the 25- to 250-kg (65- to 550-pound) bombs were strapped beneath the wings for dive-bombing missions, whatever the diving angle might be. But because Il-2 production was stepped up at several factories, all work on the PBSh-2 stopped by the end of 1940.

In some OKB documents the PBSh-2 is referred to as the MiG-6.

Specifications
Span, 8.6/12.4 m (28 ft 2.6 in/40 ft 8.2 in); length, 8.85 m (29 ft 0.4 in); height, 3.5 m (11 ft 5.8 in); wheel track, 2.97 m (9 ft 8.9 in); wheel base, 5.23 m (17 ft 1.9 in); wing area, 10.26/22.14 m² (110.44/238.3 sq ft); takeoff weight, 4,828 kg (10,641 lb); wing loading, 149 kg/m² (30.5 lb/sq ft).

Design Performance
Max speed at sea level, 426 km/h (230 kt); range, 740 km (460 mi); range with two 100-l (26.4-US gal) auxiliary tanks, 929 km (577 mi).

The DIS-200 (T) is sometimes referred to as the DIS-2AM-37 because of its power plant. The engine exhaust pipes extend above the wing's upper surface.

DIS-200 Series

DIS-200 / T

At the end of the 1930s the VVS could operate a fleet of long-range heavy bombers and tactical medium-range bombers but did not have the escort fighters needed to protect them. As early as 1940, the newly formed design bureau had tackled the development of a long-range escort fighter capable of performing high-speed reconnaissance, light bombing, and torpedoing roles. In the preliminary design it was planned to use new, efficient diesel engines developed by A. D. Charomskii, the M-40 and M-30, which would give the aircraft a much greater range because of their low specific fuel consumption.

But neither of these diesels were as yet reliable, so it was decided to equip the prototype—whose factory code letter was T—with two liquid-cooled in-line 1,030 kW (1,400 ch) AM-37s driving three-bladed variable-pitch propellers. The exhaust collectors were bent and extended above the wing upper surface. Fuel was distributed into six tanks, four in the wing center section and two behind the cockpit. Two oil coolers were located on each side of the engine cowlings as on the MiG-3. The glycol was cooled by two air scoops placed on each side of the cowlings, and the corresponding outlets were located beneath the trailing

The air scoops of the engine coolant radiators on the DIS-200 (T) are on each side of both engine nacelles. The corresponding outlets were placed just under the wing's trailing edge.

edge of the wing. The engine supercharger inlets were located in the leading edge.

The single-seat, twin-engine T prototype had a low wing, a twin fin configuration, and a structure composed primarily of wood rather than scarce light alloys. The flight tests led to the installation of slats on the wing leading edge. On the trailing edge, to complement the two-segment flaps, high lift was augmented by specially designed ailerons, which could be symmetrically lowered to 20 degrees. The latter feature was a great novelty at the time; today they are known as flaperons. The main gear (950 x 300 tires) retracted into the engine nacelles and the tail wheel into the fuselage, and both were pneumatically operated—a first in the USSR.

The cockpit was equipped for instrument flying, and the pilot had an oxygen dispenser and a radio receiver at his disposal. The sliding aft canopy could be jettisoned. The front, rear, sides, and underside of pilot's seat were protected by armored plates, and the lower part of the fuselage nose was glazed to give the pilot some downward vision.

The DIS-200 T fighter variant was especially powerful, with one 23-mm VYa cannon in a removable fairing under the nose to complement two 12.7-mm BS and four 7.62-mm ShKAS machine guns on the leading edge of the wing. The cannon fairing could be replaced by a 1,000-kg (2,200-pound) bomb or a torpedo.

The T started its taxiing tests on 15 May 1941 and made its first flight at the end of the month, with A. I. Zhukov at the controls. Flight tests went on all summer at the Khodinka airfield near Moscow and were conducted by Zhukov and V. N. Savkin, an NII VVS test pilot.

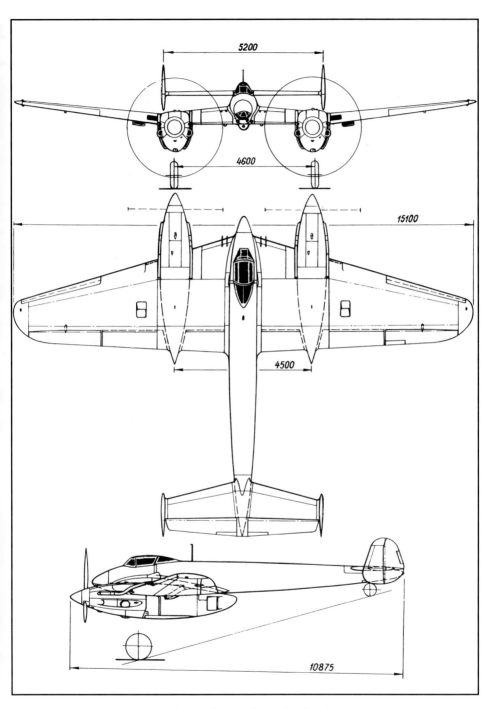

DIS-200 (T) (MiG OKB three-view drawing)

Series production of the T was to have taken place at factory no. 1, but the Germans invaded just as the first tests got under way. The aircraft was moved to Kazan, and flight tests were not continued.

In some OKB documents the fighter version is referred to as the MiG-5 and the bomber version as the MiG-2. DIS stands for *Dalniy Istrebityel Soprovozhdeniya:* long-range escort fighter. The DIS-200 had three competitors: the Taïrov Ta-3 (OKO-6bis), the Polikarpov TIS, and the Grushin Gr-1.

Specifications
Span, 15.1 m (49 ft 6.5 in); length, 10.875 m (35 ft 8.1 in); wheel track, 4.6 m (15 ft 1.1 in); wing area, 38.9 m² (418.7 sq ft); empty weight, 6,140 kg (13,530 lb); takeoff weight, 8,060 kg (17,765 lb); fuel, 1,920 kg (4,230 lb); wing loading, 207.2 kg/m² (42.4 lb/sq ft).

Performance
Max speed, 610 km/h at 6,800 m (329 kt at 22,300 ft); service ceiling, 10,800 m (35,425 ft); range, 2,280 km (1,415 mi); endurance, 5 h; climb to 5,000 m (16,400 ft) in 5.5 min.

DIS-200 / IT

Because neither Charomskii diesels nor Mikulin AM-37 engines were ready for use, the DIS-200 second prototype was powered by an M-82F, a mass-produced radial engine that could afford 1,250 kW (1,700 ch) at takeoff and 957 kW (1,300 ch) at 6,500 m (21,300 feet). The aircraft—built in Kuybyshev, where factory no. 1 and the OKB had relocated in October 1941—was first called the IT (factory code).

Both T and IT prototypes had the same structure. The only difference, except for the engines, concerned the heavier armament: four 7.62-mm ShKAS and two 12.7-mm BS machine guns plus two 23-mm VYa-23 cannons. The T could carry either a 1,000-kg (2,200-pound) bomb or a torpedo of the same weight, depending on mission requirements. The IT rolled out of the factory in January 1942 and made its first flight later that month with G. M. Shiyakov (an LII pilot) at the controls. It was then transferred from Kuybyshev to Kazan, where it was test-flown. Meanwhile, the experience of the war demonstrated that escort missions could be handled successfully by frontal fighters, making the development of a special type of aircraft unnecessary. The IT flight tests were consequently terminated.

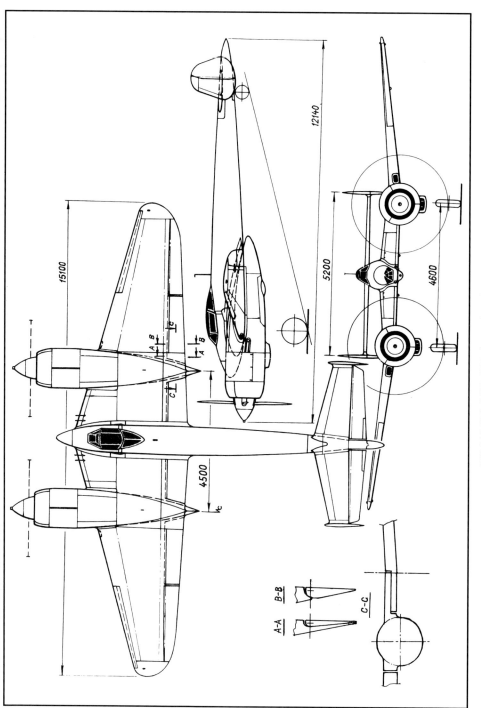

DIS-200 (IT) (MiG OKB three-view drawing)

The DIS-200 (IT) or DIS-2M-82F was the second and final prototype of this program. Because long-range escort fighters were no longer needed, flight tests were terminated in 1942.

Specifications
Span, 15.1 m (49 ft 6.5 in); length, 12.14 m (39 ft 9.9 in); height, 3.4 m (11 ft 1.8 in); wheel track, 4.6 m (15 ft 1.1 in); wing area, 38.9 m^2 (418.7 sq ft); takeoff weight, 8,000 kg (17,630 lb); fuel, 1,920 kg (4,230 lb); wing loading, 205.7 kg/m^2 (42.11 lb/sq ft).

Performance
Max speed, 604 km/h at 5,000 m (326 kt at 16,400 ft); climb to 5,000 m (16,400 ft) in 6.3 min; service ceiling, 9,800 m (32,140 ft); range, 2,500 km (1,550 mi).

A Series

I-220 / A / MiG-11

In 1942 the OKB began to build a string of high-altitude interceptor prototypes (the A series) meant to oppose the Luftwaffe's photo reconnaissance aircraft, which were able to operate with complete impunity because of their high operational ceiling. These Soviet aircraft were in fact updated remakes of the MiG-3.

The I-220 no. 01 was first powered by the AM-38F engine at 1,250 kW (1,700 ch).

The first project was assigned the letter A and the designation I-220. It was a low-wing single-seater of mixed construction. For the first time the OKB designers departed from the MiG-3 layout. All of their previous models had had the exact same span and wing area. The I-220 was different: the radiator was moved to the wing center section (with air intakes in the leading edge and a variable shutter on the wing's upper surface), the main gear legs were given a levered suspension system, its firepower was increased. Two prototypes were built. The I-220 no. 01 received an AM-38F, which was later replaced by an AM-39. The AM-38F generated 1,251 kW (1,700 ch) at takeoff and 1,104 kW (1,500 ch) at rated altitude, while the AM-39 generated 1,325 kW (1,800 ch) at takeoff and 1,104 kW (1,500 ch) at rated altitude. Its armament included two synchronized 20-mm ShVAK (SP-20) cannons above the engine with 150 rpg.

Because flight tests and the development of the AM-39 took longer than expected, the I-220 no. 02 received an engine that was not certified and could not yet be mass produced. Its armament was also different. It was the first Soviet fighter fitted with four synchronized 20-mm ShVAK (SP-20) cannons with 100 rpg and the first to have a whip antenna for its radio set.

The I-220 no. 01 with AM-38F engine was rolled out in June 1943 and made its first flight in July with A. P. Yakimov in the cockpit. Tests continued through August and involved pilot P. A. Zhuravlyev. The I-220 no. 01 with AM-39 engine was rolled out in January 1944 and

All air intakes—engine cooling, oil cooling, heat exchanger—are grouped together in the wing's leading edge of the I-220.

The I-220 no. 01 was reengined in 1943 with the AM-39 at 1,325 kW (1,800 ch).

The I-220 no. 02, powered by an AM-39, was rolled out in August 1944, but the high-altitude engine ratings were rather disappointing.

underwent flight tests between then and August. The I-220 no. 02 with AM-39 engine left the factory in August 1944 and was first flown by I. I. Shelnest in September.

Development of a suitable engine for a high-altitude interceptor intended for PVO regiments proceeded concurrently with the preliminary design of the I-220 and continued until the test flights ended in 1944. In the meantime, the notorious inadequacies of available engines at the edge of the stratosphere greatly complicated the work of the OKB. The wing loading of the I-220 was relatively moderate and marked a significant improvement over the MiG-3 in terms of both maneuverability and rate of climb. But because no existing engine could provide the needed power, the design ceiling—the raison d'être of this program—was never reached. At low and medium altitudes, however, the aircraft proved to be superior to the Yak and La "frontal" fighters then in use. For ground speed the I-220 nos. 01 and 02 were between 30 and 70 km/h (16 and 38 kt) faster than the Yak-9. At 7,000 m (22,960 feet) the no. 02 was 50 to 95 km/h (27 to 51 kt) faster than the La-5. If the I-220 had been mass-produced, it would have been called the MiG-11.

The following details refer to the I-220 no. 01 with the AM-38F engine/with the AM-39 engine.

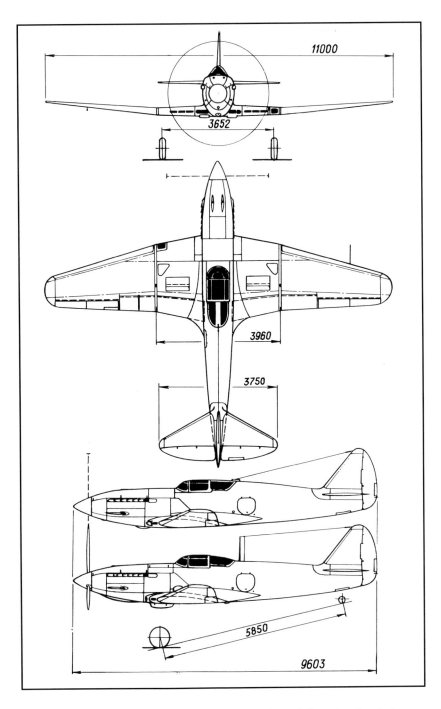

I-220 (A); *bottom,* side view of the I-225 (5A) (MiG OKB four-view drawing)

Specifications

Span, 11 m (36 ft 1 in); length, 9.603 m (31 ft 6.1 in); height, 3.16 m (10 ft 4.4 in); wheel track, 3.652 m (11 ft 11.8 in); wheel base, 5.85 m (19 ft 2.3 in); wing area, 20.38 m² (219.37 sq ft); empty weight, 2,936/3,013 kg (6,471/6,641 lb); takeoff weight, 3,574/3,835 kg (7,877/8,452 lb); fuel, 335 kg (738 lb); oil, 45 kg (99 lb); wing loading, 175.4/188.2 kg/m² (35.9/38.5 lb/sq ft).

Performance

Max speed, 630/668 km/h at 7,000 m (340/361 kt at 22,960 ft); max speed at sea level, 572/550 km/h (309/297 kt); climb to 6,000 m (19,680 ft) with AM-39 in 4.5 min; service ceiling with AM-38F, 9,500 m (31,160 ft); range, 960/630 km (595/390 mi).

The following details refer to the I-220 no. 02 with the AM-39 engine.

Specifications

Span, 11 m (36 ft 1 in); length, 9.603 m (31 ft 6.1 in); height, 3.16 m (10 ft 4.4 in); wheel track, 3.652 m (11 ft 11.8 in); wheel base, 5.85 m (19 ft 2.3 in); wing area, 20.38 m² (219.37 sq ft); empty weight, 3,101 kg (6,835 lb); takeoff weight, 3,647 kg (8,038 lb); fuel, 335 kg (738 lb); oil, 45 kg (99 lb); wing loading, 178.8 kg/m² (36.7 lb/sq ft).

Performance

Max speed, 697 km/h at 7,000 m (376 kt at 22,960 ft); max speed at sea level, 571 km/h (308 kt); climb to 6,000 m (19,680 ft) in 4.5 min; service ceiling, 11,000 m (36,080 ft); range, 630 km (390 mi).

I-225 / 5A

The I-225—last model of the A series, coded 5A—was the heaviest and most powerful experimental interceptor of the family. Its preliminary design was drawn up by a team of engineers headed by A. G. Brunov, project manager. Two I-225s were built in a back-to-basics formula: they had the same dimensions and wing area as the I-220, the family's progenitor. The I-225 no. 01 was powered by an AM-42B engine, the no. 02 by an AM-42FB. The latter power plant provided 1,472 kW (2,000 ch) at takeoff or 1,288 kW (1,750 ch) at rated altitude and also at 7,500 m (24,600 feet). There was a TK-300B exhaust-driven turbo-supercharger on its right side, and it drove a three-bladed AV-5A-22V propeller 3.6 m (11 feet, 9.7 inches) in diameter. The problems that plagued the I-222 and I-224 pressurized cockpit were solved, and the

Two I-225s were built (our photographs show no. 01). The exhaust-driven turbo-supercharger was located on the right side of the engine. These aircraft were powered by the most powerful Soviet engine: the AM-42B, delivering 1,470 kW (2,000 ch) at takeoff.

size of the heat exchanger placed under the back of the engine was reduced.

Protection for the pilot was enhanced by the addition of 9-mm thick armor plate to the back of the seat, and the front and rear parts of the canopy were fitted with 64-mm thick bulletproof glass. The cockpit was equipped for instrument flying and had an ultrashort-wave transceiver.

The armament on both prototypes comprised four synchronized 20-mm ShVAK SP-20 cannons (100 rpg), two mounted above the engine and one fitted on each side. The I-225 no. 01 built in June 1941 was a direct descendant of the I-220 and made its first flight on 21 July 1944 with A. P. Yakimov at the controls. According to design calculations the aircraft ought to have reached 729 km/h at 8,520 m (394 kt at 27,945 feet) at engine combat rating and 721 km/h at 8,850 m (389 kt at 29,030 feet) at rated power. On 7 August 1944 Yakimov clocked up 707 km/h at 8,500 m (382 kt at 27,880 feet) at rated power. Two days later, on its fifteenth flight, the plane experienced engine failure near the ground and crashed, damaged beyond repair.

A second prototype was ordered. However, by the time the I-225 no. 02 commenced flight tests on 14 March 1945 the MiG OKB had turned its attention to the I-220 (N) and its *motokompressor.* In tests the I-225 no. 02 earned the title of second-fastest Soviet piston-engine fighters at 720 km/h (389 kt)—unable to beat the record established on 19 December 1944 by the much lighter Yak-3U (2,830 kg/6,235 lb) powered by a 1,325 kW (1,800 ch) Klimov VK-108.

The I-225 ended all attempts to design a successor for the MiG-3 based on a single layout or structural design. Every member of the MiG fighter family from the I-200 to the I-225 fell victim to endless troubles with its power plant. Besides, after five years of continuous development the maximum speed of this aircraft type had increased by only 80 km/h (43 kt).

Specifications
Span, 11 m (36 ft 1 in); length, 9.603 m (31 ft 6.1 in); height in level flight position, 3.7 m (12 ft 1.7 in); wheel track, 3.652 m (11 ft 11.8 in); wheel base, 5.85 m (19 ft 2.3 in); wing area, 20.38 m² (219.37 sq ft); empty weight, 3,010 kg (6,635 lb); takeoff weight, 3,900 kg (8,595 lb); fuel, 350 kg (770 lb); oil, 41 kg (90 lb); wing loading, 191.2 kg/m² (39.18 lb/sq ft); max operating limit load factor, 8.

Performance
Max speed, 720 km/h at 8,500 m (389 kt at 27,880 ft); max speed at sea level, 590/617 km/h (319/333 kt); climb to 5,000 m (16,400 ft) in 4.5 min; service ceiling, 12,600 m (41,330 ft); landing speed, 134 km/h (72

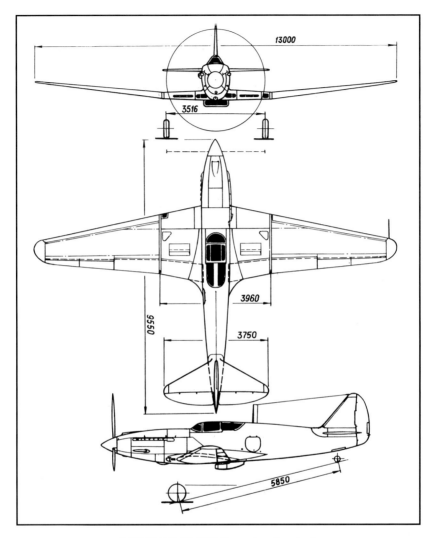

I-221 (2A) (MiG OKB three-view drawing)

kt); range, 1,300 km (810 mi); takeoff roll, 257 m (843 ft); landing roll, 450 m (1,475 ft).

I-221 / 2A

The second series-A high-altitude interceptor was assigned the provisional designation of I-221. There was not much difference between the structures of the I-220 and the I-221 beyond the wing, whose over-

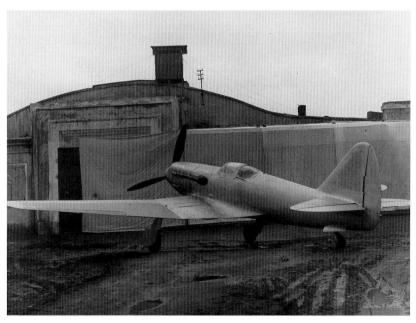

The AM-39B-1 engine of the I-222 drove at first a three-bladed propeller. The TK-300B exhaust-driven turbo-supercharger is clearly visible on the left side of the engine.

all span was raised from 11 m to 13 m (36 feet, 1 inch to 42 feet, 7.8 inches), and the rear fuselage, which on the later model was fashioned out of dural. The cockpit was not only pressurized but also air-conditioned (a first on a Soviet fighter). The heat exchanger was located under the pilot's seat and was interconnected with the air-conditioning system following a sequencing cycle.

The AM-39A offered 1,141 kW (1,550 ch) at takeoff and was fitted with two TK-2B turbo-superchargers that enabled it to maintain its maximum rated output up to 13,000 m (42,650 feet). At 5,200 m (17,055 feet) the engine put out 1,251 kW (1,700 ch). The armament on the I-221 consisted of two synchronized 20-mm ShVAK cannons flanking the engine. The prototype made its first flight on 2 December 1943 with P. A. Zhuravlyev in the cockpit, and flight tests were conducted by A. P. Yakimov, an LII pilot. The test program came to a sudden halt early on when a valve push-rod broke in flight. The engine failed, and the pilot had to bail out.

I-222 / 3A

The third series-A high-altitude interceptor received the provisional designation of I-222. Like the I-221, its cockpit was pressurized and air-conditioned. The aft-sliding canopy was fitted with an inflatable seal bead. The fuselage decking behind the cockpit was cut down to improve the pilot's rear vision. The rear fuselage and outer wing panel structure reverted to wood while the front fuselage, wing center section, flying controls, and flaps remained metallic. The engine coolant radiator and the oil cooler were both located in the leading edge of the wing center section, and the heat exchanger was placed in a ventral bath beneath the back of the engine. Flexible fuel tanks were housed in the wing center section, and the oil tank was incorporated into the wing leading edge. The tailplane area (3.34 m² [35.95 square feet]) and tail fin area (2.01 m² [21.6 square feet]) were slightly larger than those of the I-221. Tire sizes were 650 x 200 for the main gear and 350 x 125 for the tail wheel.

The I-222 was powered by an AM-39B-1 (still under development then) with an experimental exhaust-driven TK-300B turbo-supercharger on the right side of the engine. It afforded 1,288 kW (1,750 ch) at takeoff, 1,104 kW (1,500 ch) at 5,850 m (19,200 feet) and 1,052 kW (1,430 ch) at 13,200 m (43,300 feet), driving a four-bladed AV-9L-26 propeller (three-bladed at first) specially designed for high-altitude flight. The I-222 also featured an engine-driven centrifugal compressor (PTsN) to raise the engine rated altitude. The entire engine weighed

The I-222 was later fitted with a four-bladed AV-9L-26 propeller specially designed for high-altitude flights.

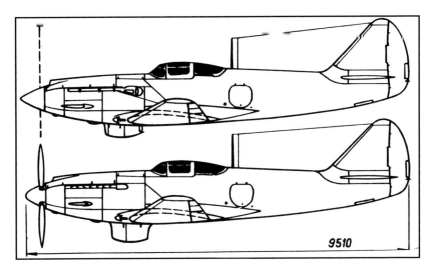

I-222 (3A) and I-224 (4A) (MiG OKB drawing)

1,845 kg (4,066 pounds): 85 kg (187 pounds) for the turbo-supercharger, 190 kg (419 pounds) for the propeller, 128 kg (282 pounds) for the radiator and its glycol, 80 kg (176 pounds) for the heat exchanger, 65 kg (143 pounds) for the oil cooler, 10.5 kg (23 pounds) for the oil tank, 12 kg (26 pounds) for the piping, 23 kg (51 pounds) for the exhaust pipes, and 30 kg (66 pounds) for the engine mount.

In the cockpit, the pilot was protected by an armored seat back, a bulletproof windscreen, and rear glazing. The aircraft's armament comprised two synchronized 20-mm ShVAK cannons (80 rpg) flanking the engine. The I-222 left the factory on 23 April 1944 and was first flown by A. I. Zhukov on 7 May. The aircraft rapidly proved to be capable of outstanding performance with a top speed of 691 km/h (373 kt) and a 14,500 m (47,560 feet) ceiling—record-setting figures at the time for this category of aircraft. But the I-222 had no future because by the time it was ready the German high-altitude reconnaissance flights over the Moscow area had stopped. The air defense regiments no longer needed this type of aircraft.

Specifications
Span, 13 m (42 ft 7.8 in); length, 9.608 m (31 ft 6.1 in); wheel track, 3.516 m (11 ft 6.4 in); wheel base, 5.85 m (19 ft 2.3 in); wing area, 22.44 m² (241.5 sq ft); empty weight, 3,167 kg (6,980 lb); takeoff weight, 3,790 kg (8,353 lb); fuel, 300 kg (660 lb); oil, 40 kg (88 lb); wing loading, 168.9 kg/m² (34.6 lb/sq ft).

Performance
Max speed, 682 km/h at 6,700 m (368 kt at 21,975 ft); 691 km/h at 12,500 m (373 kt at 41,000 ft); climb to 5,000 m (16,400 ft) in 6 min; service ceiling, 14,500 m (47,560 ft); landing speed, 169 km/h (91 kt); range, 1,000 km (620 mi).

I-224 / 4A

Considering its external layout and structure, the fourth offspring of the A family was not very different from the first, the I-220, except that the heat exchanger bath was markedly larger. The I-224 (4A) was powered by an AM-39B engine that offered 1,288 kW (1,750 ch) at takeoff and 1,052 kW (1,430 ch) at 13,100 m (42,970 feet). It was fitted with an exhaust-driven TK-300B turbo-supercharger on the right side and drove a four-bladed AV-9L-22B propeller 3.5 m (11 feet, 5.8 inches) in diameter. ("Four-paddled" is probably more accurate: the blade chord was 400 mm [15.75 inches] at its maximum breadth.) For the first time, the pressurized cockpit was constructed of welded dural sheet. It was also air-conditioned and fitted with an inflatable seal bead. Air for the cockpit was tapped from the supercharger compressor; the cockpit overpressure was 0.3 kg/cm^2 (4.27 psi). Air from the engine radiator was evacuated through four funnel-shaped variable exhausts on the upper surface of the wing. The armament was composed of two 20-mm ShVAK cannons (100 rpg) flanking the engine.

The I-224 rolled out in September 1944, and A. P. Yakimov took it for its first flight on 20 October. Its speed at high altitudes was excellent, but its range (400 km [248 miles] shorter than expected) and its ceiling (400 m [1,310 feet] below that of the I-222) were both disappointments.

Specifications
Span, 13 m (42 ft 7.8 in); length, 9.51 m (30 ft 0.2 in); height, 3.60 m (11 ft 9.7 in); wing area, 22.44 m^2 (242 sq ft); empty weight, 3,105 kg (6,843 lb); takeoff weight, 3,780 kg (8,330 lb); max takeoff weight, 3,921 kg (8,642 lb); fuel, 476 kg (1,049 lb); wing loading, 168.5/174.7 kg/m^2 (34.5/35.8 lb/sq ft); max operating limit load factor, 8.

Performance
Max speed, 693 km/h at 13,100 m (374 kt at 42,970 ft); max ground speed according to TOW, 574–601 km/h (310–325 kt); climb to 5,000

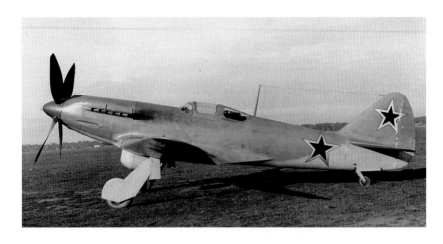

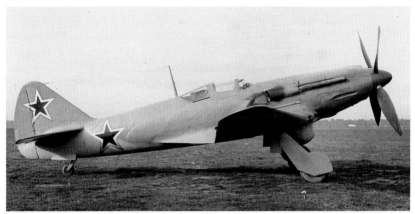

The I-224 is quite similar to the basic aircraft of the family, the I-220. The exhaust-driven turbo-supercharger was moved to the right side of the engine, and the heat exchanger was enlarged. The sizable outlets for the engine radiator on the wing's upper surface are quite noticeable.

m (16,400 ft) in 4.8 min; service ceiling, 14,100 m (46,250 ft); landing speed, 127 km/h (69 kt); range, 1,000 km (620 mi); takeoff roll, 440 m (1,445 ft).

I-250 / N

In early 1944 the GKO (state committee for defense) "gave comrades Yakovlev, Lavochkin, Mikoyan, and Sukhoi the responsibility to build jet aircraft." During the meeting held in Mikoyan's office to discuss this project, all the department managers (who knew very well that the USSR did not have a jet engine in production) decided to power their new fighter with a package unit consisting of a piston engine—a 1,214-kW (1,650-ch) VK-107A—and a special booster nozzle designed at the central institute for aeroengine construction (TsIAM) by a team led by Professor K. V. Kholshchyevnikov. In this rather odd assembly, the power delivered by the piston engine was shared between two "consumers": a propeller 3.1 m (10 feet, 2 inches) in diameter, and a compressor intended to feed a combustion chamber. This assembly was known as a VRDK (*Vozdushno-Reaktivniy-Dopolnitelniy Kompressor:* jet-propelled auxiliary compressor) or booster. For takeoff and cruise situations, the main consumer was the propeller. During these flight modes the compressor (driven by a reduction gear) ran at idle, and no fuel was pumped into the seven burners of the combustion chamber. When dash speed was needed, fuel was injected into the burners and ignited by spark plugs. The compressor was then clutched at its maximum speed, and the revolution speed of the propeller was reduced. This power package offered a total of 2,061 kW (2,500 ch) at 7,000 m (22,960 feet)—that is, 1,067 kW (1,450 ch) from the VK-107A at rated power and 994 kW (1,350 ch) from the VRDK equivalent.

It was also possible to regulate the VRDK power by means of the adjustable nozzle at the rear of the combustion chamber. This nozzle was fitted with two eyelids controlled by hydraulic jacks. The VRDK was cooled by a water-vapor system that protected the fuselage structure and the cockpit against overheating. The water tank had a capacity of 78 l (20.6 US gallons). The VRDK and the VK-107A both ran on gasoline. The VRDK operated like the compressor's second stage of a supercharged piston engine, increasing the aircraft's ceiling.

The wing had a trapezoidal planform and a thin profile. The fuel system was made up of three self-sealing bladder tanks, one in the fuselage in front of the cockpit (412 l [109 US gallons]) and two in the center section of the wing (100 l [26 US gallons] each). The main gear

The first I-250 prototype or N-1 was rolled out on 26 February 1945, just eleven months after the project received the go-ahead.

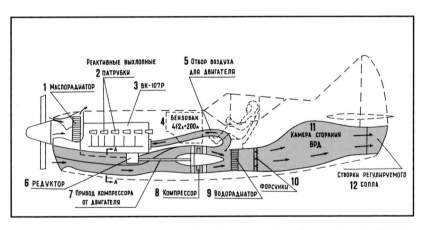

The I-250 combined power unit. (*1*) Oil cooler. (*2*) Engine exhaust pipe. (*3*) VK-107A piston engine. (*4*) Fuel tanks of 412 and 200 l. (*5*) Engine supercharger air-bleed. (*6*) Reduction gearbox. (*7*) Compressor drive shaft. (*8*) Compressor. (*9*) Water jacket. (*10*) Fuel injectors. (*11*) Combustion chamber. (*12*) Adjustable-area nozzle.

was of the levered-suspension type with lower arms leading the leg. The most original features here were the doors, which permanently closed the wheel wells except at the very moment of the gear retraction or extension (a first in the USSR).

Left to right: the VK-107A engine, the compressor drive shaft, and the VRDK compressor.

The N-1 prototype was armed with three 20-mm ShVAK (G-20) cannons—one mounted between the cylinder blocks and firing through the propeller hub, the others flanking the engine. Each of these weapons could fire 100 rounds, a figure eventually improved to 160.

The preliminary design of the I-250 was approved on 28 March 1944. At the same time, Sukhoi set to work on a similar aircraft, the Su-5 (I-107), which also had a combination power plant. But its performance was disappointing, and flight tests of the Su-5 were stopped on 15 July 1945. On 26 October 1944 the full-scale I-250 mock-up was examined in great detail and endorsed; on 30 November almost all the sets of drawings were complete. As the war continued into 1945, OKB officials elected to build not an experimental prototype but instead an operational combat fighter.

Deputy chief engineer M. I. Guryevich had this to say in the OKB house newsletter:

We are now starting work on the N, an entirely new type of aircraft that greatly departs from our previous designs. We are going to reach speeds never achieved before. We are going to use new formulas, new materials. We shall have to create new systems. There is not a single sector, a single department of our OKB that will escape the difficult problems we shall have to face. We must give up all of the old configurations and all of the systems we are

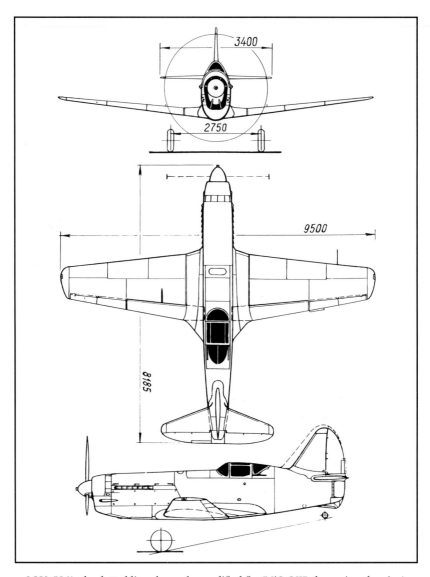

I-250 (N-1); the dotted line shows the modified fin (MiG OKB three-view drawing)

used to and create new ones. We are forging ahead in a new way, the only one enabling us to achieve levels that have never before been attained and that we could hardly have dreamed possible.

The N-1 was rolled out on 26 February 1945, less than one year after the project received the go-ahead, and the N-2 followed three months later. The initial prototype was first flown on 3 March 1945 by A. P. Deyev, and the VRDK was fired as early as the third flight. The

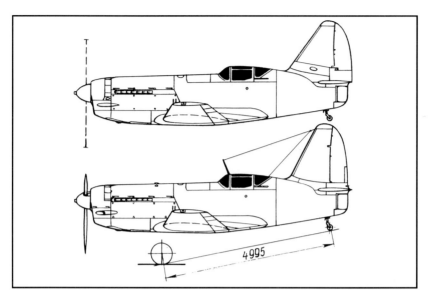

I-250 (N-2) and MiG-13 (MiG OKB drawing)

increase in speed proved to be significant: it was about 100 km/h (54 kt) faster than the best piston-engine fighter flying at the time. The only modification made involved a slight increase of the fin area.

In mid-May 1945 Deyev was killed during a flight test. The structural load factor was somehow exceeded, and the tailplane disintegrated at an altitude far too low for the pilot to bail out. The N-2 trials were conducted by a pilot from the LII, A. P. Yakimov, and later by two young OKB pilots, A. N. Chernoburov and I. T. Ivashchenko. During the first stages of the factory tests, the N-2 remained unarmed. Flight tests showed that, at top speed, the plane's yaw stability was inadequate. The fin area was increased accordingly. Takeoff roll was limited to 400 m (1,310 feet) with the VRDK in operation; but otherwise it took far too long to become airborne owing to the propeller's small diameter. The N-2 was destroyed during a crash landing, with Chernoburov at the controls.

In June 1945 the production factory was instructed to build ten I-250s in time for a Red Square flyover on 7 November in celebration of the twenty-ninth anniversary of the October Revolution. OKB test pilot I. T. Ivashchenko organized an accelerated training program for a batch of air force pilots under the command of Col. P. F. Chupikov. Nine I-250s were completed in time; unfortunately, the weather was so bad in Moscow on 7 November that the flyover had to be canceled.

Sixteen I-250s were delivered in record time to Baltic fleet aviation units at Skultye, near Riga. The first production I-250s had saber-shaped propeller blades; they were replaced later on by conventional

The I-250 N-1 in 1945 with the larger fin.

The N-2, the second I-250 prototype, was rolled out in 1945. The fin was modified once more, and the tail wheel no longer retracts. This aircraft was painted dark blue and embellished with a yellow streak.

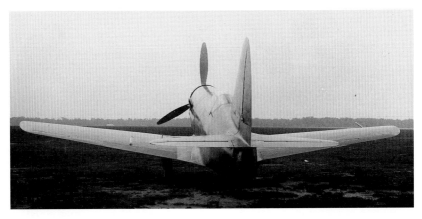

This photograph of the N-2 shows the adjustable-area nozzle, two eyelids controlled by hydraulic jacks.

The production version of the I-250 is sometimes referred to as the MiG-13. In fact, only sixteen aircraft of the type were delivered to the Baltic fleet aviation units.

The air intake duct for both the engine supercharger and the VRDK is visible below the engine mount of the MiG-13. Note also the saber-shaped blades of the propeller.

blades. The production I-250 with serial number 3810102 (381 = factory number, 01 = first series, and 02 = second machine in the series) passed its factory tests in July 1947. Its acceptance trials were carried out at the NII VVS test center between 9 October 1947 and 8 April 1948 with I. M. Sukhomlin at the controls.

All production I-250s—referred to as MiG-13s in some OKB documents—were withdrawn from service in May 1948.

The following details refer to the I-250 N-1.

Specifications
Span, 9.5 m (31 ft 1.9 in); length, 8.185 m (26 ft 10.2 in); wheel track, 2.75 m (9 ft 0.3 in); wheel base, 4.995 m (16 ft 4.6 in); wing area, 15 m^2 (161.5 sq ft); empty weight, 2,587 kg (5,702 lb); takeoff weight, 3,680 kg (8,110 lb); fuel, 450 kg (992 lb); oil, 80 kg (176 lb); water, 78 kg (172 lb); wing loading, 245 kg/m^2 (50.2 lb/sq ft); max operating limit load factor, 6.5.

Performance
Max speed, 825 km/h at 7,000 m (446 kt at 22,960 ft); max speed at sea level, 620 km/h (335 kt); climb to 5,000 m (16,400 ft) with VRDK in 3.9 min, without VRDK in 4.6 min; service ceiling with VRDK, 11,960 m (39,230 ft), without VRDK 10,500 m (34,440 ft); landing speed, 150 km/h (81 kt); range with VRDK, 920 km (570 mi), without VRDK 1,380 km (857 mi); takeoff roll with VRDK, 400 m (1,310 ft); landing roll, 515 m (1,690 ft).

The following details refer to the production I-250 (MiG-13).

Specifications
Span, 9.5 m (31 ft 1.9 in); length, 8.185 m (26 ft 10.2 in); wheel track, 2.75 m (9 ft 0.3 in); wheel base, 4.995 m (16 ft 4.6 in); wing area, 15 m^2 (161.5 sq ft); empty weight, 3,028 kg (6,674 lb); takeoff weight, 3,931 kg (8,664 lb); fuel, 590 kg (1,300 lb); oil, 80 kg (176 lb); water, 78 kg (172 lb); wing loading, 262.1 kg/m^2 (53.7 lb/sq ft).

Performance
Max speed not recorded; unstick speed, 200 km/h (108 kt); landing speed, 190–195 km/h (103–105 kt).

MiG-8 / Utka

A few months before the start of World War II, at the dawn of the jet era, members of the MiG design department started to gather and test various ideas for future aircraft. For this purpose, they decided to build an experimental prototype: an unconventional canard, or tail-first, machine (*canard* is French for duck). This is how Mikoyan and Gurye-vich justified their enterprise in a note enclosed with the preliminary design:

> The canard-tailed aircraft we have designed and that is now being built is an experimental machine intended to check the maneuverability and steadiness in flight of that type of aircraft and to verify the characteristics of highly swept wings. We have chosen the pusher-prop formula because it will enable us to check the low-speed handling with a wing that will not be blown by the propeller. This point is of special interest for aircraft pow-ered by jet engines. The Utka ["duck"] will be a useful tool with which to examine thoroughly all the problems of handling, taxi-ing, takeoff, and landing (including go-around and touch-and-go) without any propeller slipstream effects on control surfaces.

In conceptualizing the Utka or MiG-8, the OKB project engineers had in mind the installation of a jet engine on an airframe of the same layout so that its hot exhaust gases would keep away from all structural elements. The design was prepared in close cooperation with a team of TsAGI technicians. The MiG-8 had a high wing braced by V-shaped struts with a two-spar fabric-covered wooden structure that displayed a 12 percent constant thickness ratio. The wing's forward sweep angle was 20 degrees at the leading edge with a 2-degree anhedral. The fuse-lage, fins, rudders, and canard surfaces were made of wood.

The Utka was fitted with a fixed tricycle landing gear (a first for a MiG aircraft). The cabin had room for three people, with the pilot in front. Lateral and forward visibility was excellent because of the high position of the wing and the fact that the engine was in the rear. The engine bay and the fuselage were aerodynamically well matched.

The elevator was controlled by a rod and a bellcrank, while the rudder and ailerons were controlled by cables. The two duralumin fuel tanks had a total capacity of 195 l (51 US gallons) and were located in the center section of the wing. The 18-l (4.7-US gallon) oil tank was located behind the cabin. The air-cooled Shvetsov-Okromeshko M-11F radial engine offered 81 kW (110 ch) and was entirely cowled with the

On the first MiG-8 Utka the vertical tail surfaces were mounted at the wing tips, and the leading edge was fitted with protruding slots.

The vertical tail surfaces were later moved to midspan, and the slots were removed. The wing tip anhedral is noticeable.

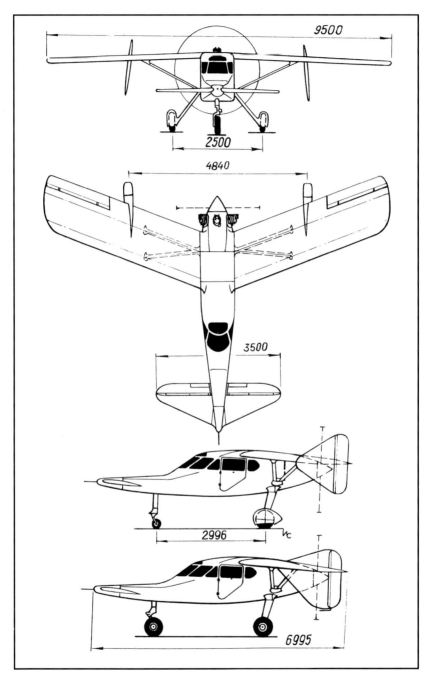

MiG-8 Utka; the two side views show the different locations of the vertical tail surfaces
(MiG OKB four-view drawing)

exception of the cylinder heads. The two-bladed, fixed-pitch propeller was made of wood and measured 2.36 m (7 feet, 8.9 inches) in diameter. Gear legs were constructed of welded metal with pneumatic shock absorbers; the front wheel (300 x 150 initially) had an oleo strut. The wheels of the main gear (500 x 150) were fitted with pneumatic brakes. Later, the wheel sizes were standardized at 500 x 150. The Utka was flown in 1945 for the first time by test pilot A. I. Grinchik. The two fins were then located at the wing tips, and the leading edge was fitted with protruding slots. The two fins were later moved to midspan. The rudders were fitted with balance weights, and the leading edge slots were removed. The OKB also tried out wing tips with a strong anhedral.

The MiG-8 was remarkable for its outstanding stability, refusing to spin even when used at great angles of attack. Many OKB pilots such as A. N. Grinchik, A. I. Zhukov, A. N. Chernoburov, and chief engineer E. F. Nashchyekin spent a great deal of time at its controls. Because of its outstanding flying qualities, safety and ease of handling, and low manufacturing costs, the OKB tried to sell the aircraft to Aeroflot, but the offer was not taken into account. The plane served as the design bureau's liaison aircraft for several years.

Specifications
Span, 9.5 m (31 ft 1.9 in); span of the canard surfaces, 3.5 m (11 ft 5.8 in); length, 6.995 m (22 ft 11.4 in); wheel track, 2.5 m (8 ft 2.4 in); wheel base, 2.996 m (9 ft 10 in); wing area, 15 m^2 (161.5 sq ft); empty weight, 642 kg (1,415 lb); takeoff weight, 997 kg (2,197 lb); max takeoff weight, 1,150 kg (2,535 lb); fuel, 140 kg (309 lb); oil, 14 kg (31 lb); wing loading, 66.5/76.7 kg/m^2 (13.6/15.72 lb/sq ft); aircraft balance, 8% MAC.

Performance
Max speed, 205 km/h (111 kt); landing speed, 77 km/h (42 kt); range, 500 km (310 mi).

MiG-9 Series

I-300 / F

At the end of World War II several OKBs—including MiG—were assigned to design fighters powered by a turbojet engine. MiG had the advantage of experience with its I-250. But above 900 km/h (486 kt) it

The I-300 or *izdeliye* F, prototype of the MiG-9, was the first Soviet jet aircraft to fly— less than three hours before the Yak-15.

was obvious that the technology of this aircraft's combined power plant was completely outdated. And at that time there was not a single homemade jet engine available in the USSR, since all research programs in this field were postponed because of the war. The Soviets had to make do with the few jets recovered either in eastern Prussia or in Germany itself. Near the end of the war one of the factories that made BMW 003 and Junkers Jumo 004 jet engines fell into the hands of Soviet troops. It was then decided to mass-produce them in the USSR. After a close examination of each type the OKB engineers chose the BMW 003, which delivered 784 daN (800 kg st) and 1,568 daN (1,600 kg st) in a twin-jet configuration. A. G. Brunov was named the chief project engineer. In the first preliminary design the future F fighter had two underwing engines like the Messerschmitt Me-262, the Gloster Meteor, and the Sukhoi Su-9, which complied with the same specifications and was first flown in August 1946.

One of Mikoyan's students, the well-known aeronautical expert A. V. Minayev, wrote in *Aircraft of the USSR:*

When Mikoyan started working on his fighter project, a lot of spadework had already been done as regards high-speed aerodynamics, aircraft aerodynamic configurations, and aeroelasticity. The more I go into this period, the more I am amazed at discovering the huge amount of R&D conducted during the war. No Soviet jet aircraft could have flown in 1946 without all this research work, which was all the more valuable to us because it was original and homemade.

Left to right, in front of the F-1, the first prototype of the I-300: A. Karyev, test engineer; G. Buchtinov, trainee mechanic; A. Grinchik, test pilot (he was killed while flying the F-1); and V. Pimyenov, field mechanic.

The preliminary designs for the project were highlighted by an innovative proposal made by Mikoyan: to place both jets side by side into the fuselage. With such an arrangement the wing remained aerodynamically clean, drag was reduced, and maneuverability improved, particularly in the event of engine failure. The preliminary design was approved in the late fall of 1945. A full-scale mock-up was built, and manufacture of the parts and systems was launched. After approval of the mock-up in January 1946 the Narkomaviaprom issued decree no. 157, assigning to the MiG OKB the task of building and flight-testing the aircraft.

The I-300, coded F at the OKB, was an all-metal midwing single-seater with a front air intake to feed the two jet engines. The two-spar straight wing had a TsAGI-1 series profile, a constant thickness ratio of 9 percent, slotted flaps, and Frise ailerons. The lower part of the rear fuselage was protected against the high-temperature exhaust gases by a heat shield (a 15-mm-thick air-gap sandwich of stainless steel plates with a corrugated core). The cantilever horizontal tail was set high on the fin to stay clear of the exhaust flow. Flying controls were standard: the stick rod-operated, the rudder pedals cable-operated. The cockpit was not pressurized.

The fuel system comprised four fuselage and six wing tanks having a total capacity of 1,635 l (432 US gallons). The engines ran on T-2 kerosene—a fuel commonly used in tractor engines—because aviation kerosene was not yet available. Each BMW 003 was fitted with a Ridel

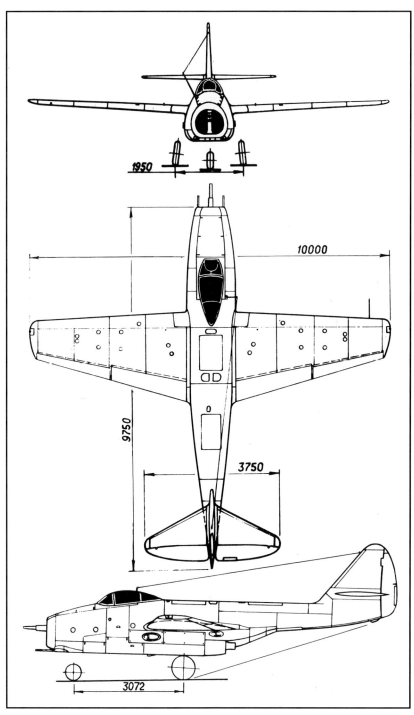

I-300 (F), MiG-9 prototype (MiG OKB three-view drawing)

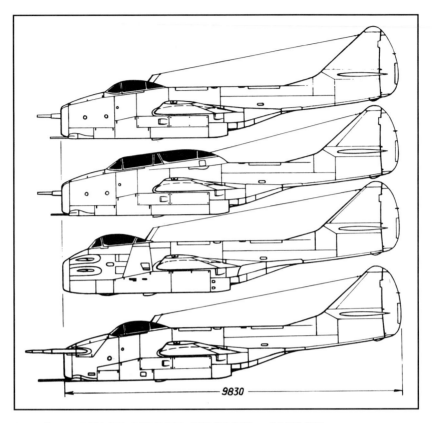

Top to bottom: I-301 (FS), I-301T (FT), MiG-9M (FR), and I-302 (FL)
(MiG OKB drawing)

combustion starter that ran on aviation gasoline. The armament of the
prototype consisted of one 57-mm N-57 (100P) cannon in the air intake
splitting wall and two 23-mm NS-23 (115P) cannons with 80 rpg at the
bottom of the air intake. This arrangement was to prove somewhat
troublesome, as will be explained later. The N-57 was eventually
replaced with a 37-mm N-37 with forty rounds.

    For the first time, a Soviet fighter was fitted with a tricycle landing
gear. The legs and wheels of the main gear retracted outward into wing
wells. The castor front wheel retracted backward into the fuselage and
was fitted with a hydraulic shimmy-damper. All of the landing gear had
levered suspension with lower arms trailing the wheel.

    On 6 March 1946 the prototype was rolled out and entrusted to the
care of the factory flight-testing team. All systems were checked, the
aircraft was weighed without and with fuel, and its engines were run
up. No one actually knew the BMW 003's time between overhauls, so a
hypothetical ten-hour TBO was adopted. On 23 March the aircraft was

Extract from Flight Log No. 19, 24 April 1946

Aircraft type	Pilot	Flight no.	Takeoff/landing time	Flight time	Test engineers
I-300 (MiG-9)	A. N. Grinchik	425	1112/1118	6 min	Poyarkov, Karyev
Yak-(3)-15	M. I. Ivanov	424	1356/1401	5 min	Konukov, Filippov

Source: LII central archives.

moved to the Ramenskoye airfield. Meanwhile, several airframes were stressed during static tests; they were progressively overloaded until they ruptured in order to determine their strength.

An experienced LII pilot, Aleksei Nikolayevich Grinchik (nick-named Lesha), was put in charge of the flight tests of the I-300 (F). In 1941–42 he flew sixty-two missions in a MiG-3 (later he flew an LaGG-3) and fought on the Kalinin front. He was wounded in the leg during a dogfight and forced to make a pancake landing. When he was released from the hospital he was appointed deputy director of flight tests at the LII. In the spring of 1946 there were no more than eleven first-class test pilots in the USSR; Grinchik was the youngest but probably the most experienced. Before the war he was a student at the Moscow Aviation Institute (MAI). When the war ended he joined the TsAGI to add to his knowledge. The flight-test team also included chief engineer A. Karyev and two field mechanics, V. V. Pimyenov and G. Bushtinov.

On 12 April 1946 Grinchik made the first ground rolls, and three days later he lifted it a few feet into the air. At more than 900 m (2,950 feet), the takeoff roll proved to be longer than expected. On 19 April the F climbed to 4 m (about 13 feet), and on 24 April it made its true first flight, which lasted twenty minutes. The first Soviet jet aircraft had flown. Three hours later the Yak-15 went up for its maiden flight.* The USSR's jet era was under way.

The second flight took place on 7 May 1946, and the third flight on the eleventh. Not until the eighth flight was severe buffet noticed around the engines. The next three flights failed to identify their cause and eliminate them. On 5 June an emergency meeting was called in Mikoyan's office to analyze the problem. After listening to the pilot and examining the flight-test data, well-known scientist M. V. Keldish proposed that the present layout of the aircraft be abandoned and the engines relocated either above or underneath the wings. He thought that the vibrations were caused by the stepping of the fuselage behind the engines. But during the twelfth flight on 7 July, after the heat shield was strengthened, the vibrations vanished inexplicably and test flights

---

*A single-engine fighter program that led to development of the Yak-15 and La-150 was launched at the same time as the twin-engine effort that produced the MiG-9 and Su-9.

were finally resumed. Tragically, during the nineteenth flight four days later Grinchik was fatally wounded while giving a flight demonstration for a group of VVS officers. The wing-to-body fairing broke off in flight, smashing the horizontal tail to pieces, and the aircraft crashed. This first I-300, the F-1, had spent a total of six hours and twenty-three minutes in the air.

Two more prototypes, the F-2 and F-3, were assembled quickly in the experimental workshop. The F-3 was flown for the first time by LII pilot M. L. Gallai on 9 August 1946, and the F-2 followed with G. M. Shiyanov at the controls two days later. On 18 August Shiyanov led a flyover in this prototype during the air force day celebrations at Tushino.

By 28 October all of the following GK NII VVS pilots had flown the I-300: A. G. Proshakov, A. M. Kripkov, A. G. Kubishkin, Yu. A. Antipov, and G. A. Sedov (who is today chief constructor at the MiG OKB). The aircraft's joint tests (factory tests plus state acceptance trials) got under way on 26 October, leading to certification. During the state trials, over two hundred aerobatics were performed—including the first spin ever attempted by a jet aircraft—without a single engine failure. The certification document contained the following statement: "Its handling characteristics have made this aircraft, on the whole, easy and pleasant to fly. Its controls are not binding and it is not hard to get accustomed to this machine." Mass production was consequently recommended and launched immediately. The aircraft entered service in the VVS with the service designation of MiG-9.

Specifications
Span, 10 m (32 ft 9.7 in); length, 9.75 m (31 ft 11.8 in); height, 3.225 m (10 ft 6.7 in); wheel track, 1.95 m (6 ft 4.8 in); wheel base, 3.072 m (10 ft 0.9 in); wing area, 18.2 m^2 (195.9 sq ft); empty weight, 3,283 kg (7,236 lb); takeoff weight, 4,860 kg (10,710 lb); fuel, 1,334 kg (2,940 lb); oil, 35 kg (77 lb); wing loading, 267 kg/m^2 (54.7 lb/sq ft); max operating limit load factor, 6.

Performance
Max speed, 910 km/h at 4,500 m (491 kt at 14,760 ft); max speed at sea level, 864 km/h (467 kt); climb to 5,000 m (16,400 ft) in 4.5 min; to 10,000 m (32,800 ft) in 14.3 min; service ceiling, 13,000 m (42,640 ft); landing speed, 170 km/h (92 kt); range, 800 km (497 mi); takeoff roll, 910 m (2,985 ft); landing roll, 735 m (2,410 ft); rate of turn, 9.73°/sec.

The first ten MiG-9s had to be delivered within seventy days; they were delivered in only fifty-five days. This explains the feverish activity in this assembly shop.

Rollout—by hand—of the premier MiG-9. The first ten aircraft were identical to the prototype.

I-301 / FS / MiG-9

During the summer of 1946, the Soviet command authorities decided that the first ten MiG-9s would take part in the flyover at Red Square on 7 November. The builders had no time to lose. The NKAP decree of 28 August 1946 stated: "Our aim being to produce the MiG-9 as soon as possible and to give the pilots time to train and get a feel for the machine, chief constructor A. I. Mikoyan and factory manager V. Ya. Litvinov are assigned the task of producing a small series of this aircraft (ten units)." By 22 October the ten aircraft were completed. They were practically handmade, without any production tooling. On the morning of 7 November, the flyover was canceled because of adverse weather conditions. These first ten machines can be regarded as preproduction aircraft and were in no way different from the prototypes.

The production aircraft I-301 (factory code FS, military designation MiG-9) was different in that RD-20 engines replaced the BMW 003s. The RD-20 was a 100-percent Soviet-made version of the BMW 003. It offered the same thrust, 784 daN (800 kg st), and its mass production was organized by D. V. Kolosov in the Kazan engine factory. The landing gear of the MiG-9 was fitted with more efficient brakes, and its fuel system was equipped with a new type of fuel cell made with a rubberized fabric developed by the VIAM (Soviet institute for aviation materials). During the test flights of the first ten MiG-9s equipped with these cells, no leaks were noted. These cells allowed the engineers to put to use all of the space available in the aircraft structure. Their capacity was of the greatest importance because the engines were so thirsty.

The armament was similar to that of the prototypes: one N-37 with forty rounds and two NS-23s with 80 rpg. The first production aircraft was rolled out on 13 October 1946 and first flown by M. L. Gallai on the twenty-sixth. The first MiG-9s were railroaded to the LII airfield, where they were taken up by GK NII VVS pilots M. L. Gallai, G. M. Shiyanov, L. M. Kushinov, Yu. A. Antipov, A. V. Proshakov, A. V. Kotshyetkov, and D. G. Pikulenko. All these men as well as a few young air force pilots had trained hard to celebrate the October Revolution.

It was not long before the first service evaluation flights revealed the aircraft's design flaws and shortcomings related to defective workmanship. Some of these could be corrected without difficulty, but others were more serious. For instance, when all three guns were fired simultaneously above 7,500 m (24,500 feet), the two jet engines frequently flamed out. It was later discovered that this phenomenon was a distinctive feature of all jet engines, and many years of research were needed worldwide to resolve this problem. It was part of the price an aircraft designer paid for doing without a propeller.

Test flights also demonstrated that jet aircraft needed airbrakes, and that above a speed of 500 km/h (270 kt) the pilot could not bail out. This led to the development of the first ejection seats. Other needs were brought to light as well, such as cockpit pressurization and fire protection in the engine bay. And soon it became obvious that a two-seat training aircraft with the same flight envelope as the single-seater had to be a priority.

The first production aircraft of the I-301 model, with its military livery. Small airbrakes (shown extended) were installed on the wing trailing edge.

This production MiG-9 was experimentally fitted with two drop tanks with a capacity of 235 l (62 US gallons) apiece.

The first jet engines were heavier than piston engines; the advantages of not having a propeller could be appreciated only at high speeds. This explains why the takeoff roll of the MiG-9 was so long: 910 m (2,985 feet), as opposed to 234 m (768 feet) for the MiG-3. And yet the primary goal—to increase flight speed—was fully achieved thanks to the jet engine.

The first two-seater, the FT-1, was not certified because of the poor visibility from the rear seat.

Specifications

Span, 10 m (32 ft 9.7 in); length, 9.83 m (32 ft 3 in); height, 3.225 m (10 ft 6.7 in); wheel track, 1.95 m (6 ft 4.8 in); wheel base, 3.072 m (10 ft 0.9 in); wing area, 18.2 m² (195.9 sq ft); empty weight, 3,420 kg (7,538 lb); takeoff weight, 4,963 kg (10,938 lb); fuel, 1,300 kg (2,865 lb); oil, 35 kg (77 lb); gas, 7 kg (15.5 lb); wing loading, 272.7 kg/m² (55.9 lb/sq ft).

Performance

Max speed, 911 km/h at 4,500 m (492 kt at 14,760 ft); max speed at sea level, 864 km/h (467 kt); climb to 5,000 m (16,400 ft) in 4.3 min; service ceiling, 13,500 m (44,280 ft); landing speed, 170 km/h (92 kt); range, 800 km (497 mi); takeoff roll, 910 m (2,985 ft); landing roll, 735 m (2,410 ft).

UTI MiG-9 / I-301T / FT-1

As mentioned above, the need to train pilots for the MiG-9 forced the OKB to design a two-seat version of the aircraft. An UTI MiG-9 (*Uchebno-trenirovochniy istrebityel:* fighter-trainer) became a priority as soon as the VVS adopted the single-seater—there was no other dedicated aircraft available.

Design of the two-seater MiG-9 was started at the OKB during the summer of 1946, and on 30 October the preliminary design was agreed upon. It was a tandem two-seater, and to make room for the

The earliest Soviet ejection seats, developed by MiG, were tested on the FT-2 by the use of mannequins at first.

second seat in the airframe one of the two fuel tanks in the fuselage had to be removed and the capacity of the other one had to be reduced by one-third.

The front student-pilot cockpit and the rear instructor cockpit were separate and had their own sliding canopies. The aircraft had dual controls, and the instructor could use an intercom system to communicate with the student. The I-301T no. 01 (or FT-1) was assembled with two German BMW 003 engines, a German K-2000 generator, and the wheels and shimmy-damper of an American Bell P-63 Kingcobra fighter.

The first ejection seats developed by the MiG OKB were due to be installed in this prototype. An emergency escape was supposed to work this way: (1) front canopy jettisoned, (2) rear canopy jettisoned, (3) rear pilot ejected, and (4) front pilot ejected. The prototype was also equipped with a new instrument, a Mach indicator (also called a Mach-meter). The two-seater had the same armament as the single-seater: one N-37 cannon whose muzzle was 1.16 m (3 feet, 4.6 inches) away from the engine air intake, and two NS-23 cannons whose muzzles were 0.5 m (1 foot, 7.7 inches) away from that spot.

The FT-1 left the factory in June 1947 and was flown by Gallai in July. In August it underwent its certification tests but failed because of the restricted view from the instructor's cockpit in the rear. The aircraft could not meet the requirement for which it was designed, pilot training. The prototype was later used for improving various MiG-9 systems and developing underwing fuel tanks.

Specifications
Span, 10 m (32 ft 9.7 in); length, 9.83 m (32 ft 3 in); height, 3.225 m (10 ft 6.7 in); wheel track, 1.95 m (6 ft 4.8 in); wheel base, 3.072 m (10 ft 0.9 in); wing area, 18.2 m² (195.9 sq ft); empty weight, 3,584 kg (7,900 lb); takeoff weight, 4,762 kg (10,495 lb); fuel, 840 kg (1,851 lb); oil, 35 kg (77 lb); gas, 7 kg (15.5 lb); wing loading, 261.7 kg/m² (53.6 lb/sq ft).

Performance
Max speed, 900 km/h at 4,500 m (486 kt at 14,760 ft); max ground speed, 830 km/h (448 kt); climb to 5,000 m (16,400 ft) in 5 min; to 10,000 m (32,800 ft) in 10 min; service ceiling, 12,500 m (41,000 ft); landing speed, 190 km/h (103 kt); endurance, 50 min; landing roll, 780 m (2,560 ft).

UTI MiG-9 / FT-2

The second UTI MiG-9, rolled out in August 1947, was a version of the FT-1 modified to answer the objections raised in its certification tests. The FT-2 was powered by two 784 daN (800 kg st) RD-20 turbojets. Other modifications included the following:

—visibility was improved for the pilot seated in the rear
—airbrakes were set in the wing trailing edge
—a FKP S-13 camera gun was positioned in the air intake lip
—wing was piped for two underslung tanks
—the bulletproof windshield panel in front was replaced by a larger one with thinner glazing
—the curvature of the lateral glazed panels was modified to improve visibility
—the glazed partition between the student and instructor cockpits was removed

The FT-2 made its first flight on 25 August 1947. During the second test phase the rear cockpit was modified to try out the first Soviet ejection seat.

After completing its factory tests the FT-2 was moved to the Shchelkovo airfield for its state acceptance trials from 4 to 17 September 1947 at the NII VVS. The aircraft made forty-seven flights and was airborne for fifteen hours and thirty-two minutes. The chief test pilot was Capt. V. G. Ivanov; he was assisted by A. S. Rozanov, captain-engineer. Other GK NII VVS pilots who took part in these tests included

After an important modification to the canopies, the FT-2 was certified as the UTI MiG-9.

Proshakov, Khomyakov, Antipov, Kuvshinov, Skupchenko, Piku-lyenko, Suprun, Teryentyev, Sedov, Alekseyenko, and Trofimov. The UTI MiG-9 no. 02 (FT-2) received its type certification. The panel also recommended the installation of airbrakes and underwing tanks on all versions of the MiG-9.

Operational MiG-9s flew at 900 km/h (486 kt) without a pilot res-cue system; to remedy that situation—and as part of the VVS experi-mental construction program approved by the council of ministers held on 11 March 1947—the MiG OKB was instructed to install an ejec-tion seat on the FT-2 and to subject it to official tests. The ejection seat was developed and installed in 1947–48. In one factory test, a man was ejected at nearly 700 km/h (378 kt).

On 29 September 1948 the FT-2 prototype equipped with the ejec-tion seat was handed over to military test pilots. The seat was placed at an angle of 22.5 degrees in the front cockpit and 18.5 degrees in the rear cockpit. It weighed 128.5 kg (238 pounds). During the first two flights, one at 596 km/h (322 kt) and the other at 695 km/h (375 kt), a mannequin was used. During the third flight on 7 October at 517 km/h (279 kt), the fourth flight on 26 October at 612 km/h (330 kt), and the fifth flight on 13 November at 695 km/h (375 kt), Flight Lt. A. V. Bistrov and his substitute, N. Ya. Gladkov, were ejected. The FT-2 was flown by Capt. V. G. Ivanov. Before these tests started, ten ejections had been carried out on the ground whose accelerations ranged from 8 to 15 g.

On the MiG-9M the location of the three guns was completely revamped. Their muzzles were set back from the air intake plane.

Specifications

Span, 10 m (32 ft 9.7 in); length, 9.83 m (32 ft 3 in); height, 3.225 m (10 ft 6.7 in); wheel track, 1.95 m (6 ft 4.8 in); wheel base, 3.072 m (10 ft 0.9 in); wing area, 18.2 m² (195.9 sq ft); empty weight, 3,460 kg (7,626 lb); takeoff weight, 4,895 kg (10,788 lb); crew, 180 kg (397 lb); fuel, 862 kg (1,900 lb); gas, 14 kg (31 lb); oil, 22 kg (49 lb); armament, 205 kg (452 lb); ammunition, 116 kg (256 lb); removable equipment, 36 kg (79 lb); wing loading, 269 kg/m² (55.1 lb/sq ft).

Performance

Max ground speed, 810 km/h (437 kt); climb to 5,000 m (16,400 ft) in 5.3 min; service ceiling, 12,000 m (39,360 ft); landing speed, 180 km/h (97 kt); range, 775 km (480 mi); design range with two 235-l (62-US gal) drop tanks, 920 km (571 mi); takeoff roll, 835 m (2,740 ft); landing roll, 775 m (2,540 ft).

MiG-9M / I-308 / FR

The engine flameout that occurred when all three cannons were fired at once puzzled OKB engineers and led them to examine the stability of the combustion process at that moment. This mystery was solved by degrees. It was thought that the solution lay in shifting the cannon

The MiG-9M's left armament bay. The lower NS-23 has been removed. The N-37 cannon was on the right side of the aircraft's nose.

muzzles behind the engine air intake plane. On the FR, all three cannons were moved aft: the N-37 was relocated to the right side of the fuselage, and both of the NS-23s were placed on the left side. This new arrangement entailed a few structural modifications of the nose section. The two RD-20s were replaced by RD-21s built at an OKB man-

The MiG-9M and the first production MiG-9s were fitted with the airbrakes tested on the UTI MiG-9.

aged by D. V. Kolosov. This was basically a "hotted-up" RD-20 rated at 980 daN (1,000 kg st).

The FR was also equipped with airbrakes first tested on a UTI MiG-9 as well as a pressurized cockpit. Five of the six original fuel tanks were retained—including a 100-l (26-US gallon) trim tank—but the total capacity remained unchanged at 1,300 kg (2,865 pounds).

The first FR was rolled out in June 1947 and flown in July by V. N. Yuganov. The data recorded during the flight tests showed that it was the first MiG to exceed M 0.8. Thanks to the greater thrust of the RD-21, the level-speed increase reached 55 km/h (30 kt). The rate of climb also improved: the MiG-9M climbed to 5,000 m (16,400 feet) in 2 minutes 42 seconds, 1 minute 36 seconds faster than any other aircraft in the same category.

The MiG-9M served as a basic model for the design of both the FL and the FN, two more powerful and sturdier versions that were built but never flown.

Specifications
Span, 10 m (32 ft 9.7 in); length, 9.83 m (32 ft 3 in); height, 3.225 m (10 ft 6.7 in); wheel track, 1.95 m (6 ft 4.8 in); wheel base, 3.072 m (10 ft 0.9 in); wing area, 18.2 m² (195.9 sq ft); empty weight, 3,356 kg

The FP marked another attempt to end the engine flameout problems that occurred when the cannons were fired simultaneously. The N-37 cannon was moved from the air intake partition wall to the left upper part of the nose.

(7,397 lb); takeoff weight, 5,069 kg (11,172 lb); fuel, 1,300 kg (2,865 lb); wing loading, 278.5 kg/m² (57.1 lb/sq ft); max operating limit load factor, 5.5.

Performance
Max speed, 965 km/h at 5,000 m (521 kt at 16,400 ft); max speed at sea level, 850 km/h (459 kt); climb to 5,000 m (16,400 ft) in 2.7 min; service ceiling, 13,000 m (42,640 ft); landing speed, 166 km/h (90 kt); range, 830 km (515 mi); takeoff roll, 830 m (2,720 ft); landing roll, 700 m (2,295 ft).

MiG-9 / I-302 / FP

The FP was designed solely to control the harmful effects of cannon fire on the combustion stability of the engine. It differed from the FS in only one point: the N-37 cannon was moved from the air intake splitting wall to the left upper part of the fuselage nose. This new arrangement did not solve the problem.

It might have become a single-engine MiG-9: the I-305 (FL) was to be powered by one Lyulka TR-1A turbojet. Unfortunately, the engine burst on its test bench, ending the FL project.

I-305 / FL

One MiG-9 airframe was to be reengined with a single 1,470-daN (1,500-kg st) TR-1A turbojet developed by A. M. Lyulka. While matching the performance level of the production model, the design takeoff weight of the FL was 350 kg (770 pounds) lower.

To fit the TR-1A—the first jet engine developed and built entirely in USSR—into the airframe in place of the two BMW 003s, the tail section of the fuselage had to be modified. Moreover, the engineers left room for an afterburner then in the works that would boost the thrust to 1,960–2,450 daN (2,000–2,500 kg st).

The I-305 was an important aircraft first and foremost because of its built-in potential. The cannon arrangement was again modified; all three arms were now on the same horizontal plane. One experimental N-37 (120P) with forty-five rounds occupied the middle space, with one NS-23 (115P) with 80 rpg on either side. The I-305 featured a pressurized cockpit and an ejection seat. To improve the aircraft's operational efficiency, most of the systems were to have been upgraded with the RSIU-10 transceiver, the Baryum-1 IFF, the NI-46 ground position indicator, and the Ton-3 direction finder.

The I-305 airframe was almost completed at the end of 1947. Unfortunately, the TR-1A turbojet burst soon afterward on the test

For the FL, the armament arrangement was modified once more. The three cannons were placed on the same horizontal plane.

bench. At that time a brand-new aircraft, the I-310 S (commonly known as the MiG-15) looked like a more promising venture. As a result the FL was discontinued.

Specifications
Span, 10 m (32 ft 9.7 in); length, 9.7 m (31 ft 9.8 in); height in level flight position, 3.2 m (10 ft 9.9 in); wheel track, 1.95 m (6 ft 4.8 in); wheel base, 3.32 m (10 ft 10.7 in); wing area, 18.2 m² (195.9 sq ft); takeoff weight, 4,570 kg (10,072 lb); fuel, 1,485 l (386 US gal); wing loading, 261 kg/m² (53.5 lb/sq ft).

Performance
Max speed, 885 km/h at 5,000 m (478 kt at 16,400 ft); max speed at sea level, 897 km/h (484 kt); climb to 5,000 m (16,400 ft) in 4.86 min; to 10,000 (32,800 ft) in 13.24 min; service ceiling, 13,400 m (43,950 ft); landing speed, 155 km/h (84 kt); range at 10,000 m (32,800 ft), 1,050 km (652 mi); takeoff roll, 815 m (2,675 ft); landing roll, 665 m (2,180 ft).

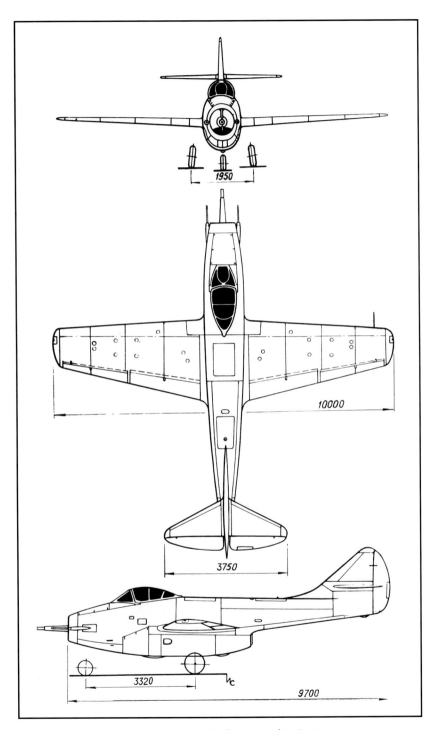

I-305 (FL) (MiG OKB three-view drawing)

I-320 / FN

The decision was made in 1947 to replace the two RD-20 engines with one of the twenty-five Rolls-Royce Nene-1 engines just purchased in Great Britain. This turbojet delivered 2,185 daN (2,230 kg st). It had a centrifugal compressor, nine separate combustion chambers, and a single-stage turbine. Later this engine developed into the RD-45 and RD-45F built in factory no. 45 (hence its designation) in the USSR with a thrust of 2,224 daN (2,270 kg st). On this version the armament arrangement was again modified. The N-37 cannon was moved to the left lower part of the nose, and its muzzle did not jut out ahead of the engine air intake. The two NS-23s flanked the fuselage nose and were also set back from the air intake plane.

Like the FL, the FN was never completed because of the promise of the I-310, whose flight tests started on 30 December 1947. By 1948 the MiG-9 was clearly obsolete.

Specifications
Span, 10 m (32 ft 9.7 in); length, 10.88 m (35 ft 8.3 in); wheel track, 1.95 m (6 ft 4.8 in); wheel base, 3.155 m (10 ft 4.2 in); wing area, 18.2 m² (195.9 sq ft).

I-307 / FF

The FF program was an update of the production MiG-9 or FS that boosted the engines' thrust and reinforced the pilot's armor protection.

The BMW 003 turbojets were retooled in factory no. 17 in Kazan. This modification was at the heart of the RD-20F, later redesignated the RD-21 and built entirely with components and accessories made in the USSR. The thrust of the hotted-up engine increased to 980 daN (1,000 kg st). The brakes were also improved, and 12-mm armor plates were installed in the front and rear of the cockpit, and the windshield front glass panel was replaced by a pane of bulletproof glass 44 mm thick. The total weight of the armor was 60 kg (132 pounds). Wheel braking was much improved. Externally very similar to the FS, the FF left the factory and made its first flight in September 1947, I. T. Ivashchenko at the controls. Tests were completed by the end of the year, and the FF was produced in small quantities with the same armament as the FS.

Specifications
Span, 10 m (32 ft 9.7 in); length, 9.83 m (32 ft 3 in); height, 3.225 m (10 ft 6.7 in); wheel track, 1.95 m (6 ft 4.8 in); wheel base, 3.072 m (10 ft

0.9 in); wing area, 18.2 m² (195.9 sq ft); empty weight, 3,471 kg (7,650 lb); takeoff weight, 5,117 kg (12,278 lb); fuel, 1,300 kg (2,865 lb); wing loading, 281.2 kg/m² (57.6 lb/sq ft).

Performance
Max speed, 950 km/h (513 kt); climb to 5,000 m (16,400 ft) in 2.9 min; service ceiling, 13,000 m (42,640 ft).

I-307 / FF with Babochkoi

Yet another attempt was made at overcoming the engine flameout problem. A large hollow plate was centered on the N-37 barrel square with the engine air intake plane. The resulting shape resembled (if one used one's imagination) a *babochka,* or butterfly.

Hot gases from the cannon's muzzle were sucked in a slot in the plate's leading edge and then vented through other slots at the top and bottom of the plate. In theory, that would neutralize the effects on the airflow of the temperature rise at the air intake level, which was disrupting the engine combustion stability, and the engines could not flame out anymore. Tests carried out at the end of 1948 (as production of the MiG-9 came to an end) showed that the "butterfly" had few positive effects—in fact, it increased the plane's drag and reduced yaw stability.

Yet another attempt to end the engine flameout when firing the guns, the "butterfly" was of doubtful effectiveness.

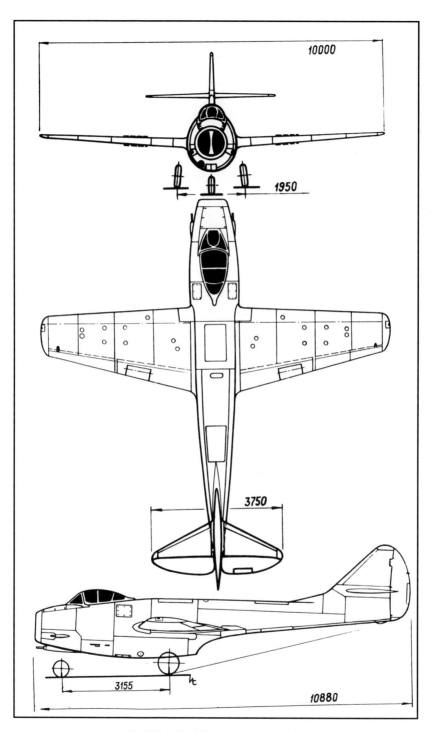

I-320 (FN) (MiG OKB three-view drawing)

103

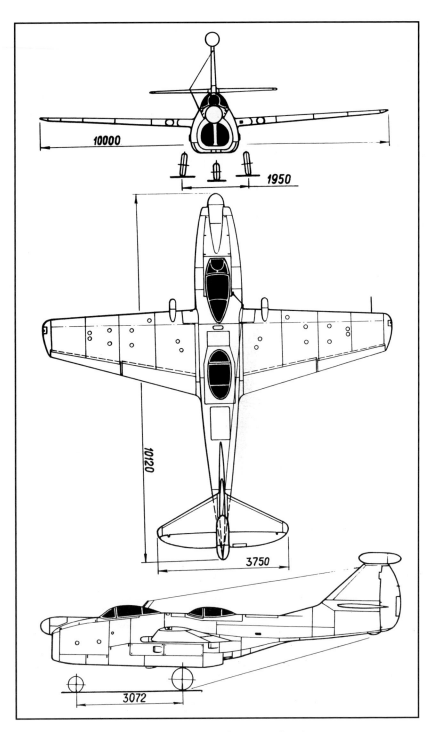

MiG-9L (FK) (MiG OKB three-view drawing)

104

The FK was a modified FS used as a two-seat test bed for the guidance system of the KS-1 missile.

The FK bristled with transmitting and receiving antennae—above the engine air intake, on the leading edge of the wing, and on top of the fin. The test engineer was seated in the rear cockpit.

The KS-1 air-to-surface missile was developed for the antiship role for use by Tu-16 bombers. A piloted model was built to assess its flight characteristics. Its landing speed was clocked at 400 km/h (216 kt).

MiG-9L / FK

This two-seat flying laboratory was basically an FS airframe that had been modified to develop the guidance system of the KS-1 Komet. This air-to-surface missile was designed to be launched from a Tu-16KS-1 bomber (one unit under each wing) in its antiship role.

The missile guidance system operator was seated in the unpressurized rear cockpit. Like the KS-1, the MiG-9L had two radar antennae. The first one above the engine air intake was used to illuminate the target, and the signals reflected back were picked up by two receivers on the wing leading edge, on either side of the cockpit. The other one, which both transmitted and received signals and was located at the top of the fin, was used to develop the guidance systems of both the launching aircraft and the missile. During the test phase, the launching aircraft was a Tu-4 bomber. This flying laboratory was reequipped in 1949 to test new radar guidance systems for four years.

Specifications
Span, 10 m (32 ft 9.7 in); length, 10.2 m (33 ft 2.4 in); wheel track, 1.95 m (6 ft 4.8 in); wheel base, 3.072 m (10 ft 0.9 in); wing area, 18.2 m^2 (195.9 sq ft).

I-270 / Zh

In the history of aviation transitional periods are always marked by attempts to develop more powerful engines. The first turbojets offered only meager thrust. For example, the first Soviet jet—the TR-1, designed and built by A. M. Lyulka—delivered power equivalent to 1,323 daN (1,250 kg st) and was not available until 1947. The wartime German jets Jumo 004 and BMW 003 had thrusts limited to 880 and 785 daN (900 and 800 kg st) respectively. Therefore, during this transitional period several aircraft manufacturers including Mikoyan and Guryevich decided to test the efficiency of rocket engines or ZhRD (*Zhidkostniy Raketniy Dvigatyel:* liquid-propellant rocket engine) for a new type of high-speed, high-altitude interceptor. At that time, only the rocket engine could meet those two requirements. One of the basic advantages of rocket engines is that their thrust is slightly subordinated to speed and altitude, two values that then depended only on the amount of combustible and oxidizer the interceptor could carry in its tanks.

The first thing that catches one's eye about the three-view drawing of the Zh is the T-tail (the stabilizer is on top of the fin). In a note dated 30 May 1946 that was included with the preliminary design, Mikoyan and Guryevich wrote: "If one reduces the effect of the wing on the stabilizer, it may be supposed that the moment characteristics will not be modified up to Mach 0.9. This is why the stabilizer has been moved upward; in relation to the wing, this displacement is equal to 1.2 MAC (mean aerodynamic chord)." Similar high-set stabilizers would appear later on transonic aircraft such as the MiG-15, MiG-17, and I-320. But in 1946 TsAGI had not yet studied the characteristics of swept-wing aircraft, and manufacturers were not yet equipped with the necessary experimental and scientific facilities. This is why both I-270 prototypes built at the end of 1946 had a straight wing. The sweep angle at the leading edge was 12 degrees. Only the stabilizer was swept back (30 degrees at the leading edge) as shown in the preliminary design.

The I-270 was an all-metal aircraft with a circular semimonocoque fuselage and a cantilever midwing. The fuselage was built in two parts and then mated (a first for MiG). The one-piece wing had a five-spar box structure and thick skin panels and was embedded in the lower part of the fuselage. Its laminar flow airfoil was relatively thin. The preliminary design called for a wing with a 20-degree quarter-chord sweep angle identical to that of the MiG-8, but as noted both prototypes received a straight wing with a modest taper to both the leading and trailing edge. The main gear had a very narrow wheel track (1.6 m

The first prototype of the I-270, or Zh-1, made its first flight without its power plant, towed behind a Tu-2. It was then released and allowed to glide to a landing.

The Zh-1 was somewhat short-lived. Test pilot Yuganov had to make a belly landing, and the aircraft was thought to be beyond repair.

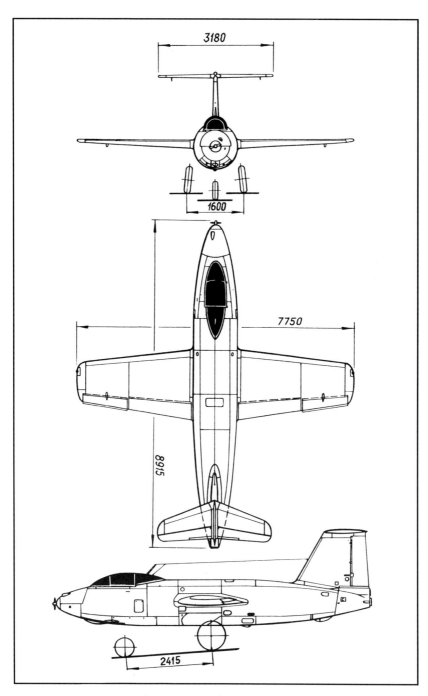

I-270 (Zh-2) (MiG OKB three-view drawing)

[5 feet, 3 inches]) and retracted inward into the wing center section. The nose gear well and the two 23-mm NS-23 cannons (40 rpg) were located under the pressurized cockpit.

It was planned to put two rocket pods (four RS-82s each) under the wing, but the idea was not accepted. The seat was equipped with a small pyrotechnic device to enable the pilot to eject in case of emergency. The power plant was a dual-chamber RD-2M-3V bipropellant rocket engine developed by L. S. Dushkin and V. P. Glushko. The combined thrust of both chambers was 1,421 daN (1,450 kg st)—that is, 1,029 daN (1,050 kg st) for the main chamber and 392 daN (400 kg st) for the cruise chamber mounted on top. The rocket engine was pump-fed by a mixture of nitric acid, kerosene, and hydrogen peroxide (80 percent). The propellants weighed 2,120 kg (4,672 pounds). All propellants were stored in three sets of tanks: 1,620 kg (3,570 pounds) of nitric acid in four tanks, 440 kg (970 pounds) of kerosene in one tank, and 60 kg (132 pounds) of hydrogen peroxide in seven tanks. The propellant turbopumps were driven by two generators: one that was part of the aircraft's electrical system; and a second, powered by the wind-milling action of a small propeller in the nose, that served as a backup.

The first prototype or Zh-1 was rolled out at the end of 1946 without its power plant and made a few flights in December 1946 towed behind a Tu-2 bomber. During these tests the bomber released it, and it glided to landings. Before these first glide descents, pilots trained on a modified Yak-9 fighter that was weighted with lead ingots to approximate the design yaw and pitch characteristics of the I-270. The RD-2M-3V rocket engine was mounted on the second prototype or Zh-2. In early 1947 an OKB test pilot, V. N. Yuganov, used the engine in flight. He handed responsibility for the test flights over to a military pilot, A. K. Pakhomov, who shortly thereafter botched a landing and destroyed the aircraft. Within weeks Yuganov belly-landed the Zh-1, and the prototype was not repaired. The I-270 never made it past its factory tests. And with the MiG-9, the Yak-15, and the first surface-to-air missiles in operation, the rocket-powered interceptor was no longer essential to the air defense forces. So work came to a halt on the I-270 and the RM-1, a similar type of interceptor designed by A. S. Moskalyev.

Specifications
Span, 7.75 m (25 ft 5.1 in); length, 8.915 m (29 ft 3 in); height, 3.08 m (10 ft 1.3 in); height in level flight position, 2.58 m (8 ft 5.6 in); wheel track, 1.6 m (5 ft 3 in); wheel base, 2.415 m (7 ft 11.10 in); wing area, 12 m^2 (129.2 sq ft); empty weight, 1,546 kg (3,407 lb); takeoff weight, 4,120 kg (9,080 lb); propellants, 2,120 kg (4,672 lb); wing loading, 343.3 kg/m^2 (70.4 lb/sq ft).

The second prototype of the I-270, or Zh-2, had a standby electrical system that was powered by the windmilling action of the small propeller in its nose.

This photograph of the Zh-2 shows the two superimposed chambers of the RD-2M-3V rocket engine.

Performance

Max speed, 900 km/h at 5,000 m (486 kt at 16,400 ft), 928 km/h at 10,000 m (501 kt at 32,800 ft), 936 km/h at 15,000 m (505 kt at 49,200 ft); climb to 10,000 m (33,800 ft) in 2.37 min; to 15,000 m (49,200 ft) in 3.03 min; service ceiling, 17,000 m (55,760 ft); landing speed, 137 km/h (74 kt); takeoff roll, 895 m (2,935 ft); landing roll, 493 m (1,617 ft); endurance with both chambers, 255 sec; with the cruise chamber only, 543 sec.

MiG-15 Series

MiG-15 / I-310 / S (S-01 and S-02)

By 1947 every avenue that promised to increase the thrust of the RD-10 turbojets had been explored. The TR-1 was not fully developed and therefore could not power a fighter prototype. A liquid-propellant rocket engine (ZhRD) like that of the I-270 (Zh) could not be used in a combat aircraft because of its short operating time. Thus there was an urgent need for a powerful and reliable turbojet.

A year earlier sixty Rolls-Royce turbojets were ordered from Great Britain. Half were Derwent Vs (1,158 daN/1,590 kg st), while the others were Nene Is (2,185 daN/2,230 kg st) and Nene IIs (2,225 daN/2,270 kg st). For their relatively lightweight fighters the Yakovlev and Lavochkin OKBs chose the Derwent V, a lighter engine (565 kg [1,245 pounds]) that would later be built in the Soviet Union as the RD-500. But for his projects A. I. Mikoyan selected the Nene I, a more powerful but also at 720 kg (1,587 pounds) a much heavier engine. It too was later produced in the Soviet Union, where it was referred to as the RD-45.

A. G. Brunov, deputy general designer, and A. A. Andreyev, chief engineer, were entrusted with the management of the program. Several TsAGI experts also took part in the preliminary research effort: S. A. Khristianovich, G. P. Svitshchev, V. V. Struminskiy, and P. M. Krassilshchikov. Several types of wing shapes—swept wing, straight wing, and even forward-swept wing—were tested in the TsAGI wind tunnels. At that time the swept wing was not favored for fast aircraft, as is shown by German and English jets designed between 1943 and 1946.

As early as March 1947 wind-tunnel tests indicated that a swept wing with fences was probably the right answer. The TsAGI engineers quickly discovered how to control the transverse stability and master the airflow breakdown. The optimum sweep angle for the wing of the

The MiG-15 at its debut was an Anglo-Soviet hybrid. This photograph shows the S-01 when it was still only the I-310.

The S-01's sliding canopy featured a thin arch in the middle.

future fighter was calculated to be 35 degrees at 25 percent chord with a 2-degree anhedral from the wing roots. The four upper-surface wing fences solved the problem of airflow straightening. But despite its obvious simplicity, the final design of the S-01 was rather unorthodox.

From the start, pilot comfort was made a high priority. The cockpit was pressurized and air-conditioned, with a canopy that offered an excellent all-around view. The aircraft was fitted with an ejection seat. The mechanical flying controls were statically and aerodynamically balanced at a time when hydraulic servo-controls did not yet exist. A high degree of serviceability was also considered important. Its structure and systems subjected to thorough research, the S aircraft was the result of a marriage of the rational and the useful. It was not by chance that the general layout of the I-310 (S)—the future MiG-15—was recognized as a classic and used for several Soviet aircraft (and even by other nations) during the 1950s. Its preliminary design allowed for future updates linked to the development of new power plants, armament, and equipment.

The I-310—founder of a great family of experimental and production machines—proved to be one of the best combat aircraft of the second postwar generation. Its top-notch performance is attributable to its optimum basic wing load, high thrust-to-weight ratio, easy-to-service armament, advanced structural technology, sturdy levered-suspension main landing gear, and reliable engine.

The main features of this all-metal aircraft included a 35-degree swept wing with four fences, a pressurized and air-conditioned cockpit, an ejection seat (the canopy was jettisoned first), and a two-section fuselage. The armament included three cannons: one N-37D and two NS-23s arranged at first like those of the I-305 (FL) with all three muzzles on the same horizontal plane near the engine air intake. For the first time on a Soviet fighter, fire warning and extinguishing systems were standard. Also for the first time on a fighter, the aircraft was fitted with an OSP-48 instrument landing system that included an ARK-5 automatic direction finder with a range of 200 km (125 miles), an RV-2 two-level radio-altimeter, and an MRP-48 marker receiver. Mating the two sections of the fuselage at the no. 13 bulkhead allowed for easy access to the engine, its accessories, and its exhaust nozzle, facilitating engine removal or installation. Mating the fuselage to the wings by means of attachment fittings meant that the aircraft could be assembled or disassembled quickly in field maintenance conditions and that, once taken apart, it could be transported in containers carried by ship, train, or another aircraft.

Assembly of the S-01 was stopped without notice as unexplained flameouts continued to hamper the development of the MiG-9. Engineer N. I. Volkov, with the cooperation of MiG armament specialists,

The second prototype or S-02 was equipped with small rocket engines beneath the wing to counter any spin, intentional or not, which could prove risky during test flights.

A close-up of the antispin rocket used for the MiG-15 tests.

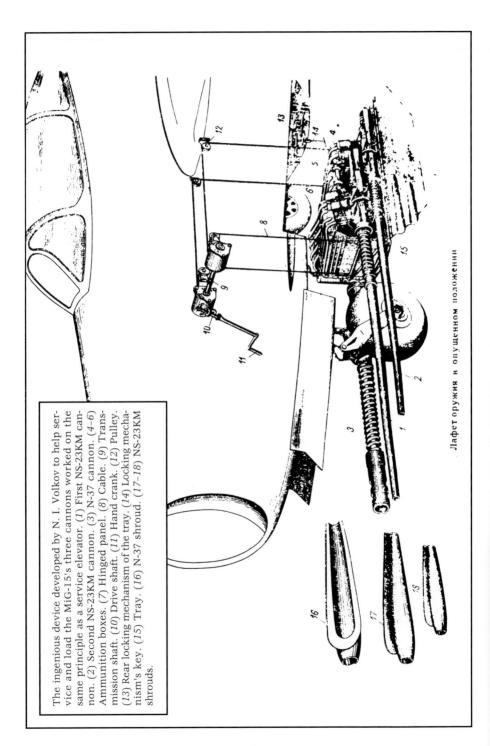

The ingenious device developed by N. I. Volkov to help service and load the MiG-15's three cannons worked on the same principle as a service elevator. (1) First NS-23KM cannon. (2) Second NS-23KM cannon. (3) N-37 cannon. (4-6) Ammunition boxes. (7) Hinged panel. (8) Cable. (9) Transmission shaft. (10) Drive shaft. (11) Hand crank. (12) Pulley. (13) Rear locking mechanism of the tray. (14) Locking mechanism's key. (15) Tray. (16) N-37 shroud. (17-18) NS-23KM shrouds.

Лафет оружия и опущенном положении

116

Slipper tanks of various sizes were tested on the S-02.

The great simplicity of the I-310 instrument panel is noticeable.

proposed a revolutionary rearrangement of the cannons. He built a single tray for the three cannons, ammunition boxes, cartridge cases, and link outlet ports. This tray was embedded under the nose and could be lifted or lowered by four cables controlled by a hand crank, a drive shaft, and four pulleys—like a small service elevator. The idea seemed so inspired that it was immediately approved for use on the I-310. The system made the cannons easier to load and service and also reduced the aircraft's turnaround time when missions had to be flown at close intervals.

The front part of the S-01 was modified to accommodate this tray. Finished at last, the S-01 was rolled out on 27 November 1947 and made its maiden flight on 30 December with test pilot V. N. Yuganov at the controls. But right away sizable losses of thrust were recorded, and all flights had to be canceled. To remedy these losses, TsAGI and TsIAM engineers proposed reducing the length of the fuselage and the exhaust pipe slightly. This change necessitated modifications to the ailerons, the wing chord, and the sweep angles of the tail unit (which were increased). The well-known silhouette of the MiG-15 was not created in one pass.

The second prototype or S-02 joined the test program before long and flew for the first time on 27 May 1948, powered by a 2,225 daN (2,270 kg st) Rolls-Royce Nene II. The state trials of the S-01 and S-02 were carried out at the GK NII VVS in two stages, from 27 May to 25 August and from 4 November to 3 December. The report concluded, "The I-310 has passed its state acceptance trials; its performance was in accordance with calculations; and the preparation of the preliminary design for a two-seat version for pilot training [the UTI MiG-15] is recommended." Test pilots who flew the I-310 were unanimous in their praise of the aircraft's handling characteristics while taking off, climbing, and landing as well as its steadiness in flight and its maneuverability. In August 1948 the council of ministers of the USSR decided to order the I-310 for the VVS. It was given the military designation MiG-15.

The following details refer to the S-01.

Specifications
Span, 10.08 m (33 ft 1 in); overall length, 10.102 m (33 ft 1.7 in); fuselage length, 8.125 m (26 ft 7.9 in); wheel track, 3.852 m (12 ft 7.6 in); wheel base, 3.075 m (10 ft 1.1 in); wing area, 20.6 m² (221.7 sq ft); empty weight, 3,380 kg (7,450 lb); takeoff weight, 4,820 kg (10,623 lb); fuel, 1,210 kg (2,667 lb); oil, 35 kg (77 lb); wing loading, 234 kg/m² (48 lb/sq ft); operational limit load factor, 8.

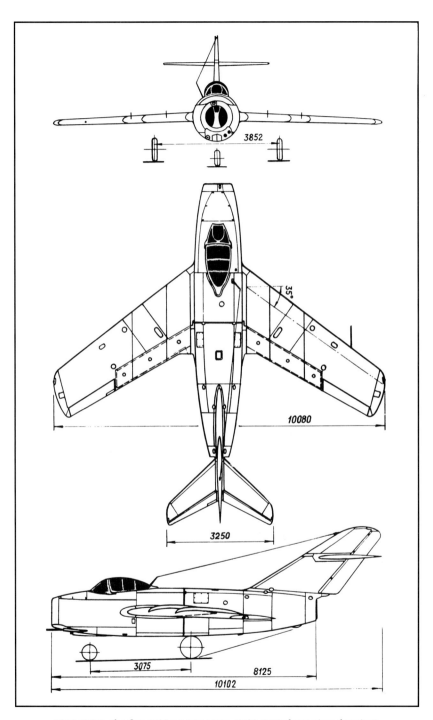

I-310 (S-01), the first MiG-15 prototype (MiG OKB three-view drawing)

Performance
Max speed, 1,042 km/h at 3,000 m (563 kt at 9,840 ft); max speed at
sea level, 905 km/h (489 kt); climb to 5,000 m (16,400 ft) in 2.3 min; to
10,000 m (32,800 ft) in 7.1 min; landing speed, 160 km/h (86 kt); service ceiling, 15,200 m (49,855 ft); endurance at 10,000 m (32,800 ft),
2.01 h; range, 1,395 km at 12,000 m (866 mi at 39,360 ft); takeoff roll,
725 m (2,380 ft); landing roll, 765 m (2,510 ft).

MiG-15 / I-310 / S-03

The S-03 prototype was built in March 1948 within the context of the
test program. Nearly all of the shortcomings found in the first two prototypes were eliminated on the S-03 under the supervision of chief
engineer A. A. Andreyev, who was in charge of the program. Like the
S-02, the S-03 was powered by a Nene II. But it differed from the S-02 in
many other respects:

—it was equipped with hydraulically powered airbrakes hinged on
  the fuselage tail section (the rear structure had to be strengthened
  for this purpose)
—the stabilizer was moved 150 millimeters (5.9 inches) aft to
  improve its efficiency (this change necessitated a modification of
  the tail fin)
—the elevator was fitted with balance weights
—the canopy was attached by a new latch mechanism
—capacity of the no. 1 and no. 3 fuel tanks was reduced, limiting
  total fuel capacity to 1,450 l (383 US gallons) from 1,538 l (406 US
  gallons)
—two store points were added beneath the wing for auxiliary fuel
  tanks or bombs (FAB-100s, FAB-50s, or AO-25s)
—removal of the cannon fairings was simplified
—new equipment was introduced, from an ASP-1N gunsight and an
  S-13 camera gun to a fire extinguishing system

The most serious challenge to be met with the S-03 was giving the
wing structure adequate strength to comply with 1947 standards. This
is why the new V-95 alloy was widely used for the wing structure (in
place of D-16 duralumin) and 30-KhGSA chromansil steel for the spar
webs and flanges. This structural reinforcement added 180 kg (397
pounds) to the weight of the wing. There were twenty-three ribs
instead of twenty, and the skin was 1.8 mm thick instead of 1.5 mm.
The efficiency of the aileron was improved by increasing the area from
0.96 m² (10.3 square feet) to 1.17 m² (12.6 square feet). The span of the

The I-310 S-03 was the master aircraft selected by the VVS. But to be safe the air force also ordered a small number of La-15s, a competing fighter created by the Lavochkin OKB.

The S-03 was equipped with airbrakes hinged on the rear fuselage. Their modest area of 0.52 m² (5.6 square feet) had to be increased on production aircraft.

flaps was reduced slightly, but their chord was increased. The gear, air-brakes, and flaps were controlled by the hydraulic system with a no-load running valve. Switching on no-load running was automatic.

The S-03 was the first MiG aircraft that used its flaps for takeoff. That reduced takeoff roll to 695 m (2,280 feet) from the S-02's 810 m (2,655 feet). But the airbrakes reduced the landing roll by only 30–35 m (98–115 feet). The prototype left the factory in March 1948 and was first flown on 17 June by I. T. Ivashchenko. The factory tests ended on 15 October (LII test pilot S. N. Anokhin came to assist Ivashchenko). The S-03 made a total of forty-eight flights, and all its flaws were eradi-cated one after the other. During one of these flights it reached a top speed of Mach 0.934. On 1 November the S-03 was sent to Saki, in the Crimean branch of the GK NII VVS, for another series of tests carried out by two military pilots, Yu. A. Antipov and V. G. Ivanov. They made thirty-five flights and spent fifteen hours and twenty-one minutes in the air, wrapping up their examination on 3 December. On 23 Decem-ber Marshall Vershinin, commander-in-chief of the VVS, ratified the "acceptance trials report of the frontal MiG-15 single-seat fighter." These are some of the report's conclusions:

> Considering its performance, we recommend choosing this air-craft to equip our squadrons, to prepare its series production and its availability for issue in compliance with the VVS standards.
>
> —this aircraft can be operated from rough strips
> —dogfight tests have not yet been carried out, but because of its high maneuverability it will be possible to involve the aircraft in fierce close combats
> —it can be flown inverted
> —because of its handling characteristics, it can be flown by average pilots

The outstanding performance of the I-310 S-03 (as well as the S-01 and S-02) in test flights was undoubtedly the reason that mass produc-tion was ordered by the Soviet government.

Specifications
Span, 10.085 m (33 ft 1 in); overall length, 10.102 m (33 ft 1.7 in); fuse-lage length, 8.125 m (26 ft 7.9 in); wheel track, 3.852 m (12 ft 7.6 in); wheel base, 3.075 m (10 ft 1.1 in); wing area, 20.6 m² (221.7 sq ft); empty weight, 2,955 kg (6,513 lb); takeoff weight, 4,806 kg (10,592 lb); pilot, 97 kg (214 lb); fuel, 1,210 kg (2,667 lb); oil, 40 kg (90 lb); ammu-nition, 117 kg (258 lb); removable equipment, 35 kg (77 lb); wing load-ing, 233.3 kg/m² (47.8 lb/sq ft); max operating limit load factor, 8.02.

Performance

Max speed, 1,031 km/h at 3,000 m (557 kt at 9,840 ft), 983 km/h at 10,000 m (531 kt at 33,800 ft); max speed at sea level, 905 km/h (489 kt); climb to 5,000 m (16,400 ft) in 2.3 min; to 10,000 m (32,800 ft) in 7.1 min; landing speed, 160 km/h (86 kt); service ceiling, 15,200 m (49,850 ft); range of S-03 at 1,000 m (3,280 ft), 660 km (410 mi); at 5,000 m (16,400 ft), 908 km (564 mi); at 12,000 m (39,360 ft), 1,433 km (890 mi); range of S-02 at 1,000 m (3,280 ft), 695 km (432 mi); at 5,000 m (16,400 ft), 955 km (593 mi); at 12,000 m (39,360 ft), 1,530 km (950 mi); takeoff roll, 695 m (2,280 ft); landing roll, 710 m (2,330 ft).

MiG-15 / SV

While the S-03 was chosen as the master aircraft, a few engineering modifications were still necessary before the production standard—the SV—was ready.

The Nene II engine was replaced by a RD-45F. In reality it was the same engine, but manufactured in factory no. 45 in Moscow. Several parts of the aircraft's structure were strengthened once more: wing spar flanges, fuselage rear frames, wing top skin, and the skin of the airbrakes (in the latter, duralumin was replaced by EI-100N steel). A tab was added to the left aileron, and the wing was equipped with an additional flutter damper. The outlet port for the links and cartridge cases of the three cannons' ammunition belts was modified to prevent jams when firing. Compared to the S-03, the production MiG-15 differed in many particulars:

—an engine-start switchboard was added in the cockpit, and 12-A-30 batteries were replaced by 12-SAM-25s (the self-starter worked only on the ground)
—NS-23 cannons were replaced by NR-23s
—bothersome glints on the canopy were eliminated
—the efficiency of the ailerons was improved with the first hydraulic servo-control unit ever installed on a MiG aircraft, in this case a B-7 model developed by TsAGI
—stick forces caused by the elevator or the ailerons were reduced
—vibrations experienced while the N-37D cannon was fired were eliminated
—the nose-up attitude caused by airbrake deployment was compensated for
—a homemade GS-3000 generator starter was installed

The SV was the first-series production MiG-15, the "soldier aircraft" whose fame dates from the Korean War and whose success earned its manufacturer a worldwide reputation.

124

—the wing was piped for two 496-l (131-US gallon) underslung tanks
—the nose gear leg was fitted with a new shock absorber
—fuel tanks were kept under pressure by bleeding air from the engine compressor
—the NR-23 cannon mounts were modified, and their hydraulic dampers were removed
—the newer ASP-3N optical gunsight replaced the ASP-1
—a newer IFF interrogator was installed

As the MiG-15 was mass-produced in several factories, its structure, armament, and equipment were continuously updated. During an acceptance test, the engine flamed out when the pilot started to fly upside down. Other aerobatic maneuvers such as barrel rolls also seemed to cause flameouts. To solve this problem OKB engineers developed a small tank inside the fuel system that could feed the engine in all negative-g situations. This feeder tank was fitted with a fuel connector that swiveled according to gravitational acceleration (g) and provided a continuous flow to the engine for up to ten seconds whatever the aircraft's attitude in space (including zero-g or negative-g conditions). After special tests, this tank was installed on all MiG-15s on the assembly line and retrofitted on those that had left the factories.

A certain roll instability experienced at high speed was cured by increasing the stiffness of the wing and its control surfaces at the trailing edge. The canopy's ice and mist problems were solved as well: a new canopy, molded in one piece, was kept clear by engine air bleed. Two other systems were also developed to fight icing: one generated heat electrically, and another employed an alcohol-based deicing fluid. Many other improvements were introduced gradually as production continued:

—the efficiency of the airbrakes was improved by increasing their area from 0.52 m^2 (5.6 square feet) to 0.88 m^2 (9.5 square feet) with the following operating limits: a 0.7 design Mach number during a 16.8-second vertical dive and 1.03 during a 45-degree dive
—to increase the survivability of the aircraft in combat, a standby cable-operated elevator control was added
—to improve the pilot's rear view, a TS-23A periscope was fitted on the front arch of the canopy
—the protective armor in the cockpit was strengthened, and the pilot's life support equipment was improved
—to optimize the accelerations the pilot was subjected to while ejecting, a variety of pyrotechnic cartridges were chosen (there were winter and summer versions); also, the left armrest of the pilot's

A MiG-15 (SV) with its flaps down and set for landing.

seat was equipped with an emergency handle to trigger the ejection procedure in case the pilot's right hand was wounded
—in 1951 the wing was piped for two 300-l (79-US gallon) standardized drop tanks (subsequent MiG-15 bis and MiG-15R bis received 600-l [158-US gallon] tanks); these more streamlined tanks enabled the MiG-15 to fly at speeds up to 900 km/h (486 kt) or Mach 0.9 and were able to withstand a load factor of 5 when filled or 6.5 when empty
—to improve both operational safety and fire protection, the duralumin used in the aircraft's pipes was replaced by steel
—the AGK-47B artificial horizon was replaced by an AGI-51 plus an EUP-46 standby horizon
—for night landings, a powerful headlight was inserted in the air intake partition
—the RD-45F turbojet was continuously improved by the V. Ya. Klimov OKB

V. A. Romodin, MiG deputy chief constructor, coordinated the mass production of the MiG-15 in several factories and their introduction in fighter regiments. On 20 May 1949 the council of ministers ordered the mass production of the MiG-15. The aircraft was deemed

Exploded view of the MiG-15 (MiG OKB document)

127

so important that series production of four other aircraft models—the La-15, Yak-17, Yak-23, and Li-2—was discontinued in order to clear the assembly lines of four factories for the MiG-15. The first production aircraft, MiG-15 no. 101003, was built in factory no. 1. State acceptance tests began on 13 June 1949 but were interrupted because of a faulty cannon. Tests resumed on 26 October and ended on 7 January 1950.

GK NII VVS test pilots Kuvshinov, Blagoveshchyenskii, Kochyetkov, Sedov, Dzyuba, Ivanov, and Pikulenko made fifty-nine flights for a total of forty hours and fifty-five minutes of air time. Eight in-flight engine relights were carried out successfully. In early 1950 several MiG-15s were taken away for static tests as well as armament and equipment tests. Simultaneously, twenty MiG-15s of the fourth and fifth series passed their military acceptance tests with fighter regiments, making 2,067 flights in a total of 872 hours and 47 minutes. Once those tests were completed, the instruction manual for the MiG-15 was issued for VVS and PVO pilots. The *samolyot soldat* (soldier aircraft) was born.

The first MiG-15s were delivered to operational units during the winter of 1949–50. A short time later, on 25 June 1950, war broke out in Korea. Over the next three years the MiG-15 would make a name for itself in that conflict. In 1952 the VVS and the PVO called a joint meeting to discuss their operational experience with the aircraft. Pilots were unanimous in praising its performance, its versatility, and its superiority in combat at medium and high altitudes up to 15,000 m (49,200 feet), in clouds, at night, and in the worst weather conditions. Their assessment is hard to dispute.

The MiG-15 became the first MiG built under license, first in Czechoslovakia and then in Poland. The first Czechoslovakian MiG-15 (built by Aero) took off on 13 April 1953. The factory built 853 machines of the type, referred to as the S-102. Production started in 1954 in Poland, where the aircraft was called the LIM-1.

Specifications
Span, 10.085 m (33 ft 1 in); overall length, 10.102 m (33 ft 1.7 in); fuselage length, 8.125 m (33 ft 7.9 in); wheel track, 3.852 m (12 ft 7.6 in); wheel base, 3.23 m (10 ft 7.2 in); wing area, 20.6 m^2 (221.7 sq ft); empty weight, 3,253 kg (7,170 lb); takeoff weight, 4,963 kg (10,938 lb); max takeoff weight, 5,405 kg (11,913 lb); fuel, 1,225 kg (2,700 lb); wing loading, 240.9–262.4 kg/m^2 (49.4–53.8 lb/sq ft).

Performance
Max speed, 1,031 km/h at 5,000 m (557 kt at 16,400 ft); max speed at sea level, 1,050 km/h (567 kt); climb to 5,000 m (16,400 ft) in 2.5 min; to 8,000 m (26,240 ft) in 5 min; to 10,000 m (32,800 ft) in 7.1 min;

The I-312 (ST), prototype of the UTI MiG-15, was developed from a MiG-15 airframe and powered by the RD-45F.

climb with two 248-l (65-US gal) auxiliary tanks to 5,000 m (16,400 ft) in 3.5 min; to 8,000 m (26,240 ft) in 7 min; to 10,000 m (32,800 ft) in 10.5 min; service ceiling, 15,200 m (49,850 ft); landing speed, 160 km/h (86 kt); range, 1,175 km at 10,000 m (730 mi at 32,800 ft); range with auxiliary tanks, 1,650 km at 12,000 m (1,025 mi at 39,360 ft); take-off roll, 630 m (2,065 ft); takeoff roll with auxiliary tanks, 765 m (2,510 ft); landing roll, 720 m (2,360 ft).

UTI MiG-15 / I-312 / ST

In the MiG-15 test report the NII VVS experts had pointed to the need for a two-seat training version of the aircraft. Because the MiG-15 was mass-produced for the VVS, the PVO, and the VMF, a "jet-flying class" had to be developed for the many pilots who had flown only piston-engine aircraft. Decrees from the council of ministers on 6 April 1949 and the ministry of aircraft production on 13 April ordered the MiG OKB to design and construct a UTI (*Uchebno-trenirovich istrebityel:* formation and training fighter) by 15 May—that is, in just five weeks! Once the test flights were completed, an entire factory would be dedicated to the sole production of the UTI MiG-15.

Armament of the prototype and first-series UTI MiG-15s consisted of one NR-23 cannon and one 12.7-mm UBK-E machine gun.

The aircraft was a straight two-seat modification of the MiG-15 powered by the same RD-45F engine. Both the student in the front seat and the instructor in the rear were in a pressurized cockpit. A reduced armament suite—one machine gun and one cannon—was placed on a removable rack. Both flight stations had fully instrumented panels and first-generation ejection seats (the same as those in the MiG-15). The front-seat instrument panel and equipment were identical to those of the MiG-15 so that the student pilot could become fully acquainted with his environment in preparation for his first solo flight in the single-seater. With the dual controls, the instructor could always control the aircraft and counteract the student's mistakes. The front canopy was starboard-hinged, and the rear canopy was rearward-hinged; both could be jettisoned in case of emergency. The rear-seat occupant was ejected first.

For gunnery exercises, the first-series UTI MiG-15s had a 23-mm NR-23 cannon with 80 rounds and a 12.7-mm UBK-E machine gun with 150 rounds. Starting with the sixth series, and after tests with the ST-2, the cannon was replaced on the removable rack by an OSP-48 instrument landing system. The student cockpit was equipped with an ASP-1N gunsight. In the front, two armor plates were bolted on frame no. 4—one to protect the crew, and the other to shield the ammunition

The UTI MiG-15 toured fighter regiments to familiarize pilots with operation of the ejection seat.

boxes and equipment placed on the machine gun rack. The ejection seats were fitted with armored headrests. Two store stations were provided beneath the wing for 50-kg (110-pound) and 100-kg (220-pound) bombs. An S-13 camera gun was added to the upper lip of the engine air intake. Communication equipment included an RSI-6 transceiver, but most UTI MiG-15s were fitted with an RSIU-3 VHF transceiver, an SRO-1 IFF transponder, and an AFA-1M wide-angle camera. From aircraft no. 10444 onward armament was reduced to a single UBK-E 12.7, and the ASP-1N gunsight was replaced by an ASP-3N; but the fire control and bombing systems were identical to previous versions. All UTI MiG-15s also had electrically operated flare launchers, one for each of four colors: red, green, white, and yellow.

The ST was built in the OKB experimental workshop between March and May 1949, using a MiG-15 airframe built in Kuybyshev factory no. 1. It underwent factory tests from 23 May to 20 August 1949 under the supervision of three pilots: I. T. Ivashchenko, K. K. Kokkinaki, and A. N. Chernoburov. State acceptance trials took place at the GK NII VVS from 27 August to 25 September, and mass production was approved. The prototype was sent to a fighter regiment in Kubinka from October 1949 to April 1950, when it returned to the OKB for the adjustments needed to eliminate the shortcomings recorded in operation. On 15 May it passed the final factory tests administered by Chernoburov and S. Amet-Khan of the LII. In a little over a year the prototype made 601 flights—33 in factory tests, 58 in state trials, 502 in a fighter regiment, and 8 more back at the factory.

As early as 1952 there were in practice four UTI MiG-15s in every fighter regiment equipped with MiG-15s. Besides their normal role as pilot trainers, they were also used to make weather reconnaissance flights and to teach pilots how to fly at night and under adverse weather conditions. To this end, the front cockpit was equipped with a curtain so that pilots could train for blind flying in the daytime. The aircraft was also used to train pilots for dive-bombing and photo reconnaissance missions.

From the outset, MiG-15s equipped with the older ejection seats were not popular: many pilots were afraid to risk serious injury by using the seat in emergencies. After all, most accidents happen when flying near the ground at minimum control speeds or when landing—and in those situations use of the first-generation seats was risky. This fear presented a psychological obstacle that had to be overcome in order to restore the airmen's confidence. So it was decided that some of the UTI MiG-15s would be used as ejection trainers. Military instructor parachutists went to VVS and PVO fighter regiments and demonstrated ejections above each unit airfield. Then they asked for volunteers among the regiment's pilots. These volunteers were ejected from the aircraft's rear seat. The seat carried a reduced pyrotechnic charge, but one still powerful enough to catapult the seat—and its contents—above the aircraft's fin.

The UTI MiG-15 was mass-produced in the USSR, Czechoslovakia, Poland, and China. Czechoslovakia alone built 2,012 of them (referred to as the CS-102). They have been operational for more than thirty years and have trained all MiG-15, MiG-17, and MiG-19 pilots. In the early 1970s nearly all of the UTI MiG-15s still active in the VVS and the PVO were handed over to the DOSAAF (the voluntary association for the support of the army, the air force, and the fleet). That kept the two-seaters in the air for a few more years with pilots from various Soviet flying clubs in the cockpit.

Specifications
Span, 10.085 m (33 ft 1 in); overall length, 10.11 m (33 ft 2 in); fuselage length, 8.08 m (26 ft 6.1 in); height, 3.7 m (12 ft 1.7 in); wheel track, 3.81 m (12 ft 6 in); wheel base, 3.175 m (10 ft 5 in); wing area, 20.6 m² (221.7 sq ft); empty weight, 3,724 kg (8,208 lb); takeoff weight, 4,850 kg (10,690 lb); takeoff weight with two 260-l (69-US gal) auxiliary tanks, 5,320 kg (11,725 lb); takeoff weight with two 300-l (79-US gal) auxiliary tanks, 5,400 kg (11,900 lb); fuel, 900 kg (1,984 lb); useful load (fuel, ammunition, crew), 1,588 kg (3,500 lb); wing loading, 235.4–262.1 kg/m² (48.3–53.7 lb/sq ft); max operating limit load factor, 8.

Performance
Max speed, 1,015 km/h at 3,000 m (548 kt at 9,840 ft); 720 km/h at 13,000 m (389 kt at 42,640 ft); max Mach number, 0.894; climb to

The UTI MiG-15P (ST-7) was used to train pilots to operate the RP-1 Izumrud radar.

3,000 m (9,840 ft) in 1.5 min; to 5,000 m (16,400 ft) in 2.6 min; to 10,000 m (32,800 ft) in 6.8 min; service ceiling, 14,625 m (47,970 ft); landing speed, 172 km/h (93 kt); range, 680 km at 5,000 m (422 mi at 16,400 ft); 950 km at 10,000 m (590 mi at 32,800 ft); range with two 300-l (79-US gal) auxiliary tanks, 1,054 km at 5,000 m (655 mi at 16,400 ft), 1,500 km at 10,000 m (930 mi at 32,800 ft); takeoff roll, 510 m (1,675 ft); landing roll, 740 m (2,425 ft).

UTI MiG-15 / ST-2

The need to train pilots to fly at night or in adverse weather conditions led to several engineering modifications of the UTI MiG-15. The most noteworthy upgrading was the installation of the OSP-48 instrument landing system; to make room, the NR-23 cannon was removed and the capacity of fuel tank no. 1 was reduced. The instrument panel was completed with a KI-11 additional compass, and the cockpit pressurization system was fitted with a filter. On the other hand, the outlet port for spent links and cartridge cases from the UBK-E machine gun was changed to prevent it from jamming when firing. A newer gunsight, the ASP-3N (the same one used in the MiG-15), replaced the ASP-1 of the ST prototype. These modifications marked the birth of the ST-2.

The two RP-1 antennae were housed in the air intake's partition and upper lip.

After certification tests at the GK NII VVS, the ST-2 became the new master aircraft on the UTI MiG-15 production line.

UTI MiG-15P / ST-7

The VVS needed a two-seater to familiarize pilots with the operation of the RP-1 Izumrud radar. For this purpose the nose of the UTI MiG-15 was modified to resemble that of the MiG-15P bis (SP-5). The instrument panel of the student's cockpit in front was identical to that of the single-seat fighter.

The armament on this trainer was limited to one 12.7-mm UBK-E machine gun. The ST-7 passed its acceptance tests in 1952, but its production run was limited. Its flight performance did not differ significantly from that of the UTI-15.

MiG-15 / SU

From the inception of fighter aircraft in World War I, pilots aimed at objects in the air or on the ground by pointing the aircraft so that the target appeared in the gunsight's cross hairs. The fighter's armament

The MiG-15 no. 935 was modified to be equipped with an experimental weapons system having a limited slew angle.

was fixed, making it difficult to direct (especially at high speed). Pilots had little time to aim and fire at their targets.

This is why aircraft manufacturers and armament specialists joined forces to develop rotating gun systems to simplify aiming and firing sequences for the fighter pilot and thereby to guarantee a decisive tactical advantage in dogfights. In late 1949 OKB engineers and armament experts decided to design an experimental weapon system with a limited slew angle and to test it on a MiG-15. It was one of the very first installations of the kind in the USSR.

First, cannons of a new type—23-mm Shpitalniy Sh-3s—were installed in a MiG-15 bis (ISh). They had standard mountings and passed their firing tests. On 14 September 1950 the council of ministers ordered MiG-15 no. 109035, built in factory no. 1 at Kuybyshev, to be sent to the OKB's experimental workshop and equipped with the limited slew angle V-1-25-Sh-3 weapon system, which consisted of two experimental 23-mm Sh-3 short-tube cannons with 115 rpg. The specification called for the weapons to rotate in the vertical plane (11 degrees upward, 7 degrees downward) concurrently with a synchronous displacement of the gunsight in the cockpit. The aim of the cannons was remotely controlled by two switch knobs—one on the throttle (RUD) and one on the stick (RUS). Either knob could be used.

A front view of the V-1-25-Sh-3 weapons system with its two 23-mm Sh-3 cannons.

The specifications and performance of the SU were virtually identical to those of the MiG-15 with the RD-45F engine. The MiG-15 (SU) no. 109035 was tested by Yu. A. Antipov and sent to the NII VVS on 20 June 1951 for state trials, which took place from 30 June to 10 August. The aircraft was put through its paces by military test pilots such as Trofimov, Makhalin, Dzyuba, Lukin, Kotlov, Tupitsin, and Filippov. They made sixty-three flights with a total of forty-two hours and forty-six minutes of flying time. The rotating elements of the new weapon system functioned for fifty-two hours, with the cannon pivoting for six and one-half hours.

Firing tests in flight gave prominence to the tactical advantages of the SU over the production MiG-15, but these advantages were tempered by the relatively small angular movement of the cannons and by the limited possibilities of the ASP-3N production gunsight. Combat simulations were staged against an Il-28 and a MiG-15 bis. These proved that the V-1-25-Sh-3 could widen the possibilities of head-on attacks without a risk of collision. At a distance of 800 m (2,600 feet) and identical load factor for both the target and the attacker, the SU could fire at a heading angle 7 to 13 degrees wider than that of a conventionally armed fighter. The pivoting Sh-3 also made possible a much longer burst. The trial attacks on the Il-28 were launched from the rear from quadrants two and three, while the combat with the MiG-15 bis took place at 15,000 m (49,200 feet).

Tests showed that fifteen to twenty flights were sufficient to train pilots to operate the new system; compared to the three-cannon standard armament, the new layout was more straightforward. In-flight exercises also demonstrated that firing at extreme slew angles did not affect the aircraft's speed and trim at 5,000 m (16,400 feet) but did

Close-up of one of the two Sh-3 cannons, which rotated only in the vertical plane between +11 and -7 degrees.

detract from the lateral stability somewhat. Moreover, because of a minor buffeting produced by rudder deflection, undamped oscillations were generated at Mach 0.845 on the longitudinal and vertical axes. The aircraft's ground handling deteriorated because of the much larger turning radius entailed by the V-1-25-Sh-3 installation. Its ranging display system also proved to be too slow.

These first tests with limited slew angle cannons proved that it was essential to widen the angles—to 25–30 degrees upward and 10–15 degrees downward—and to use an automatic gunsight. A mobile gunsight was tested on the experimental MiG-15 to assess its simplicity of operation. Another attempt of this kind was made in 1953–54 with the SN, an experimental member of the MiG-17 family.

The two drop tanks of the MiG-15 bis—capacity 250 l (66 US gallons)—were braced.

MiG-15 bis / SD

In 1946 the Klimov OKB developed the VK-1, a more powerful version
of the RD-45F boosted to 2,645 daN (2,700 kg st). Because this engine
was practically the same size and weight as the RD-45F it could be
installed in the MiG-15 without many modifications, confirming the
hopes pinned on the aircraft's growth potential. This is how the MiG-15
gave way to the MiG-15 bis.

Its silhouette did not differ much from that of the MiG-15, but it
offered better performance. The wing structure was strengthened, the
pitch trim was increased to 22 percent, and the shape of the elevator
and rudder noses was modified. The upper surface of the wing was fitted with a long, squared blade called a *nozh* (knife) to retard any stall
tendency. The airbrakes, whose set switch was on the pilot's stick,
were redesigned. But the most successful innovation was the use of a
BU-1 servo-control—with its own hydraulic system—on the aileron control. The hydraulic system for the flaps was backed up by a pneumatic
system.

The SD armament comprised one N-37 cannon with 40 rounds,
two NS-23KMs with 160 rounds. The gunsight was of the ASP-3N type.

The greater thrust of the VK-1 improved the performance of the MiG-15 bis significantly.

With four store stations under the wing, the aircraft could carry two 50-kg (110-pound) or 100-kg (220-pound) bombs and two 250-l (66-US gallon) drop tanks. New tactical methods were tested with the MiG-15 bis. For instance, the aircraft was to be able to drop bombs upon invading bombers at altitudes of up to 12,000 m (39,360 feet) and speeds of up to 700 km/h (378 kt). For this unusual assignment the MiG-15 bis carried special 100-kg (220-pound) OFAB-100M and PROSAB-100 bombs that were fused at the command of the squadron leader. The cockpit pressurization system was improved, the pilot had a warming system for his legs, and the windscreen front panel was made of 64-mm-thick bulletproof glass.

The MiG-15 bis was built in two versions: one that was equipped with an OSP-48 instrument landing system and one that was not and therefore was limited to daytime missions. The first series had an RSI-6 VHF transceiver (later replaced by an RSIU-3), and installation of the first SRO IFF transponders was already planned.

In 1952 a few of these planes were fitted with a 15-m^2 (161.5–square foot) brake chute to make it possible for them to use small airfields. It is worth noting that because of the aircraft's strength, handling, and flutter characteristics, Mikoyan had limited its speed to 1,070 km/h (578 kt) IAS or Mach 0.92. For the first time, pilots wore

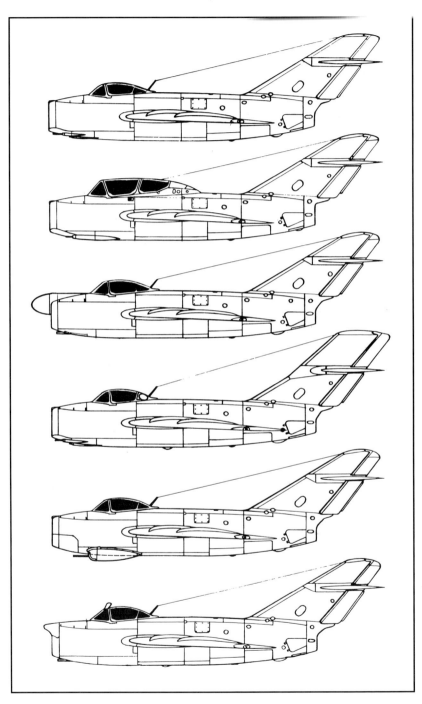

Top to bottom: MiG-15 bis, UTI MiG-15 (ST), MiG-15 bis (SP-1), MiG-15 bis (SYe), MiG-15 (SU), and MiG-15 bis (SP-5) (MiG OKB drawing)

the PPK-1 g-suit aboard the MiG-15 bis. It increased the pilot's resistance to the effects of gravitational accelerations experienced at high
altitudes and worked best at between 1.75 and 8 g.

Also in 1952 a TS-23 periscope was installed on MiG-15 bis no. 235
to help the pilot look backward in combat or while taxiing. It was developed by the Vavilov State Optical Institute, a branch of the defense
ministry. The optical head of the periscope was located on the canopy's
windshield arch, and its mirror was hung on it. A heating coil kept the
glass clear. The field of view scanned by the TS-23 was 16 degrees. But
it was not certified, and its development was halted. From 1 June 1952
a new model, the TS-25, was tested; it was a one-piece periscope placed
on the canopy and offering a much wider field of view. The TS-25
allowed pilots to watch the sky behind them and spot aircraft approaching from that sector without having to focus solely on the periscope. It
covered between 50 and 55 degrees on each side of the aircraft's longitudinal axis and between 20 and 25 degrees vertically. This periscope
was approved for use in Soviet aircraft. The MiG-15 bis—as well as the
MiG-17 and all its variants—was equipped with either the TS-25 or the
improved TS-27.

The MiG-15 bis was built under license in Czechoslovakia (620
units referred to as S 103s) and in Poland (LIM-2s).

*The following details refer to the MiG-15 bis equipped with an OSP-48
instrument landing system weighing 84 kg (185 pounds).*

Specifications
Span, 10.085 m (33 ft 1 in); overall length, 10.102 m (33 ft 1.7 in); fuselage length, 8.125 m (26 ft 7.9 in); wheel track, 3.852 m (12 ft 7.6 in);
wheel base, 3.23 m (10 ft 7.2 in); wing area, 20.6 m² (221.7 sq ft);
empty weight, 3,681 kg (8,113 lb); takeoff weight, 5,044 kg (11,117 lb);
max takeoff weight clean, 5,380 kg (11,857 lb); with two 260-l (69-US
gal) drop tanks, 5,508 kg (12,140 lb); with two 300-l (79-US gal) drop
tanks, 5,574 kg (12,285 lb); with two 600-l (158-US gal) drop tanks,
6,106 kg (13,458 lb); fuel, 1,173 kg (2,585 lb); wing loading, 244.9–296.4
kg/m² (50.2–60.8 lb/sq ft).

Performance
Max speed, 1,107 km/h at 3,000 m (598 kt at 9,840 ft); 1,014 km/h at
5,000 m (548 kt at 16,400 ft); max speed at sea level, 1,076 km/h (591
kt); climb to 5,000 m (16,400 ft) in 1.95 min; to 10,000 m (32,800 ft) in
4.9 min; service ceiling, 15,500 m (50,840 ft); landing speed, 178 km/h
(96 kt); range, 1,130 km at 12,000 m (702 mi at 39,360 ft); with two
260-l (69-US gal) drop tanks, 1,860 km (1,155 mi); with two 300-l (79-US
gal) drop tanks, 1,975 km (1,227 mi); with two 600-l (158-US gal) drop
tanks, 2,520 km (1,565 mi); endurance at 12,000 m (39,360 ft), 2 h 6

min; with two 260-l (69-US gal) drop tanks, 2 h 57 min; with two 300-l (79-US gal) drop tanks, 3 h 9 min; with two 600-l (158-US gal) drop tanks, 3 h 52 min; takeoff roll, 475 m (1,560 ft); landing roll, 670 m (2,200 ft).

MiG-15S bis / SD-UPB   MiG-15R bis / SR

The MiG-15S bis and MiG-15R bis were both direct derivatives of the MiG-15 bis. The first was an escort fighter, the second a frontline photo-reconnaissance aircraft—two roles that demand long-range capabilities. The main difference between the MiG-15 bis and these two versions (other than the AFA-40 photo equipment on the MiG-15R bis) was the addition of two 600-l (158-US gallon) drop tanks beneath the wing.

Considering the greater endurance made possible by those tanks, an additional 2-l (0.53-US gallon) oxygen bottle was installed in the nose section, bringing the total oxygen reserve to 8 l (2.11 US gallons). The additional takeoff weight also led the engineers to increase the tire pressure from 7 kg/cm^2 (100 psi) to 8$^{+0.5}$ kg/cm^2 (113.8 psi) and to add restrictions to the flight protocols. Pilots were not allowed:

—to fly the aircraft under a negative load factor with full drop tanks
—to make a long side-slip with full drop tanks (since a steady fuel draining could not be assured)
—to fly for a long time at speed limits
—to land with full drop tanks (they had to be jettisoned first)

The specifications and performance of the MiG-15S bis and MiG-15R bis did not differ much from those of the MiG-15 bis, and both aircraft were held to the same speed and altitude limits as their precursor.

MiG-15 bis / SYe / LL (Flying Laboratory)

From 1947 to 1952 intensive research on wing profiles and aerodynamic design for supersonic speeds was carried out at TsAGI, LII, and other science centers. The top speed of the production MiG-15 was limited to Mach 0.92—above that, the aircraft's transverse stability deteriorated. Two LII engineers, I. M. Pashkovskiy and D. I. Mazurskiy, proposed to relieve the ailerons and to increase the rudder area. To test their ideas, two SYe prototypes were built using MiG-15 bis SD airframes at factory no. 1 in Kuybyshev.

The rudder of the SYe was taller and thus larger than that of the SD; the fin had to be modified accordingly. V. P. Yatsenko engineered these changes and took charge of the prototypes. (Yatsenko had acquired some fame as the designer of the I-28 fighter in 1938. He joined the MiG OKB in July 1941.) The fin was enlarged along all of its chords and made taller in order to position the upper hinge fitting of the rudder. The original drawings did not impel the builders to construct an entirely new tail fin but only to match it with the new rudder dimensions—hence the break of the fin's leading edge (see the side drawing of the SYe).

To reduce the wing divergence at high speed that resulted in a wing dropping, strengthening panels were installed on the upper surface near the wing roots, above the wheel wells. Moreover, the span of the ailerons was increased. Thus the wing area increased—since the wing tips, instead of being rounded, were now angular at the trailing edge—but the wing span remained unchanged. At first there were no aileron servo-controls, but the enlarged rudder and the much improved wing stiffness allowed the aircraft to fly at higher speeds while still handling well.

The MiG-15 bis SYe was flight-tested at the LII by D. M. Tyuteryev. On 21 September 1949 he reached Mach 0.985 at 12,000 m (39,360 feet) by climbing to the aircraft's service ceiling and then going into a shallow dive at maximum engine thrust. The first BU-1 hydraulic servo-controls appeared at the end of 1948, and the SYe prototype was fitted with one. On 18 October 1949 Tyuteryev broke the sound barrier in the SYe.

The test flights of the SYe contributed a great deal to the research work on supersonic speed then under way in the USSR. They also had a hand in the achievement of I. T. Ivashchenko, who in February 1950 reached Mach 1.03 on a production MiG-17.

MiG-15P bis / SP-1

The development of an all-weather fighter for the PVO had become a necessity. When the first Soviet airborne radars appeared at the end of the 1940s, it was decided that the MiG-15 bis would receive this advanced equipment. But first a number of questions had to be answered regarding the capabilities and efficiency of both the radar control unit when engaging enemy aircraft and the sighting system in blind flying (at night or in clouds).

The council of ministers called for development work on both the airborne radar and the aircraft on 7 December 1948. The first ranging

Installation of the Toriy radar on the MiG-15P bis necessitated alteration of the fuselage nose structure up to the no. 8 frame.

radar, the Toriy, was developed by A. B. Slepushkin, pioneer of Soviet
radar technology. It was a peculiar system: its one antenna both transmitted and received signals. It was housed in a small radome made of a
specially developed dielectric material. The Toriy was not easy for the
pilot to control while trying to intercept enemy aircraft because it
could not track targets automatically.

From the start MiG OKB engineers were determined not to let the
efficiency and performance of the MiG-15 bis suffer because of the
addition of radar. Production MiG-15 bis no. 3810102 built at factory
no. 1 was sent to the OKB workshop and modified, becoming the SP-1.
The two guns on the left of the fuselage were removed; only the N-37D
with forty-five rounds was retained. The ASP-3N gunsight was replaced
by a new model, and the S-13 camera gun usually placed above the air
intake was moved to the right side of the fuselage. Most important, a
radar display was set into the instrument panel. With that display the
pilot was able to track an invader, bring his aircraft into line with it,
and measure its distance before firing.

As it turned out, many other modifications had to be made, mainly
structural ones:

—because of the radar installation and armament removal, the fuselage nose section was made over up to the no. 8 frame and lengthened by 120 mm (4.7 inches)
—the area of the airbrakes was increased, and their shape and axis of
   rotation were altered (22 degrees in relation to the vertical)

—the cockpit windshield was fitted with 64-mm-thick bulletproof glass, and the shape of the windshield and the canopy was changed in order to retain a good forward view despite the nose modifications

—the wing anhedral was increased from 2 to 3 degrees

—the front leg of the landing gear had to be moved 80 mm (3.5 inches) forward to bring the NR-37 cannon axis as close as possible to the aircraft datum line

—the wheel fork was replaced by a half-fork, and the double gear doors were replaced by a single door

—the elevator control was fitted with a BU-1 servo-control unit

The SP-1 prototype was equipped with an ARK-5 automatic direction finder and an MRP-48 marker receiver. After the factory flight tests conducted in December 1949 by A. N. Chernoburov and G. A. Sedov, the aircraft was transferred to the NII VVS on 31 January 1950 for its state trials. They ended on 20 May 1950.

The test report noted a number of defects. The pitching stability was too scanty at landing, and compared with the MiG-15 (SV) the dynamic stability margin had decreased. In straight level flight, the aircraft tended to bank to the left and then side-slip at 940–950 km/h (508–513 kt). Poor aileron efficiency limited the bank angle to 5 degrees.

The report concluded that the SP-1 could not be used as an all-weather interceptor because its Toriy ranging radar did not work properly. The all-weather radar tests were conducted by Suprun, Kalachev, Pibulyenko, Blagoveshchenkiy, Antipov, Dzyuba, and Ivanov, all military pilots. Several passes were made in attempts to locate Il-28 and Tu-4 bombers. The SP-1 was not certified because it was too difficult for a pilot to fly his aircraft and operate the radar at the same time—and moreover, the Toriy was not very reliable. Its manufacturer upgraded the unit, which then became known as the Toriy A and was installed on the MiG-17 (SP-2). But its most serious shortcoming was not addressed: the Toriy A still could only track incoming aircraft manually.

In 1951 five SP-1s equipped with RP-1M radars were assembled at factory no. 1. On 25 November one was sent to the NII VVS for trials, but the aircraft and its upgraded radar unit still failed to earn certification. Like the MiG-15 bis, the MiG-15P bis (SP-1) was powered by a 2,645-daN (2,700-kg st) VK-1 turbojet.

Specifications
Span, 10.085 m (33 ft 1 in); overall length, 10.222 m (33 ft 6.5 in); wheel track, 3.852 m (12 ft 7.6 in); wheel base, 3.075 m (10 ft 1.1 in);

On the MiG-15P bis the two NR-23 cannons usually found at the lower left of the MiG-15 bis front fuselage had to be removed.

The wing anhedral of the MiG-15P bis was increased slightly, as was the area of the airbrakes.

The SD-21 was a MiG-15 bis used for testing S-21 rockets, hence its designation.

wing area, 20.6 m² (221.7 sq ft); empty weight, 3,760 kg (8,287 lb); takeoff weight, 5,080 kg (11,196 lb); fuel, 1,168 kg (2,574 lb); wing loading, 246.6 kg/m² (50.55 lb/sq ft).

Performance
Max speed, 1,022 km/h at 5,000 m (552 kt at 16,400 ft); 979 km/h at 10,000 m (529 kt at 32,800 ft); climb to 5,000 m (16,400 ft) in 2.15 min; to 10,000 m (32,800 ft) in 5.35 min; service ceiling, 14,700 m (48,200 ft); range, 1,115 km at 10,000 m (692 mi at 32,800 ft); takeoff roll, 510 m (1,670 ft).

MiG-15 bis Burlaki

Burlaki is a nickname that means "towed." The purpose of this experimental prototype was to assess the feasibility of a rather odd idea. In the early 1950s, the Dalnyaya Aviatsiya (DA) or long-range bomber force still used Tu-4s because no jet bomber was yet available, and therefore it needed escort fighters. It was to meet this need that the Yakovlev OKB designed the Burlaki system for the MiG-15 bis.

This SD-21 has, in addition to its rockets, two 250-l (66-US gallon) slipper tanks.

A boom fitted with a hooking device (called a *garpun* or harpoon) and a steel wire was set in front of the engine air intake. The hook was sent outward by a pneumatic cylinder at the pilot's order. When the bomber reached its flight level, the fighter closed in and lined up outside the bomber's wake vortex. The fighter pilot then threw his harpoon toward the end of a cable trailing behind the Tu-4. Once the docking was made, the fighter pilot shut off the engine and was towed like a glider. In case of emergency the pilot could restart the engine and separate from the bomber. He could engage enemy aircraft and return to dock once more, a procedure that could be repeated several times.

With the Burlaki system the range of an escort fighter could be almost doubled. In tests at the OKB and with a DA bomber unit, however, the system's drawbacks quickly became evident. For example, once the engine was shut down the pressurization or heating systems cut out as well. At altitudes between 8,000 m (26,240 feet) and 10,000 m (32,800 feet) pilots could breathe through their oxygen masks, but few could withstand a long ride in an unpressurized and icy cockpit. This was enough to seal the fate of the system, and it is no wonder that the prototype was not certified. In any case, the Tu-4s were soon replaced by Tu-16 jet bombers.

The SD-57 was a production MiG-15 bis used for testing automatic 57-mm rocket pods, hence its designation.

MiG-15 bis / SD-21

In addition to its usual armament suite, the experimental SD-21 was fitted with two wing store stations for 210-mm S-21 unguided air-to-surface rockets. This same prototype was tested in overload conditions with two S-21 rockets and two 250-l (66-US gallon) drop tanks.

MiG-15 bis / SD-57

This modified production MiG-15 bis was developed for tests of two automatic rocket pods (each carrying twelve 57-mm ARS-57 rockets), which were attached to the store stations usually reserved for drop tanks.

A small batch of the MiG-15 bis (ISh) was built, fitted with long wing pylons. "ISh" stands for *Istrebityel Shturmovik:* fighter attack plane.

MiG-15 bis / ISh

This version of the MiG-15 bis was equipped with two wing pylons that could accept either heavy rockets, automatic rocket pods (six or twelve rockets per pod), or drop tanks. Twelve of these planes were built and flight-tested at the NII VVS.

Flight-Refueled MiG-15 bis

The limited range of first- and second-generation jet fighters posed nightmarish problems for their operators. The first turbojets were quite thirsty, and auxiliary tanks of various types and sizes did not provide the long "legs" that the aircraft's mission demanded. In-flight refueling was the best answer because it increased the range in direct proportion to the amount of fuel transferred.

To study the feasibility and capabilities of such a system, three production MiG-15 bis's were modified; in total, five aircraft were involved in the development process. The equipment required at both ends of

This flight refueling system, tested with a Tu-4 as the tanker aircraft and two MiG-15 bis's, was developed by the Yakovlev OKB.

such an operation was developed by the Yakovlev OKB. The refueling process unfolded this way. From the wing tips of the tanker aircraft (in this case a Tu-4 bomber) flexible hoses were released. At the end of each hose was a funnel-shaped device called a drogue. The MiG-15s were fitted with a probe in the left upper nose of the fuselage. To refuel, the fighter pulled up to one of the drogues. Once the connection between the probe and the drogue was secure and the ball joint locked in place, the refueling operator aboard the Tu-4 activated a motor-pump that sent fuel down the hose to the fighter. The tanker could refuel two fighters simultaneously.

The first test flights helped to clear up three important points:

1. New homing equipment was needed to simplify the rendezvous of the tanker and the fighters in midair
2. Pumps with faster delivery rates would have to be developed in order to shorten the refueling process as much as possible
3. Very precise rules were required to govern the movements of both tankers and fighters during the refueling process

As the tests continued, several unfortunate phenomena came to light and complicated the procedure. Immediately after the fighter broke the link with the drogue, for instance, the fuel that remained in the tanker's hose spilled into the fighter's engine air intake or over its

Two flexible hoses fitted with drogues were unreeled from the Tu-4's wing tips.

canopy. The engine did not flame out because the VK-1 was far better in terms of combustion stability than its predecessors; but kerosene vapors did enter the cockpit via the pressurization conduit, and the pilot had no choice but to inhale them until the next air-conditioning blowout cycle. This situation was remedied by fitting the drogue with an electromagnetic shutoff valve controlled by the tanker's refueling operator.

The MiG-15 bis in-flight refueling tests were never completed, since the coupling process required very highly trained pilots. The two men in charge of those tests conducted in 1953 were two LII pilots, S. A. Anokhin and V. Pronyakin.

MiG-15 bis / SDK-5/SDK-5s/SDK-7

In order to train its fighter pilots for combat, the PVO needed a large number of aircraft with the same performance characteristics as combat aircraft—especially speed. Towed windsocks or plywood gliders were no longer sufficient for target practice. Customized mobile targets were needed, targets capable of maneuvering or changing their speed, heading, and altitude. Then it occurred to someone that MiG-15 bis's

This MiG-15 bis, the SP-5, was used as a test bed for the RP-1 Izumrud radar. A new type of periscope then under development is mounted on the windshield arch.

The RP-1 antennae on the SP-5 were housed in the air intake's partition and upper lip.

that had outlasted their prescribed operational life span could be used to meet that need. Their ejection seats were replaced by remote control equipment so that they could be flown either from the ground or from another aircraft. This is how the SDK-5 was born.

In 1955 the MiG OKB used MiG-15 and MiG-15 bis airframes to test the unmanned SDK-5s and SDK-7, which were in essence remotely controlled bombs that could be used against ground targets.

MiG-15P bis / SP-5

This MiG-15 bis was used as a test bed for the RP-1 Izumrud ("emerald") ranging radar developed by V. V. Tikhomirov in 1950. This radar had two antennae, one that scanned the sky for its target and one that could track enemy aircraft automatically. The scanning antenna was housed in a dielectric radome at the top of the engine air intake; the tracking antenna was set into a small, streamlined compartment in the air intake partition. The radar rack was in the equipment bay, in front of the cockpit. Radar control and target acquisition were automatic, but only the pilot could determine the distance and open fire. Like the SP-1, the SP-5 had only one cannon. Because the Izumrud proved to be easier to operate and much more reliable than the Toriy, series production was recommended. It would later be used for both the MiG-17 and the MiG-19.

MiG-15 bis / SO/SA-1/SA-3/SA-4

These four designations were given to production MiG-15 bis's used as test beds for various equipment and systems. They did not differ externally.

I-320 / R-1/R-2/R-3

Toward the end of the 1940s a specification was laid down calling for a cover interceptor, a fighter whose role would be to oppose any invading aircraft as far as possible from its target and under any weather conditions. Several manufacturers put in a tender for the program:

The I-320 (R-1) was designed for the cover-interceptor mission and was powered by two RD-45F engines.

Sukhoi, Mikoyan, Lavochkin, and (later) Yakovlev. This is how a number of famous aircraft were created, including the Su-15 (the first of two), La-200, and La-200B. In this instance the result was the I-320, an all-weather fighter proposed and built by Mikoyan in 1949.

This twin-jet had a cantilever midwing with a 40-degree sweep angle at the leading edge, and its tail unit was also swept back. Because this aircraft was designed at the same time as the MiG-15 and MiG-17, the family resemblance will come as no surprise. However, the I-320 outdid both of them in size as well as weight. It differed from them in its side-by-side cockpit layout (one captain and one pilot/radar operator) and its unusual power plant arrangement (two tandem-mounted turbojets in the fuselage). Its fuselage was 1.9 m (6 feet, 2.8 inches) in diameter with a maximum cross section of 2.83 m² (30.4 square feet). The crew had dual controls and was equipped with two radar scopes. This certainly made the pilot's job easier in combat, since the second crew member could scan the invaders or even fly the aircraft during the long defensive patrol flights. Each pilot had his own oxygen supply, and the overall reserve amounted to 6 l (1.6 US gallons).

Both bladder fuel tanks—capacities 1,670 l (441 US gallons) and 1,630 l (430 US gallons)—were placed behind the cockpit. The rear tank included a 45-l (12-US gallon) antigravity feeder tank that supplied fuel

The R-1 had two tandem-mounted turbojets. The front engine's exhaust nozzle emerged from the underside of the fuselage, while the rear engine's nozzle came out under the tail.

The R-2 differed from the R-1 by its engines—two VK-1 turbojets—and its strengthened armament.

The R-2 cockpit canopy was deeper than that of the R-1.

to the engines during inverted flights. It was also planned to equip the aircraft with two 750-l (198-US gallon) drop tanks beneath the wings.

The front engine was located beneath the crew compartment, its exhaust nozzle emerging from the underside of the fuselage. The other engine was placed in the conventional position at the rear of the fuselage, its straight exhaust nozzle emerging from under the fin. The airbrakes flanked the tail section (area per unit, 1.08 m^2 [11.6 square feet]; maximum deployment angle, 45 degrees).

The trailing edge of the wing was fully occupied in the inner section by Fowler flaps designed by TsAGI (span, 3.18 m [10 feet, 5.2 inches]; area per unit, 3.1 m^2 [33.4 square feet]; takeoff setting, 22 degrees; landing setting, 56 degrees) and in the outer section by internally balanced ailerons (span, 2.497 m [8 feet, 2.3 inches]; area per unit, 1.47 m^2 [15.8 square feet]). Both R-1 and R-2 prototypes had four fences on the upper surface of the wing. To compensate for the retroaction of the rudder and the transverse instability it caused, 900-mm (2-foot, 11.5-inch) spoilers were installed on the wing's lower surface. Operated by electric actuators, they could be extended 40 millimeters downward. Spoiler extension was automatic whenever the rudder deflection exceeded 2 degrees. The stabilizer had a sweep angle of 40 degrees at the leading edge, while the tail fin had a sweep angle of 59 degrees, 27

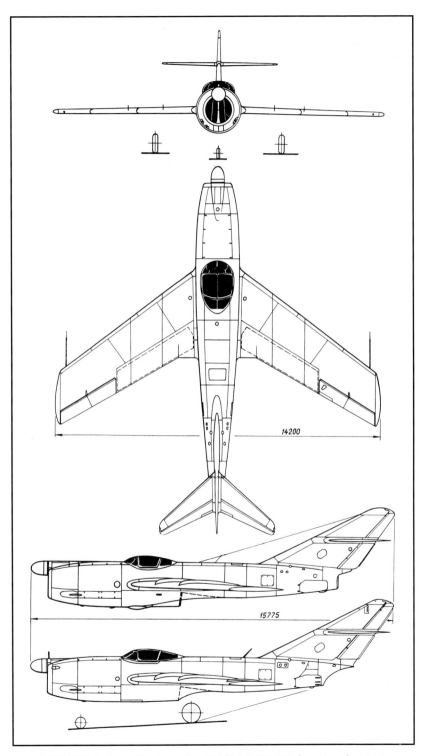

I-320 (R-1); *bottom,* side view of the R-2/R-3 (MiG OKB four-view drawing)

minutes at the leading edge. The maximum deflection angle of the elevator was 33 degrees upward and 17 degrees downward. The maximum deflection angle of the rudder was plus or minus 24 degrees, 48 minutes.

The tricycle landing gear was hydraulically controlled, with air-and-oil shock absorbers on the legs of the main gear. They retracted into the wing, and their wheels were fitted with double brake shoes and 900 x 275 tires. The front leg and its wheel (520 x 240 tires, no brakes) retracted forward. The wheel doors, flaps, airbrakes, and booster cylinders for the aileron and the elevator were also hydraulically controlled. The hydraulic reservoir had a capacity of 35 l (9 US gallons). The dual pneumatic system consisted of a main circuit that controlled the wheel brakes and the cannon loading plus a standby system that governed the gear, flaps, and wheel brakes. Fire control was electrical, activated by one button on the captain's stick.

The first prototype, or R-1, was powered with two 2,225-daN (2,270-kg st) RD-45F turbojets. The R-2 and R-3 (in fact, a modified R-2) featured 2,645-daN (2,700-kg st) VK-1 turbojets. The air intake was divided into three ducts: the one in the middle fed the front jet, and the other two channeled air to the rear jet. The I-320 could fly and even take off on either of its two engines. The polystyrene dome that housed the Toriy-A radar designed by A. B. Slepushkin was located in the upper lip of the air intake. The armament comprised two N-37 cannons flanking the air intake.

The R-1 was rolled out in April 1949 and made its first flight on 16 April with Ya. I. Vernikov and S. Amet-Khan at the controls. Factory tests continued until 18 January 1950 under two pilots, A. N. Chernoburov and I. Y. Ivashchenko. The R-1 was also flown by four LII pilots—Ya. I. Vernikov, S. Amet-Khan, S. N. Anokhin, and M. L. Gallai—and by pilots of the PVO, a potential customer. Lt. Gen. Ye. Ya. Savitskiy, commanding officer of the PVO's fighter regiments, made the following comments after flying the I-320: "The aircraft handles well at takeoff, in flight, and while landing. It has no tendencies to yawing or swinging. Being easy to handle, it can be flown by average pilots." The official test report added:

The aircraft has excellent in-flight steadiness on its three axes. Given the aircraft's layout and the location of its fuel tanks, there is no need to use the elevator's tab in a flight envelope ranging from takeoff speed to 700 km/h [378 kt]. Gear extension and retraction do not modify the aircraft trim. When performing a tight turn or a combat half-flick roll, the I-320 handling characteristics remain safe. The airframe was initially stressed with a load factor of up to 5.9 for an aircraft weight of 8,530 kg [18,800

The R-2 radome, like that of the R-1, housed a Toriy-A radar. Note the partitions in the air intake, to supply air to both engines.

pounds]. The load factor was later increased to 8. To test the radar's performance fourteen flights were carried out, nine of which involved targeting a Tu-2, an Li-2, a B-17, or a Tu-4. While trying to intercept a Boeing B-17 Flying Fortress, the I-320 was caught in the propeller slipstream of the bomber, causing the fighter to make a spectacular pirouette.

Yu. A. Antipov, M. L. Gallai, N. P. Zakharov, and G. T. Beregovoy were four of the pilots who took part in the combat tests. The R-1 was not certified because of its transverse instability in a narrow speed range between Mach 0.89 and 0.9, and also because of its wing dropping between 930 and 940 km/h (502 and 508 kt).

The I-320 (R-2) was modified after an accident, becoming the R-3. Note the third wing fence.

The VK-1 engine of the R-2 prototype boosted the maximum speed by only 3 percent—1,090 km/h (589 kt) IAS versus 1,040 km/h (562 kt) for the R-1—considering the severe limitations imposed by the stiffness problems of a thin, high-aspect-ratio swept wing. Except for its new engines, the R-2 was not greatly modified. The crew's all-around visibility was improved, and the canopy was fitted with a more reliable emergency release system. The wing and the stabilizer were equipped with deicers, and the air intake ducts were electrically warmed. Its armament was supplemented by another cannon so that the prototype fielded a total of three N-37s, one on the left and two more on the right of the lower nose section.

The R-2 received its Toriy-A radar at the beginning of its test program. It was later replaced by a Korshun, also developed by Slepushkin. Neither radar was able to track targets automatically. The R-2 was equipped with an RV-2 radio-altimeter, an RSIU-6 VHF transceiver, and a Bariy (barium) IFF system.

This second prototype was rolled out in early November 1949. During its factory tests from December 1949 to September 1950 the aircraft made 100 flights, executed a steep spin, jettisoned its canopy in flight, performed several aerobatic maneuvers under negative gravity, flew at night, and dropped its auxiliary tanks. Between 13 and 30 March all test flights had to be suspended after a shell exploded in an ammunition belt and damaged the aircraft's nose. The OKB took advantage of the repair time to make a few modifications. The wing anhedral was reduced to 1.5 degrees from 3 degrees, the span of the spoilers was increased to compensate for the transverse instability at high speeds,

Slipper tanks were tested on the I-320 (R-3).

The exhaust nozzle breaches and fin of the R-2/R-3 were somewhat different from those of the R-1.

an automatic airbrake deployment system was installed, and two fences were added on the upper surface of the wing. This repaired and modified R-2 became the R-3.

The first flight of the new version took place on 31 March. The test pilot noted that the wing anhedral modification had changed the transverse stability/yaw stability ratio. To deal with that problem a provisional ventral fin was added under the tail section. Moreover, the spoilers were mechanically linked to the ailerons. The tests were resumed on 13 April and ended on 23 April 1951. During the state trials sixty flights were made, and the R-3 logged forty-five hours and fifty-five minutes in the air. All of these tests were carried out within certain operational limitations: speed, 1,000 km/h (540 kt); Mach, 0.95; load factor, 7.5; maximum speed with underslung tanks, 800 km/h (432 kt); load factor with underslung tanks, 3.5. The VK-1–powered I-320R-3 was not certified either—nor was its competitor, the Lavochkin La-200. As the saying goes, opportunity makes the thief; it was a third manufacturer, Yakovlev, that—despite its late entry into the competition—gathered the fruits of much hard labor. His Yak-25M equipped with RP-6 Sokol radar was selected for mass production. The R-1 and R-2/R-3 were used for a long time as test beds for new equipment; for example, from 13 July to 31 August 1950 LII test pilot Sultan Amet-Khan made thirty-one flights to develop the Materik and Magniy-M instrument landing systems.

The following details refer to the I-320R-1.

Specifications
Span, 14.2 m (46 ft 7 in); length, 15.775 m (51 ft 9 in); fuselage length without radome, 12.31 m (40 ft 6.6 in); wheel track, 5.444 m (17 ft 10.3 in); wheel base, 4.754 m (15 ft 7.2 in); wing area, 41.2 m² (443.5 sq ft); empty weight, 7,367 kg (16,237 lb); takeoff weight, 10,265 kg (22,625 lb); fuel, 2,700 kg (5,950 lb); wing loading, 249.2 kg/m² (51.1 lb/sq ft); max operating limit load factor, 8.

Performance
Max speed, 994 km/h at 10,000 m (537 kt at 32,800 ft); max speed at sea level, 1,040 km/h (562 kt); climb to 5,000 m (16,400 ft) in 2.3 min; to 10,000 m (32,800 ft) in 5.65 min; service ceiling, 15,000 m (49,200 ft); range, 1,100 km (683 mi); takeoff roll, 610 m (2,000 ft); landing roll, 770 m (2,525 ft).

The following details refer to the I-320R-2/R-3.

Specifications
Dimensions and area identical to R-1; takeoff weight, 10,725 kg (23,638 lb); max takeoff weight, 12,095 kg (26,657 lb); fuel, 2,700 kg (5,950 lb); fuel with two 750-l (198-US gal) underslung tanks, 3,950 kg (8,705 lb); wing loading, 260.3–293.6 kg/m^2 (53.4–60.2 lb/sq ft).

Performance
Max speed, 1,090 km/h at 1,000 m (589 kt at 3,280 ft); max Mach, 0.9; service ceiling, 15,500 m (50,840 ft); range, 1,205 km at 10,000 m (748 mi at 32,800 ft); range with two 750-l (198-US gal) underslung tanks, 1,940 km (1,205 mi).

MiG-17 Series

MiG-17 / I-330 / SI / MiG-15 bis 45° / SI-2/SI-02/SI-01

The next major challenge faced by MiG designers was to increase the maximum speed of a fighter solely by improving its aerodynamic factor—that is, without giving it a single additional pound of thrust. Both in silhouette and in structure, the SI-2 (prototype of the MiG-17) and its double the SI-3 looked very much like the MiG-15. (The SI-1 was set aside for static tests and never left the ground.) But there were many important differences:

—the wing sweepback C/4 (at quarter chord) was 45 degrees from the root to midspan (hence the aircraft's designation) and 42 degrees beyond that, creating a sweepback on the leading edge of 49 degrees and 45 degrees, 30 minutes (this compound sweep came about not only because of trim considerations but also because the wing root rib had to be bolted to a section of the fuselage inherited from the MiG-15)
—the wing area was enlarged by 2 m^2 (21.5 square feet)
—the anhedral was increased to 3 degrees with a 1-degree wing incidence

The SI-2, the second I-330 built, was the first prototype of the MiG-17 program to fly.

Note the leading-edge compound sweepback—45 and 42 degrees at quarter chord—of the SI-2, plus its six wing fences.

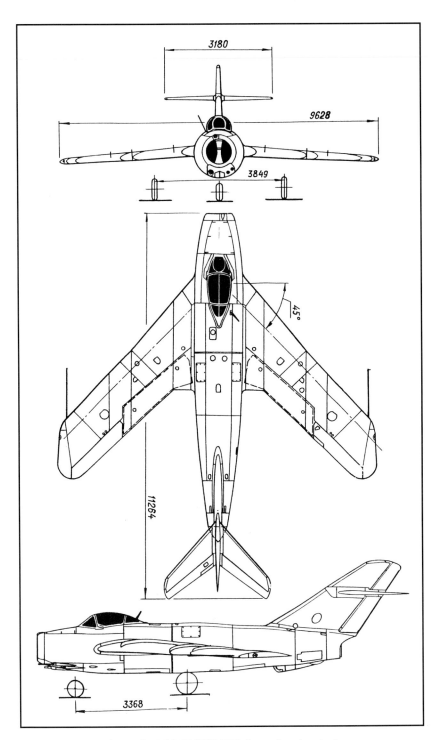

First-series MiG-17 (MiG OKB three-view drawing)

166

—the wing had six fences
—the wing profile was thinner, with a TsAGI S-12s at the root and a
  TsAGI SR-11 at the tip (wing aspect ratio, 4.08; taper ratio, 1.23;
  mean aerodynamic chord, 2.19 m [7 feet, 2.2 inches])
—the wing-to-fuselage junction was improved near the trailing edge
—the fuselage was lengthened by 900 mm (2 feet, 11.5 inches) in
  proportion to the sweepback increase
—the area of the airbrakes was increased to 1.76 m^2 (18.9 square
  feet)

The semimonocoque fuselage was built in two parts that joined at the main wing–fuselage splice fittings to facilitate engine removal and replacement. The cockpit was pressurized and air-conditioned. The hood was made of a 64-mm-thick bulletproof glass windshield and a sliding canopy. In the lower part of the fuselage beneath the cockpit, an inspection panel allowed for easy access to the cannon tray. The cockpit was equipped with a first-generation ejection seat controlled by handles on both armrests.

The monospar wing, reinforced with stiffeners and a stressed skin structure, had a 12-percent thickness ratio. The ailerons each had a span of 1.512 m (4 feet, 11.5 inches), an area of 0.8 m^2 (8.6 square feet), and a maximum deflection angle of plus or minus 18 degrees, and they were balanced internally. The aileron on the right side was fitted with a tab measuring 0.034 m^2 (0.37 square feet), and the aileron control was boosted by a hydraulic servo-control. The Fowler-type flaps (span, 4 m [13 feet, 1.5 inches]; area, 2.86 m^2 [30.8 square feet]) had two settings: 20 degrees for takeoff and 60 degrees for landing. The fin had a sweep angle of 55 degrees, 41 minutes at the leading edge and a total area of 4.26 m^2 (45.85 square feet), including 0.947 m^2 (10.2 square feet) for the rudder. The horizontal tail—an ASA-M airfoil—had a sweep angle of 45 degrees at the leading edge and a total area of 3.1 m^2 (33.37 square feet), including 0.884 m^2 (9.5 square feet) for the elevator.

Two bladder tanks with capacities of 1,250 l (330 US gallons) and 150 l (40 US gallons) were located behind the cockpit. They could be placed in the fuselage via the cannon access port, and because of their location their contents had no effect on the aircraft's trimming. Two store stations beneath the wing could receive either two drop tanks or two 100-kg (220-pound) or 250-kg (550-pound) bombs. Other equipment included the RSIU-3 Klen ("maple") VHF and the SRO-1 Bariy-M IFF transponder as well as the OSP-48 ILS, which included the ARK-5 Amur automatic direction finder, the MRP-48 Khrizantema ("chrysanthemum") marker receiver, and the RV-2 Kristall low-altitude radio-altimeter. Its armament was identical to that of the MiG-15: one N-37D with forty rounds and two NR-23s with 80 rpg.

Armament on the SI-2 is identical to that on the MiG-15 and MiG-15 bis. The landing light in the air intake partition was later moved.

The SI-02: third prototype built, second to fly, and first production-line MiG-17.

The SI-02 with airbrakes deployed and flaps lowered at the landing setting.

The SI-2 was flight-tested by I. T. Ivashchenko. An in-depth test sequence was planned, including even the most difficult aerobatic maneuvers. From the start Ivashchenko noted that the SI-2 was 40 km/h (22 kt) faster than the MiG-15 bis, and on 1 February 1950 he reached 1,114 km/h (602 kt) in level flight at 2,200 m (7,215 ft). But 20 March proved to be a fateful day for the aircraft and its pilot. After completing the day's exercises at 11,000 m (36,080 ft), Ivashchenko started to descend normally when suddenly the aircraft dived and crashed.

After the tragic death of the pilot, it took more than a year to establish the causes of this accident, remedy them, and build a new prototype, the SI-02. (The SI-01, whose assembly was delayed, rolled out of the factory after the SI-02 and was thus the fourth prototype.) To test this machine Mikoyan called on a military test pilot, G. A. Sedov. The SI-02 passed its factory tests and state acceptance trials in short order. In decree no. 851 of 1 September 1951 the GK NII VVS and the ministry of aircraft production ordered mass production of the aircraft in no fewer than six factories (recall that the MiG-15 was built in eight factories).

The MiG-17 was able to carry out the most complicated aerobatic maneuvers, but it was necessary to impart a greater deflection to the control surfaces than was the case with the MiG-15 bis. Moreover, its

The SI-01—fourth prototype built, third to fly—was rolled out after the SI-02, delayed by production problems. It carried two 400-l (106-US gallon) slipper tanks.

acceleration after takeoff had deteriorated somewhat. With the help of its airbrakes the MiG-17 was able to roll at any speed or altitude up to 14,000 m (45,920 feet). The area of the airbrakes was increased slightly on the assembly line starting in September 1952.

At high altitudes the MiG-17 was a very stable aircraft. It was even possible to roll in at its operational ceiling without losing much altitude. However, at 270–280 km/h (146–151 kt) its gliding descent speed was significantly faster than that of the MiG-15 bis. Several modifications were made on the assembly line to improve the aircraft's structure and safety. For instance, the ejection seat was fitted with a tight strap slinged to the pilot's bucket. With that safety device the pilot was much better secured. And like the MiG-15, the left armrest of the pilot's seat was equipped with a backup activator for the ejection seat in case the pilot's right hand was wounded in flight. As early as the end of 1953, all MiG-17s were equipped with a more modern curtain-type ejection seat that could be used safely at various speeds. This seat— developed by MiG—protected the pilot's face from the relative wind, was equipped with stabilizing panels meant to prevent a disorderly free-fall, and secured the pilot's legs. The OKB also developed for the MiG-17 a one-piece canopy (without a rear arch) that improved the pilot's field of vision in the rear to between 24 and 27 degrees on both

Performance Comparison of the MiG-17 and the Mystère IV

Aircraft	MiG-17	Mystère IV
Takeoff weight	5,200 kg (11,460 lb)	7,400 kg (16,310 lb)
Engine thrust	2,645 daN (2,700 kg st)	3,430 daN (3,500 kg st)
Maximum speed	1,114 km/h (602 kt)	1,090 km/h (589 kt)
Flight endurance	1 h 50 min	1 h 10 min
Armament	1 x 37 mm and 2 x 23 mm	2 x 30 mm

Source: MiG OKB.

sides. Nevertheless, the rearward view—so important in a fighter—was not totally panoramic. That is why periscopes, the only piece of equipment that offered a 360-degree view, soon made their appearance.

The VK-1 turbojet generated the equivalent of 12,000 shaft horsepower (shp), ten times the power of the AM-35A piston engine of the first MiG aircraft. Before long the MiG-17 was equipped with the VK-1A, which also delivered 2,645 daN (2,700 kg st) but had a much longer service life. The MiG-17 as well as the last series of the MiG-15 bis was fitted with a new starting system that could use either a ground power unit or an airborne storage battery, making the aircraft self-contained. However, the start cycle was longer in the self-contained mode (forty-four seconds versus thirty seconds). The fuel system was greatly improved by inserting a pressure relief valve in the drop tanks' pressurization pipes to ensure a regular flow of fuel at all operational speeds. Finally, the landing light was moved from the engine air intake to a place under the wing.

The SI-02 tests revealed a few structural shortcomings. For example, during one flight strong vibrations were felt in the elevator. By yanking and throttling back immediately Sedov thought he could stop the vibrations, but the elevator had already partly disintegrated. By trimming the aircraft on its glide path solely with the throttle, Sedov managed to return to the airfield and land. The cause of the vibrations was identified, and the aircraft was modified accordingly. Ever since, pilots have had nothing but praise for the performance of the MiG-17.

In tests two LII pilots, S. A. Anokhin and P. I. Kazmin, achieved Mach 1.14, a speed that was never used operationally.

At different times, nearly forty countries on three continents chose the MiG-17 for their air forces: Albania, Algeria, Afghanistan, Angola, Bulgaria, Cambodia, China, the Congo, Cuba, Czechoslovakia, East Germany, Egypt, Ethiopia, Finland, Guinea, Hungary, India, Indonesia, Iraq, Madagascar, Mali, Mongolia, Morocco, Mozambique, Nigeria, North Korea, Pakistan, Poland, Rumania, Sri Lanka, Somalia, the Sudan, Syria, Uganda, Vietnam, Yemen, and Yugoslavia. The aircraft has proven itself in combat, first in Egypt in the fall of 1956 against French Mystère IVs as well as British Vampires and Meteors. It is interesting to compare the characteristics of the MiG-17 with those of a contemporary fighter, the Dassault Mystère IV (see table). In the early 1960s MiG-17s twice engaged American F-105 Thunderchiefs and F-4 Phantoms over North Vietnam.

From this basic airframe the MiG OKB developed a full range of tactical fighters for specific missions, endowing this aircraft with a great versatility. The MiG-17 has also been used as a test bed for a number of systems intended for the next generation of fighters.

Specifications
Span, 9.628 m (31 ft 7 in); overall length, 11.264 m (36 ft 11.5 in); fuselage length, 9.206 m (30 ft 2.4 in); height with depressed shock absorbers, 3.8 m (12 ft 5.6 in); wheel track, 3.849 m (12 ft 7.5 in); wheel base, 3.368 m (11 ft 0.6 in); wing area, 22.64 m^2 (243.7 sq ft); empty weight, 3,798 kg (8,371 lb); takeoff weight, 5,200 kg (11,460 lb); max takeoff weight, 5,929 kg (13,068 lb); fuel + oil, 1,173 kg (2,585 lb); wing loading, 230–262.3 kg/m^2 (47.2–53.8 lb/sq ft); max operating limit load factor, 8.

Performance
Max speed, 1,114 km/h at 2,000 m (602 kt at 6,560 ft); max speed at sea level, 1,060 km/h (572 kt); Mach limit (fluttering conditions), 1.03; climb to 5,000 m (16,400 ft) in 2 min; to 10,000 m (32,800 ft) in 5.1 min; service ceiling, 15,600 m (51,170 ft); landing speed, 170–190 km/h (92–103 kt); range, 1,295 km at 12,000 m (805 mi at 39,360 ft); 1,185 km at 10,000 m (735 mi at 32,800 ft); range with two 400-l (106-US gal) drop tanks, 2,150 km at 12,000 m (1,335 mi at 39,360 ft); 1,907 km at 10,000 m (1,184 mi at 32,800 ft); takeoff roll, 535 m (1,755 ft); landing roll, 825 m (2,700 ft).

MiG-17F / SF

Toward the end of the 1940s aircraft manufacturers began hunting for more powerful turbojets. The way the engines were positioned on the MiG-15 and MiG-17 posed an obstacle to any increase in thrust. It had become impossible to augment the pressure ratio with a centrifugal compressor distributing air to separated combustion chambers. On the other hand, it seemed impossible to increase the thrust of axial flow engines because the turbine blade temperature had already reached its upper limit.

The service life of the VK-1F afterburner was initially limited on the MiG-17F.

With the help of the TsIAM, MiG engineers tried to obtain this thrust increase by burning fuel downstream from the turbine, an idea that proved to be the simplest and most efficient way to augment the nozzle exit impulse. The first homemade afterburner—together with its flame holder, its controlled fuel-air mixture light-up, and its adjustable nozzle—was designed and tested in the MiG OKB by a team managed by A. I. Komossarov and G. Ye. Lozino-Lozinskiy, who is presently responsible for development of the Buran space shuttle. At the time, reheat systems were not in use anywhere else in the world.

This afterburner consisted of a diffuser, the nozzle itself, and a variable exhaust nozzle breech with two open positions, 540 and 624 mm (21.26 and 24.57 inches). The basic nozzle subassembly consisted of the flame holder (a U-shaped ring) and the burner manifold. Trials and adjustments of the afterburner were carried out on TsIAM test benches, and the final product boosted engine thrust by 25 percent. The VK-1A with its afterburner was renamed the VK-1F. Its maximum dry thrust was 2,250 daN (2,600 kg st), a figure that rose to 3,310 daN (3,380 kg st) with reheat. The afterburner was internally cooled by forced convection of a part of the airflow from the engine intake ducts. The first production MiG-17 equipped with the VK-1F was no. 850. A few minor modifications had to be made in the engine bay to install the afterburner. The fuel system piping also had to be modified to take into account the significant increase in fuel flow and consumption caused by the reheat system.

MiG-17 no. 850 was reengined with the first VK-1F, thereby becoming a MiG-17F. Note the sizable fairing of the airbrake lever.

Factory tests on the SF started on 29 September 1951 with A. N. Chernoburov at the controls. Other OKB pilots such as G. A. Sedov and K. K. Kokkinaki also took part, and the tests ended on 16 February 1952. The SF was then passed to the GK NII VVS for state trials. Mass production of what became known as the MiG-17F was launched at the end of 1952. At first, use of the afterburner was limited to just three minutes at altitudes up to 7,000 m (22,960 feet) and ten minutes above that. Equipment included the R-800, RSIU-3M, or RSIU-4V VHF; the SRO-1 IFF transponder; the OSP-48 ILS with the ARK-5 ADF, the MRP-48P marker receiver, and the RV-2 radio-altimeter; the ASP-4NM gunsight; the FKP-2 monitoring camera; the S-13 camera gun; the KSR-46 flare launcher; the GSR-3000 generator; the 12 SAM25 accumulator battery; and the RD-2ZhM pressure control unit. During its state trials, a number of difficult maneuvers were carried out with the afterburner in full operation.

In the course of their service life, the MiG-17Fs underwent many modifications. In November 1953 the first turbine cooler unit fitted with an automatic temperature regulator was installed in the aircraft to improve the pilot's working conditions. Drop tanks with a capacity of 600 l (158 US gallons) were considered, but only a few were built. In early 1953 production MiG-17Fs were fitted with a collector tank to

The exhaust nozzle breech and fully deployed airbrakes of the MiG-17F.

feed the engine in negative-g flight conditions despite the fuel flow-rate increase when operating the afterburner. To reduce the pressure drop, this tank was fitted with six additional nonreturn valves. This delivered a reliable supply of fuel to the engine and the afterburner in inverted flight for at least five seconds.

The armament of the MiG-17F included one N-37D cannon with forty rounds and two NR-23s with 80 rpg. For its ground-attack role the aircraft could carry under its wing four 190-mm TRS 190 or two 212-mm ARS 212 air-to-surface rockets; or two rocket pods; or two 50-kg (110-pound), 100-kg (220-pound), or 250-kg (550-pound) bombs.

The MiG-17F could nearly break the sound barrier in level flight. Its revolutionary engine thrust augmentation device, the afterburner, would soon be adopted the world over. This model was also built in Poland, where it was called the LIM-5M.

Specifications

Span, 9.628 m (31 ft 7 in); overall length, 11.264 m (36 ft 11.5 in); fuselage length, 9.206 m (30 ft 2.4 in); height with depressed shock absorbers, 3.8 m (12 ft 5.6 in); wheel track, 3.849 m (12 ft 7.5 in); wheel base, 3.368 m (11 ft 0.6 in); wing area, 22.6 m^2 (243.3 sq ft); takeoff weight, 5,340 kg (11,770 lb); max takeoff weight, 6,069 kg (13,375 lb);

fuel, 1,170 kg (2,578 lb); max landing weight, 4,164 kg (9,177 lb); wing loading, 263.3–268.5 kg/m² (54–55 lb/sq ft); max operating limit load factor, 8.

Performance

Max speed, 1,100 km/h at 3,000 m (594 kt at 9,840 ft); with reheat, 1,145 km/h at 3,000 m (618 kt at 9,840 ft), 1,071 km/h at 10,000 m (578 kt at 32,800 ft); max speed at sea level, 1,100 km/h (594 kt); max speed with two 400-l (106-US gal) drop tanks, 900 km/h (486 kt); max permissible Mach, 1.03 (increased in 1954 to 1.15 at altitudes above 7,000 m [22,960 ft]); climb rate at sea level, 65 m/sec (12,800 ft/min); climb to 5,000 m (16,400 ft) in 2.4 min (2.1 min with reheat); climb to 10,000 m (32,800 ft) in 6.2 min (3.7 min with reheat); climb to 14,000 m (45,920 ft) in 14 min (6.3 min with reheat); takeoff speed, 235 km/h (127 kt); landing speed, 170–190 km/h (92–103 kt); range at 12,000 m (39,360 ft) with reheat operating to reach 3,000 m (9,840 ft), 1,160 km (720 mi); range at 12,000 m (39,360 ft) with two 400-l (106-US gal) drop tanks, 2,020 km (1,255 mi); range at 12,000 m (39,360 ft) with two 400-l (106-US gal) drop tanks and reheat operating to reach 3,000 m (9,840 ft), 940 km (584 mi); flight endurance at 12,000 m (39,360 ft), 1 h 52 min (1 h 40 min with reheat); flight endurance at 12,000 m (39,360 ft) with two 400-l (106-US gal) drop tanks, 3 h; service ceiling with reheat, 16,600 m (54,450 ft) (at that altitude, prototype no. 850 still had a climb rate of 3.6 m/sec [710 ft/min]); service ceiling without reheat, 15,100 m (49,530 ft); takeoff roll, 590 m (1,935 ft); landing roll, 820–850 m (2,690–2,790 ft).

MiG-17F / SP-2

The focus of this program was the performance of the RP-3 Korshun ("kite") ranging radar designed by the Slepushkin OKB. The Korshun-equipped MiG-17F was built in response to two directives from the USSR council of ministers, dated 10 June 1950 and 10 August 1951. These directives called for the development of a fighter "to intercept and destroy, day or night and whatever the weather conditions, any enemy bomber, reconnaissance aircraft, or escort fighter." The Korshun ranging radar used a Cartesian-coordinate scanning mode.

The SP-2, derived from the SI-2, differed from the basic MiG-17F in the following ways:

1. The fuselage nose section had to be modified
2. The N-37D cannon was removed, but the ammunition reserve for

The SP-2 was an experimental variant of the MiG-17F intended to assess the performance of the Korshun radar.

the remaining NR-23 guns was increased to 90 and 120 rounds, respectively
3. Several systems were relocated
4. The airbrakes opened automatically at M 1.03 and retracted as soon as the aircraft's speed dropped to M 0.97 (and they could still be operated manually)
5. The camera gun was moved to the right side of the engine air intake
6. The capacity of the rear fuselage tank was increased to 250 l (66 US gallons) from 195 l (51 US gallons)

G. A. Sedov conducted the factory tests between March and November 1951. State acceptance trials were carried out from 28 November to 29 December by NII VVS and PVO military pilots such as A. P. Suprun, Yu. A. Antipov, V. G. Ivanov, Ye. I. Dziuba, Ye. Ya. Savitskiy, and R. N. Sereda. Experiments with the Korshun radar proceeded in July and August 1951 on an I-320 fighter prototype. The Korshun was basically a modified rendition of the Toriy-A, which was tested from February to May 1950 in an SP-1. Neither radar could operate in an automatic tracking mode.

The SM-1 is in fact the MiG-15 bis 45° reengined with two Mikulin AM-5As. These turbojets were later fitted with afterburners and renamed AM-5F.

Specifications and Performance of the SP-2

Specifications and performance	Design requirements	SP-2
Takeoff weight	—	5,320 kg (11,725 lb)
Fuel capacity	—	1,510 l (399 US gal)
Maximum speed		
At 3,000 m (9,840 ft)	—	1,109 km/h (599 kt)
At 5,000 m (16,400 ft)	1,094 km/h (591 kt)	1,097 (592)
At 10,000 m (32,800 ft)	1,042 (563)	1,046 (565)
At 12,000 m (39,360 ft)	1,022 (552)	1,020 (551)
Maximum Mach number	—	1.03
Climb		
To 5,000 m (16,400 ft)	2 min	2 min
To 10,000 m (32,800 ft)	5.1	5.2
Service ceiling	15,600 m (51,170 ft)	15,200 m (49,860 ft)
Range at 12,000 m (39,360 ft)		
Without drop tanks	1,300 km (805 mi)	1,375 km (855 mi)
With drop tanks	2,500 (1,550)	2,510 (1,560)
Armament	2 x NR-23	2 x NR-23
Ammunition		90 + 120 rounds

Source: MiG OKB.

The following extract from the final test report is especially noteworthy:

1. The SP-2's performance meets the targets stipulated in the directives of the USSR council of ministers
2. The SP-2's performance data are almost identical to those of the MiG-17F
3. The aircraft's combat capabilities are limited because it is difficult for the pilot of a single-seater to follow up the Korshun data (searching for, approaching, and sighting the target) for several reasons: first, it is impossible to determine with sufficient accuracy the distance between the fighter and its target; second, it is impossible to reduce the aircraft's speed quickly when approaching the target because of the poor efficiency of the airbrakes; and third, the radar is not very reliable
4. Taxiing and taking off with two 600-l [158-US gallon] drop tanks is rather tricky because of the aircraft's inertia

For these and other reasons the NII VVS put an end to the Korshun research program. The table above is quite noteworthy because it allows us to compare the design performance data outlined in the government directives with those of the SP-2.

MiG-17 / I-340 / SM-1

When they began work on the preliminary design of a fighter capable of breaking the sound barrier in level flight in 1950, the OKB engineers decided to power it with a new, smaller Mikulin turbojet. At that time Mikulin, the engine manufacturer and academician, had just developed a big and powerful turbojet, the AM-3, to power the Tu-16 bomber. Rated at 8,575 daN (8,750 kg st), it was probably the most powerful jet engine in the world. Of course, it was much too large to use in a fighter. So Mikulin hit upon the idea of developing an engine with the same layout, operating cycle, and architecture as the AM-3 but on a scale one-third as large.

On 30 June 1950 Khrunichev, minister of the aviation industry, Mikoyan, Yakovlev, and Mikulin were called to the Kremlin to discuss the plans for the engine that, by decree of the USSR council of ministers, would power the new Yakovlev and Mikoyan fighters. This engine, referred to as the AM (Aleksandr Mikulin)-5, was not an immediate success. Numerous adjustments proved to be necessary, and it was obvious that they could be performed best on a flying test bed rather than a factory test bench. Mikoyan, who was very interested in the new engine, offered to install two AM-5s side-by-side in a MiG-15, a proven aircraft. For his part, Yakovlev proposed arranging them in pods under the wing of his new fighter, the Yak-25.

In the end the first two AM-5s replaced the single VK-1 of the MiG-15 bis 45° (the experimental aircraft that had led the way to the MiG-17). This modification was approved on 20 April 1951 by the council of ministers and renamed the SM-1. The prototype rolled out of the factory at the end of 1951 and was put into the hands of test pilot G. A. Sedov. The goals of the SM-1 tests were to improve on the performance of the MiG-17 with a minimum of modifications and to bring the AM-5A to the required level of reliability and fuel efficiency.

The AM-5A had no afterburner, and its maximum rating was 1,960 daN (2,000 kg st). But the thrust of the two engines together was greater than that of a single VK-1F with reheat. Moreover, the two AM-5As weighed 88 kg (194 pounds) less than one VK-1F. Yet it quickly became apparent that the thrust of the AM-5A was inadequate to meet the design specifications. Mikulin then decided to add an afterburner to the engine, which thus became the AM-5F and was rated at a maximum dry thrust of 2,015 daN (2,150 kg st) and a reheated thrust of 2,645 daN (2,700 kg st). Both fuel tanks—with capacities of 1,220 l (322 US gallons) and 330 l (87 US gallons)—were located in the fuselage behind the cockpit. To accommodate the required increase in airflow, the engine air intake ducts were widened. A canister for a 15-m²

The typical shape of the SM-1's dual exhaust nozzles. The aircraft was used as a test bed for the AM-5 engine.

(161–square foot) tail chute was attached to the fuselage under the tail section.

The AM-5F development flights with the SM-1 and later the SM-2 convinced Mikoyan, Mikulin, and other experts that the thrust of this engine was still inadequate for the next generation of Soviet aircraft. Mikulin embarked immediately on the creation of a new afterburner and increased the engine compressor output from 37 to 43.3 kg/sec. Out of this came a much more powerful turbojet, the AM-9, later renamed the RD-9B. The top speed of the compressor's first stage was already supersonic, and with the afterburner the thrust reached 3,185 daN (3,250 kg st). This was the engine that the MiG OKB counted on for its new supersonic interceptor.

Specifications
Span, 9.628 m (31 ft 7 in); overall length, 11.264 m (36 ft 11.5 in); fuselage length, 8.603 m (28 ft 2.7 in); wheel track, 3.849 m (12 ft 7.5 in); wheel base, 3.368 m (11 ft 0.6 in); wing area, 22.6 m^2 (243.3 sq ft); empty weight, 3,705 kg (8,166 lb); takeoff weight, 5,210 kg (11,483 lb); wing loading, 230.5 kg/m^2 (47.2 lb/sq ft).

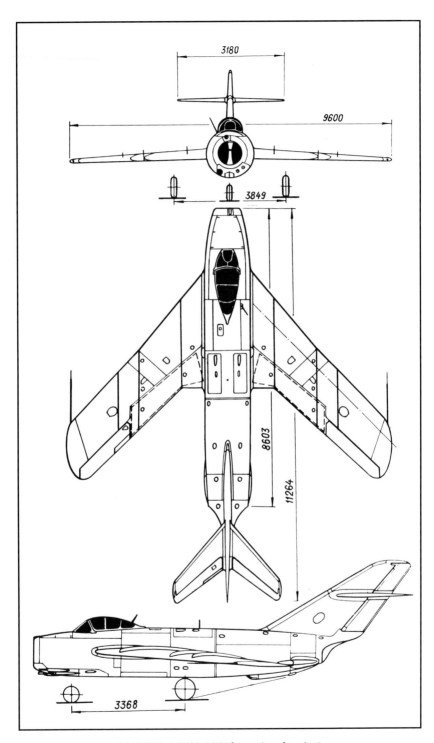

I-340 (SM-1) (MiG OKB three-view drawing)

Performance

Max speed, 1,193 km/h at 1,000 m (644 kt at 3,280 ft); 1,154 km/h at 5,000 m (623 kt at 16,400 ft); climb to 1,000 m (3,280 ft) in 0.16 min; to 5,000 m (16,400 ft) in 0.94 min; to 10,000 m (32,800 ft) in 2.85 min; to 15,000 m (49,200 ft) in 6.1 min; service ceiling, 15,600 m (51,170 ft); range, 920 km at 5,000 m (570 mi at 16,400 ft); 1,475 km at 10,000 m (915 mi at 32,800 ft); 1,965 km at 15,000 m (1,220 mi at 49,200 ft); take-off roll, 335 m (1,100 ft); landing roll, 568 m (1,863 ft).

MiG-17P / SP-7

The purpose of this program was to convert the MiG-17 day fighter into an all-weather night fighter. The radar developed for the new aircraft was supposed to provide target scanning and fire control capabilities day and night as well as in clouds.

The SP-7, powered by a VK-1A rated at 2,645 daN (2,700 kg st), differed from the MiG-17 in the nose section, which was modified to accommodate the RP-1 Izumrud radar designed by V. V. Tikhomirov. This modification led engineers to redesign the cockpit windshield and to rearrange the armament. The N-37D cannon of the MiG-17 was replaced by another NR-23, for a total of three NR-23 cannons with 100 rpg. Protection for the pilot included a bulletproof windshield, an armor plate in front of the cockpit, an armored headrest, and an armored seat back.

The SP-1 radar was combined with an ASP-3N gunsight and had two antennae: one (for scanning) housed in the upper lip of the engine air intake, and one (for ranging and fire control) housed in the air intake partition. Once the target was within 2 km (1.24 miles) the fire control antenna activated automatically to sharpen the pilot's aim. In clear weather the radar was disconnected, and the pilot used the gunsight. With the exception of the aileron controls, which were boosted by a BU-1U servo-control unit, all systems were identical to those of the MiG-17.

G. A. Sedov was the first pilot to fly the SP-7; it passed its tests in the summer of 1952. After certification as the MiG-17P, it was mass-produced for the PVO and land-based naval aviation. Approval was expedited by the fact that the RP-1 radar had been installed beforehand on the SP-5, a modified MiG-15 bis. Development of the RP-1 continued in 1953 on the ST-7, a version of the UTI MiG-15, in various weather conditions.

The MiG-17P was flown only by above-average pilots. It was the first radar-equipped lightweight interceptor ever built in the USSR.

Specifications
Span, 9.628 m (31 ft 7 in); length, 11.680 m (38 ft 3.9 in); height, 3.085 m (12 ft 1 in); wheel track, 3.849 m (12 ft 7.5 in); wheel base, 3.44 m (11 ft 3.4 in); wing area, 22.6 m² (243.3 sq ft); empty weight, 4,154 kg (9,155 lb); takeoff weight, 5,550 kg (12,232 lb); max takeoff weight with two 400-l (106-US gal) drop tanks, 6,280 kg (13,841 lb); wing loading, 245.6–277.9 kg/m² (50.3–57 lb/sq ft); max operating limit load factor, 8.

Performance
Max speed, 1,115 km/h at 3,000 m (602 kt at 9,840 ft); max speed at sea level, 1,060 km/h (572 kt); climb to 5,000 m (16,400 ft) in 2.5 min; to 10,000 m (32,800 ft) in 6.6 min; to 14,000 m (45,920 ft) in 16.2 min; climb rate at sea level, 37 m/sec (7,280 ft/min); landing speed, 180–200 km/h (97–108 kt); range, 1,290 km at 12,000 m (800 mi at 39,360 ft); with two 400-l (106-US gal) drop tanks, 2,060 km (1,280 mi); flight endurance, 1 h 53 min at 12,000 m (39,360 ft); with two 400-l (106-US gal) drop tanks, 2 h 58 min; takeoff roll, 630 m (2,065 ft); landing roll, 860 m (2,820 ft).

MiG-17PF / SP-7F

The purpose of this project was to combine the combat resources of the MiG-17P and the MiG-17F into a single aircraft (hence the equation MiG-17P + MiG-17F = MiG-17PF). Rolled out in 1952, the MiG-17PF marked a new stage in the history of the MiG-17. It was powered by the same engine as the MiG-17F, a VK-1F rated at 2,595 daN (2,650 kg st) dry thrust and 3,310 daN (3,380 kg st) reheated thrust. It carried three NR-23 cannons, just like the MiG-17P. Its fire control radar was the RP-1 Izumrud. But the plans for this aircraft contained a number of structural and equipment modifications:

— the armament array and other equipment in the nose of the fuse-lage were repositioned
— because of the size of the afterburner duct, the exhaust pipe had to be redesigned
— a cooling shroud was set between the aircraft's skin and the after-burner to protect some structurally significant items (SSI) of the fuselage
— additional hydraulic actuators were added to the afterburner control
— the GSR-3000 generator was replaced by the more sophisticated GSR-6000

—early versions of a radar warning receiver (nicknamed Sirena-2) and a ground position indicator (NI-50B) were installed

In terms of performance, the MiG-17F and the MiG-17PF were virtually identical. Despite the added takeoff weight the MiG-17PF did not differ much from the basic model except for its 360-degree turn time, which rose to 85 seconds (62 seconds with reheat), and its climb rate, which dropped to 55 meters per second (10,800 feet per minute). The MiG-17PF served in PVO units for several years before a complete reappraisal of its armament was ordered. All cannons were then removed and replaced by four radar-guided air-to-air missiles, and the MiG-17PFU was born.

The MiG-17PF was built in Poland as the LIM-5P and in Czechoslovakia as the S-104.

Specifications
Span, 9.628 m (31 ft 7 in); length, 11.68 m (38 ft 3.9 in); height, 3.8 m (12 ft 5.6 in); wheel track, 3.849 m (12 ft 7.5 in); wheel base, 3.44 m (11 ft 3.4 in); wing area, 22.6 m² (243.3 sq ft); empty weight, 4,150 kg (9,147 lb); takeoff weight, 5,620 kg (12,386 lb); max takeoff weight, 6,280 kg (13,841 lb); fuel, 1,143 kg (2,519 lb); wing loading, 245.6–277.9 kg/m² (50.3–57 lb/sq ft).

Performance
Max speed, 1,121 km/h at 4,000 m (605 kt at 13,120 ft); initial climb rate, 55 m/sec (10,800 ft/min); climb to 5,000 m (16,400 ft) in 2.5 min; to 10,000 m (32,800 ft) in 4.5 min; takeoff roll with reheat, 600 m (1,970 ft); landing roll with flaps set at 60 degrees, 830 m (2,720 ft).

MiG-17R / SR-2   SR-2s/MiG-17F

The SR-2, derived from a MiG-17F airframe, was designed to study the feasibility of a frontline photo-reconnaissance aircraft powered by the new VK-5F turbojet. The preliminary designs of the SR-2 and VK-5F were ordered by decree no. 2817-1338 signed on 3 August 1951 by the USSR council of ministers. Strangely, the aircraft's performance data were not described in the council's specifications. The structure of the cannon tray was now riveted; new equipment included a MAG-9 tape recorder and a special AFA-BA-21s camera capable of taking oblique, vertical, or double-corridor-wide photos. A hydraulic servo-control was added to the elevator control, and the instrument panel was rearranged once more.

The VK-5F was rated at 2,940 daN (3,000 kg st) maximum dry thrust and 3,775 daN (3,850 kg st) with reheat. Its afterburner was much more efficient than that of the VK-1F thanks to an increase in turbine inlet temperature, the use (for the first time) of refractory alloys resistant to thermal stress for turbine blades, and better cooling. Because of this new technology the reheated thrust was 460 daN (470 kg st) greater than that of the VK-1F—but the size, weight, and specific fuel consumption of the two engines were identical. The VK-5F was certainly an unqualified success for the Klimov OKB.

To counterbalance the weight of the SR-2 camera set, the N-37D cannon was removed. The two NR-23s with 100 rpg were retained. The pilot could use the tape recorder to note all of his observations while flying a mission, saving him the bother of taking notes on his plotting chart or remembering a lot of details until debriefing time. The camera was attached to a tilting tray that permitted to take either single- or double-corridor vertical or oblique photos with a 30-degree setting angle in relation to the horizontal (on the left side of the flight path). Small protective flaps that opened automatically before each shot were flushed into the skin of the fuselage just in front of the camera lens. While the aircraft was on the ground these flaps remained closed, protecting the lens against foreign objects. The AFA-BA-21s camera could be replaced by the more sophisticated AFA-BA-40R. The SR-2 was equipped with a curtain-type ejection seat fitted with stabilizing panels.

The aircraft was rolled out in May 1952 and made its first flight under A. N. Chernoburov in June. Factory tests continued for quite a long time—until January 1954. The state acceptance trials were conducted concurrently, lasting from July 1952 until 10 August 1954. They were carried out by two military pilots, S. A. Mikoyan and P. N. Belyasnik, both wing commanders. The state test report concluded: "1. The SR-2, powered by a VK-5F turbojet, has passed its state acceptance trials. 2. Entry into service of a VK-5F-powered MiG-17R has no justification, since its performance is not very different from that of the VK-1F-powered MiG-17F. 3. On the other hand, we recommend the production of the MiG-17R powered by the VK-1F and with the same camera installation." Thus, while the SR-2 met the state's specifications, it was not recommended for air force units. But the SR-2s—which carried the same photographic equipment as the SR-2 but was powered by a VK-1F engine—was accepted after a series of tests as the air force's frontline daytime photo-reconnaissance aircraft and, surprisingly, named the MiG-17F. In this case the "F" stood for *fotografia* (photography) and not *forsirovanie* (reheat), as is commonly presumed.

The following details refer to the SR-2.

The MiG-17PFU was equipped with the RP-1 Izumrud radar and could carry two slip-per tanks with a capacity of 400 l (106 US gallons) apiece.

Specifications
Span, 9.628 m (31 ft 7 in); length, 11.36 m (37 ft 3.2 in); height, 3.8 m (12 ft 5.6 in); wheel track, 3.849 m (12 ft 7.5 in); wheel base, 3.368 m (11 ft 0.6 in); wing area, 22.6 m² (243.3 sq ft); takeoff weight, 5,350 kg (11,790 lb); wing loading, 236.72 kg/m² (48.5 lb/sq ft).

Performance
Max speed, 1,132 km/h (611 kt); service ceiling, 16,800 m (55,100 ft); range with two 600-l (158-US gal) drop tanks, 2,115 km (1,313 mi); climb to 5,000 m (16,400 ft) in 2 min.

MiG-17PFU / SP-6

The SP-6 was basically a MiG-17PF fitted with an RP-1 Izumrud radar and four missile-launcher pylons under the wing. The air-to-air K-5 mis-sile (one per pylon) was renamed the RS-2U after its acceptance by the VVS. Its semiactive radar seeker operated with the aircraft's RP-1 radar. The experimental SP-6 retained one NR-23 cannon on the right side of

The SP-6—a MiG-17PF equipped with launch rails for four missiles—later became the MiG-17PFU.

the nose, but on the assembly line this last cannon was removed. The MiG-17PFU thus became the first missile-only MiG fighter.

MiG-17 / SN

The SN project marked the OKB's second attempt to develop rotating cannons in the vertical plane. But unlike the experimental SU (a modified MiG-15), in which the cannons' angular movement was limited by their position under the engine air intake, the rotating cannons on the SN were housed in the aircraft's nose. This arrangement led to a complete reshaping of the front of the fuselage:

—because the axial engine air intake had to be replaced by two intakes, one on each side of the fuselage, the structure of the nose was modified up to frame no. 13, making the fuselage 1.069 m (3 feet, 6.1 inches) longer

—the main gear was fitted with KT-23 wheels for better braking, and the doors were moved to the sides of the air intake ducts

—the cockpit canopy was enlarged to improve the pilot's view

—the fuel capacity was increased by 50 l (13 US gallons)

—the instrument panel was rearranged and topped by special sighting equipment

On paper, the SV-25–MiG-17 system was supposed to give the aircraft a decisive advantage. Pointing the fighter toward an intruder is a maneuver that costs a pilot many precious seconds. If he makes even the slightest error or if his adversary proves to be more agile, he has no choice but to withdraw from the engagement. If he chases an enemy aircraft in a curved trajectory, the fighter pilot has to point his aircraft toward a point in space ahead of the intruder (a process called target correction). But if the fighter's angular velocity is too low, its pilot will once more be forced to withdraw. Rotating guns are more accurate and can be pointed toward a predetermined point; moreover, they give the fighter pilot a far better chance to aim and shoot first.

The SV-25–MiG-17 system consisted of three 23-mm TKB 495 rotating cannons. The angular displacement of the weapons in the vertical plane (27 degrees, 26 minutes upward, and 9 degrees, 48 minutes downward) was electrically controlled. These experimental guns, developed in Tula by two famed armorers, Afanasyev and Makarov, had a rate of fire of 250 rounds per minute—a record for a single cannon at the time. The whole unit weighed 469 kg (1,034 pounds); the rotating support mount by itself, 142.4 kg (314 pounds); the ammuni-

The SN marked the MiG OKB's second attempt to develop rotating guns.

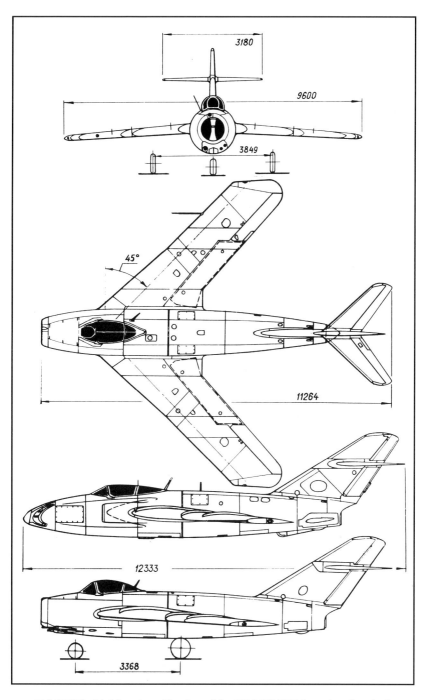

SI-2 (I-330); *third from top,* side view of the SN (MiG OKB four-view drawing)

The SV-25-MiG-17 weapons system consisted of three 23-mm cannons, two on the left and one on the right of the fuselage.

tion, 139.7 kg (308 pounds); and miscellaneous equipment, 70 kg (154 pounds).

The experimental SN prototype was the first MiG jet fighter to have lateral air intakes. From the MiG-25 forward, this became the standard arrangement on all MiGs.

The factory tests were conducted by G. K. Mosolov in 1953, and state trials began on 15 February 1954 under GK NII VVS pilots Yu. A. Antipov, A. P. Molotkov, N. P. Zakharov, S. A. Mikoyan, V. N. Makhalin, A. S. Saladovnikov, and V. G. Ivanov. They completed a total of 130 flights, mainly on a specially modified Ilyushin Il-28 twin-jet bomber; only three of the test flights involved the SN. Thirteen flights were dedicated to firing exercises against ground targets. Altogether, the pilots fired 15,000 rounds with the SV-25–MiG-17.

The results of those tests were far from satisfactory to N. I. Volkov, MiG's program manager. The SN's maximum speed proved to be 60 km/h (32 kt) slower than that of the production MiG-17. Its climb rate had also suffered: 0.4 additional minutes were needed to climb to 5,000 m (16,400 feet), and 1.5 additional minutes to climb to 10,000 m (32,800 feet). The aircraft's service ceiling was almost 500 m (1,640 feet) lower. And to top it all off, the aircraft's maneuverability had deteriorated. For instance, a tight 360-degree turn could be completed in 77 seconds at best—15 seconds slower than was possible in a production MiG-17.

When the guns were fired, other unpleasant surprises occurred. For example, firing in gusts with the three weapons rotated upward or downward altered the aircraft's flight path in the opposite direction. It was impossible to fire the cannons at all when the weapons' slew angle exceeded 10 degrees upward unless special equipment was used to balance the angular momentum of their recoil. The setbacks suffered while experimenting with rotating weaponry on both the MiG-15 and the MiG-17 convinced the OKB once and for all that any such system would be useless if installed too far from the center of gravity in single-seat fighters. All research work in that direction was subsequently abandoned.

The SN was powered by a VK-1A with a rated thrust of 2,645 daN (2,700 kg st).

Specifications
Span, 9.628 m (31 ft 7 in); length, 12.333 m (40 ft 5.5 in); height, 3.8 m (12 ft 5.6 in); wheel track, 3.849 m (12 ft 7.5 in); wheel base, 3.368 m (11 ft 0.6 in); wing area, 22.6 m² (243.3 sq ft); empty weight, 4,152 kg (9,150 lb); takeoff weight, 5,620 kg (12,385 lb); fuel, 1,455 kg (3,207 lb); wing loading, 248.7 kg/m² (51 lb/sq ft).

Performance
Max speed, 1,047 km/h at 2,000 m (565 kt at 6,560 ft); 1,058 km/h at 5,000 m (571 kt at 16,400 ft); 1,027 km/h at 10,000 m (555 kt at 32,800 ft); 986 km/h at 12,000 m (532 kt at 39,360 ft); climb to 5,000 m (16,400 ft) in 2.54 min; to 10,000 m (32,800 ft) in 6.9 min; service ceiling, 14,500 m (47,560 ft); range, not recorded.

MiG-17 / SI-10

This experimental version was developed to improve the handling of the MiG-17 by modifying its lift devices and adding a variable incidence stabilizer. This effort was made in the context of a plan drawn up by the ministry of aircraft production between 26 and 30 March 1953 to eliminate some of the shortcomings revealed during flight tests. The SI-10 was built with the MiG-17 no. 214 airframe. It differed from the production machine in several ways:

—the Fowler-type wing flaps were replaced by split flaps with take-off and landing deflections of 16 and 25 degrees, respectively

Three technical innovations on MiG-17 no. 214: automatic slats on the leading edge of the wing, spoilers, and variable incidence stabilizer with elevator.

—67 percent of the wing's leading edge was taken up by automatic slats (maximum extension angle 12 degrees)
—the standard stabilizer was replaced by a variable incidence stabilizer with elevator (surface deflection -5 to +3 degrees)
—the SI-10 received the first spoilers ever installed on a MiG fighter, with their operation linked to that of the ailerons; located on the lower surface of the wing, they extended 55 millimeters (2.16 inches) downward when the aileron displacement was greater than 6 degrees
—the wing fences were removed

The SI-10 was probably the most technically advanced fighter of its time. All of these modifications added weight to the aircraft, however: the new stabilator, 28 kg (62 pounds); spoilers, 14 kg (31 pounds); slats and flaps, 120 kg (265 pounds); and the balance weight, 70 kg (154 pounds).

The aircraft was rolled out at the end of 1954. It was powered by a VK-1A turbojet rated at 2,645 daN (2,700 kg st). Armament consisted of one N-37D and two NR-23 cannons. The factory tests took place in early 1955 with G. K. Mosolov, G. A. Sedov, and A. N. Chernoburov at the controls. State trials were completed in July. Four GK NII VVS

Close-up of the SI-10's deep-chord leading edge slats.

The SI-10 wing with lift augmentation at its best: automatic slats at their maximum extension angle of 12 degrees, and split flaps at their maximum deflection of 25 degrees.

Comparison of the Takeoff and Landing Performance of the SI-10 and SI-02

Aircraft	SI-10	SI-02
Takeoff roll	560 m (1,835 ft)	535 m (1,755 ft)
Takeoff speed	232 km/h (125 kt)	232 km/h (125 kt)
Runway needed for takeoff	1,070 m (3,510 ft)	1,260 m (4,130 ft)
Landing roll	1,095 m (3,590 ft)	825 m (2,700 ft)
Landing speed	194 km/h (105 kt)	190 km/h (103 kt)
Runway needed for landing	1,650 m (5,410 ft)	1,460 m (4,790 ft)
Angle of flaps at takeoff/landing	16°/25°	20°/60°

Source: MiG OKB.

pilots participated in these tests: S. A. Mikoyan, A. P. Molotkov, V. N. Makhalin, and N. A. Korovin. They made forty-seven flights and spent thirty-two hours and ten minutes in the air. The tests proved that the variable incidence stabilizer and spoilers' action on the pitch control significantly improved the aircraft's handling characteristics, especially at high speeds and altitudes. However, the addition of the slats and the modification of the flaps did not seem to have any effect.

Specifications
Span, 9.628 m (31 ft 7 in); length, 11.264 m (36 ft 11.5 in); height, 3.8 m (12 ft 5.6 in); wheel track, 3.849 m (12 ft 7.5 in); wing area, 22.6 m² (243.3 sq ft); empty weight, 4,140 kg (9,125 lb); takeoff weight, 5,490 kg (12,100 lb); fuel, 1,128 kg (2,486 lb); wing loading, 242.9 kg/m² (49.8 lb/sq ft).

Performance
Except as noted in the table above, the SI-10's performance data were almost identical to those of the MiG-17 (SI-02).

MiG-17PF / SP-10

In 1955 MiG-17PF no. 627 was chosen as the test bed for a new twin-barrel cannon with a high rate of fire. Two of these guns were installed on the standard armament tray, which could be lowered or lifted with the help of hoists and cables. The pilot used buttons on the stick handle to fire with one of the cannons or to launch a combined salvo. This experimental weapon failed its certification tests and was never mass-produced.

The SP-10 was a production MiG-17PF modified for tests of a new twin-barrel cannon.

The SP-10 had a support mount for its two twin-barrel cannons which resembled that of the MiG-15.

The SDK-5, as well as the MiG-9L, was used to test the guidance equipment of an air-to-surface missile.

The SDK-5 experimental prototype was unarmed.

Guidance from the mother aircraft was received by two antennae, one in the SDK-5's nose and the other on top of the fin.

MiG-17 / SDK-5

The SDK-5 was a MiG-17 modified to simulate the flight track of a winged air-to-surface missile and to test the airborne guidance equipment of the whole system (mother aircraft plus missile). On this prototype all cannons were removed. Above the air intake, a small radome housed an electromagnetic detector to vector the missile/aircraft toward its target; another small radome on top of the fin housed the antenna that received radio signals—transmitted by the mother aircraft—coordinating the missile's/aircraft's flight path. Similar equipment was mounted to a MiG-9L during development of the KS-1 cruise missile.

MiG-17 / Experimental Versions with Various Equipment and Armament

The SP-8, SP-9, SP-11, SI-05, SI-07, SI-16, SI-19, SI-21, SI-21m, and SI-91 are all MiG-17s equipped with various weapon systems—basically,

This MiG-17 (SI-16) was used to test short-tube rocket pods for eight 57-mm rockets.

This MiG-17 (SI-19) was used to test 190-mm TRS-190 heavy unguided rockets fired from tubes.

This MiG-17 (SP-8) was used to test a new ranging radar dubbed "Grad" and housed in this bay in front of the cockpit. Left, the dielectric lip for the antenna.

unguided rockets fired from pylons, pods, or tubes. On the SI-16 and SI-19, for instance, firing experiments were conducted from short rocket pods. The SI-19 was also used for experiments with heavy-caliber TRS-190 (190 mm) rockets fired from rails or tubes under the wing. Several types of bomb racks were also tested, and much attention was paid to whether the rockets interfered with underslung fuel tanks.

In 1953 the SP-9 was used to test rocket pods attached under long pylons. An automatic ZP-6-Sh device allowed the rockets to be fired one after the other. That prototype had no cannons. A new ranging radar, the Grad ("hail"), was also tested on the SP-8 that year.

The standard MiG-17 gunsight, the ASP-3M, was replaced by an experimental head-up display nicknamed Sneg ("snow") that displayed the distance from the target and the collimator reticle on the windshield front glass panel. Because of development problems, the Grad radar was abandoned and replaced by the Kvant ("quantum"), which passed its tests and was recommended for mass production. All MiG-17s on the assembly lines were equipped with this new radar, and the equipment was retrofitted on all MiG-17s already in service.

The I-350, which was supposed to be a frontline supersonic fighter, became a victim of its power plant and was flown only seven times.

I-350 / M

In 1947 A. M. Lyulka, a renowned engine specialist, began to develop a new turbojet that would produce a thrust of 4,410 daN (4,500 kg st). It was the VRD-5, later called the TR-3. This engine, based on a seven-stage axial compressor, was developed years later and built as the AL-5 and then the AL-7 with thrust limited at first to 5,095 daN (5,200 kg st). But in the late 1940s the only engine available to power the new MiG supersonic fighter was the TR-3A. The program was officially launched by a decree of the council of ministers on 10 June 1950.

The I-350—known at MiG as the M project—was a single-engine, all-metal, midwing fighter. The wing sweep at the leading edge was 60 degrees, and the wing aspect ratio was 8.6. The wing had been designed at TsAGI by a team of aerodynamicists managed by V. S. Struminski and G. S. Byushgens. The M was equipped with an RP-1 Izumrud radar whose antennae were housed in the upper lip and partition

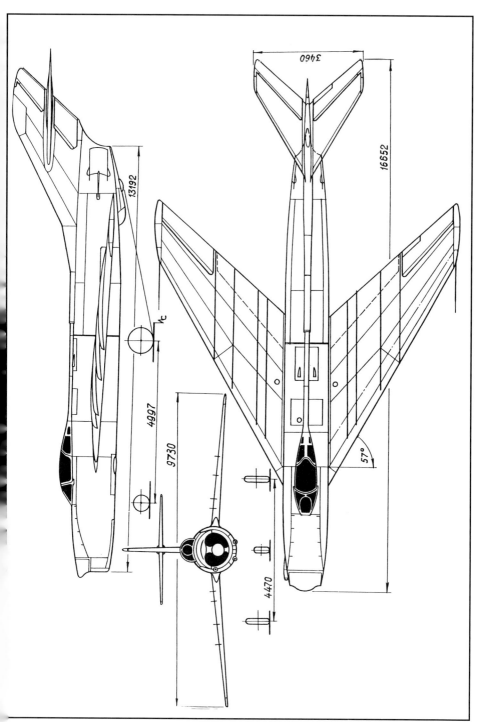

I-350 (M) (MiG OKB three-view drawing)

of the air intake, just like the MiG-17P, MiG-17PF, and MiG-17PFU. The M wing had eight fences on the upper surface. Under the rear part of the fuselage was a small ventral fin with a bumper, and the nozzle throat was flanked by two airbrakes. The I-350 was the first MiG fighter to be fitted with a tail chute. Armament was identical to that of the MiG-15 and the MiG-17: one N-37D and two NR-23 cannons mounted on a removable tray.

The flight tests were conducted by G. A. Sedov, and the first flight took place on 16 June 1951. This flight nearly ended in disaster: as Sedov throttled back to prepare to land the TR-3A shut down abruptly at an altitude of only 2,000 m (6,560 feet). Only the sangfroid, experience, and decisiveness of the pilot could save the aircraft. The engine shutdown caused a pressure drop in the hydraulic system, and the stick force increased suddenly. When Sedov tried to extend the gear with the usual hydraulically controlled lever, the right main gear strut refused to extend. The ground staff immediately notified Sedov over the radio, but the aircraft was now only 20 to 30 meters off the ground. At the last second Sedov actuated the emergency gear lever, which was pneumatically controlled. Eyewitnesses reported that the gear strut snapped into place just as the wheel touched the runway.

This first flight lasted nine minutes and ended in what was probably the first power-off landing for a supersonic fighter in the USSR. If Sedov had used the occasion to demonstrate his prowess, the TR-3A (AL-5) had proved that it was not nearly reliable enough to power a single-engine fighter. Six more flights took place before development efforts were abandoned. If the I-350 (M) program had continued, two other versions would have appeared: the M-2, equipped with Korshun radar, and the MT, a tandem-seat trainer.

Specifications
Span, 9.73 m (31 ft 11 in); overall length, 16.652 m (57 ft 7.6 in); fuselage length, 13.192 m (43 ft 3.4 in); wheel track, 4.47 m (14 ft 8 in); wheel base, 4.997 m (16 ft 7 in); wing area, 36 m² (387.5 sq ft); empty weight, 6,125 kg (13,500 lb); takeoff weight, 8,000 kg (17,630 lb); max takeoff weight, 8,710 kg (19,197 lb); wing loading, 222.2–241.9 kg/m² (45.5–49.6 lb/sq ft).

Performance
Max speed, 1,266 km/h at 10,000 m (684 kt at 32,800 ft); max speed at sea level, 1,240 km/h (670 kt); climb to 5,000 m (16,400 ft) in 1.1 min; to 10,000 m (32,800 ft) in 2.6 min; service ceiling, 16,600 m (54,450 ft); range, 1,120 km (695 mi); range with 800-l (211-US gal) drop tank, 1,620 km (1,006 mi).

MiG-19 Series

The First Soviet Supersonic Fighter

The MiG-19 was taken for its premier flight on 5 January 1954 by G. A. Sedov, now the chief constructor at the Mikoyan OKB. It is no secret that the transition to supersonic speed was lengthy, tricky, and bloody. Ivashchenko died in a MiG-17, and many pilots were lost in other OKBs and aircraft manufacturers all the world over.

Some pilots succeeded in reaching and even surpassing Mach 1 for a short while, but for the true supersonic effect one had to maintain that speed for a long time in level flight. The SM-2, the first prototype of the MiG-19, seemed to have all of the prerequisites for supersonic flight: a thin wing with a high sweep angle (57 degrees), a reduced master cross-section, and a pair of AM-5A engines, a new type of compact and efficient turbojet. But the problem proved to be far more intricate than expected. More than one person would have lost heart, but all concerned clenched their teeth and committed themselves deeply.

Work on the engine got under way when it was decided to add an afterburner to the axial flow AM-5A turbojet. Mikulin knew how to make a success of this afterburner with an efficient flame holder that did not reduce the gas rate of flow in the combustion chamber. The armorer N. I. Volkov moved two of the three cannons and their ammunition into the leading edge of the wing near the root. In this manner, empty space in the wing was filled and some much-sought-after room was made in the fuselage for new equipment.

For their part, A. G. Brunov, deputy chief constructor, and R. A. Belyakov, department manager, developed a new servodyne-powered flight control unit. The variable incidence stabilizer was replaced by a stabilator, a single pivoted tailplane (without elevator) for pitch control (also called a slab tailplane). All flight control systems were duplicated to guard against failure of the main unit, and the stabilator was fitted with a booster control and an artificial feel unit. (As explained by Bill Gunston in *Jane's Aerospace Dictionary,* "In aircraft control system artificial feel can be explained by forces generated within system and fed to cockpit controls to oppose pilot demand. In fully powered or boosted system there would otherwise be no feedback and no 'feel' of how hard any surface was working.") The engine flameout problems that occurred during cannon tests with the MiG-9 had not been forgotten, and everything was done to dodge the difficulty. The ejection procedures were also improved to protect the pilot at much higher speeds.

A lot of useful information was collected during the SM-2 flights. Unexpected spins occurred due to the blanketing effect of the wing on the stabilator at great angles of attack (AOA). The aircraft had to be returned to the wind tunnel, and tests there led engineers to move the stabilator from the top to the base of the fin. Moreover, the location of the wing fences was modified. This is how the SM-2 became the SM-9.

At this time the North American F-100 Super Sabre could not exceed Mach 1.09. From the start Sedov reached Mach 1.3 or 1,400 km/h (756 kt) in the MiG-19 and thereby beat—unofficially—the world speed record. But there remained many youthful inadequacies to cure. The stretch of the turbine blades at high rotation speeds ceased to be a problem once new heat-resistant steel was used to make the blades. The inadequate roll handling was improved by placing spoilers ahead of the ailerons. The longitudinal swings noticed at high speeds vanished thanks to the new artificial feel system. The pressure surges felt on the rudder pedals at transonic speed were remedied by initiating a vortex flow—or burbling—on the rear of the fuselage. All of this was done step by step.

Only fourteen months after the SM-9's first flight, two production MiG-19s were delivered to a fighter regiment. The MiG-19 was mass-produced and operated in many countries.

I-360 / SM-2/SM-2A/SM-2B

To develop a fighter capable of supersonic speeds in level flight, many requirements had to be met:

- —the layout had to have the smallest possible master cross-section to reduce drag
- —the drag of the wing and the tail assembly had to be reduced by increasing their sweep angle at the leading edge
- —a series of intricate technical problems had to be resolved in designing duplicate flying controls, artificial feel systems, supersonic air intakes, and the like
- —the engines and fuel systems had to be positioned to prevent flameouts during maneuvers within the aircraft's speed and altitude range, including when firing the cannons

The SM-2 became the flying laboratory that allowed engineers to explore ways to get beyond the sound barrier.

The SM-2 was designed in record time under the supervision of A. G. Brunov, deputy chief constructor, and R. A. Belyakov, who was then

The SM-2 no. 01 before being modified with its T-tail.

The same SM-2 after modification of its tail unit. The stabilizer was lowered to the base of the fin to avoid the wing-blanketing effect.

The SM-2 no. 02, built at the same time as no. 01, also had a stabilizer set high on the fin.

chief of the general affairs brigade. A. A. Chumachenko took care of the aerodynamic design while V. M. Yezuitov studied pilotage and handling problems. Engineer A. V. Minayev played a great part in the development of the SM-2. G. Ye. Lozino-Lozinskiy was put in charge of the power unit. The stress analysis was placed under the management of D. N. Kurguzov, who had worked with N. N. Polikarpov before World War II.

The first SM-2 was a midwing, T-tail, twin-jet fighter. The wing sweep back C/4 was 55 degrees with a 4-degree, 30-minute anhedral. The sweep of the stabilizer and fin leading edges was 55 and 56 degrees, respectively. The wing structure was identical to that of the I-350 (M) except that there were only two fences on the wing's upper surface. Armament consisted of two N-37D cannons located in the leading edge, near the wing roots. Rolled out in April 1952, the SM-2 made its first flight, with G. A. Sedov in the cockpit, on 24 May.

It soon became obvious that the aircraft could not really exceed Mach 1 in level flight. It did reach Mach 1.19—but in a shallow dive. At 3,920 daN (4,000 kg st) the cumulative thrust of the two first-series AM-5A turbojets was not sufficient because they lacked an afterburner. The engines were replaced by reheated AM-5Fs—first developed for the SM-1—rated at 2,645 daN (2,700 kg st). Other faults were noted in the aircraft's aerodynamic qualities and fuel control system.

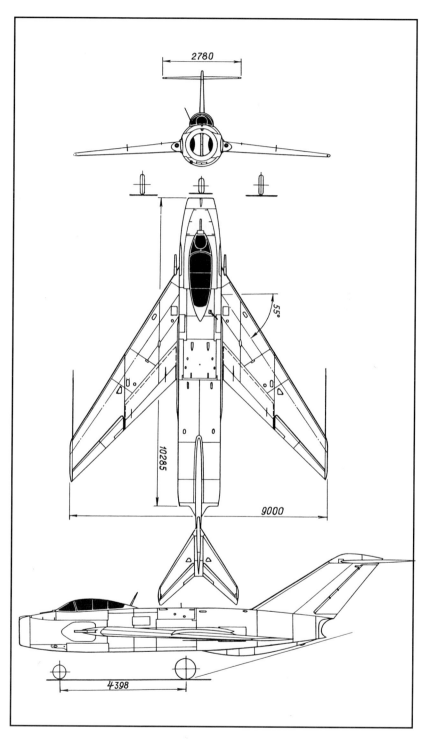

I-360 (SM-2) (MiG OKB three-view drawing)

The rear of SM-2 no. 02, with its airbrakes set at their maximum deployment.

The spin problem was solved by moving the stabilizer to the base of the fin and modifying the location of the wing fences. Various other changes put a stop to engine flameouts and surges. After completing its factory tests, the SM-2 commenced its state trials in early 1953. They proceeded normally until V. G. Ivanov, a military pilot, discovered a serious shortcoming: a pitch instability caused by diminution of the stabilizer's efficiency at high speeds. The flight tests were canceled, and the prototype was returned to the factory for modifications. The stabilizer was lowered once more and positioned on the rear section of the fuselage. The tailplanes on MiG fighters have remained on the fuselage and "abandoned" the fin ever since. Moreover, to suppress the buffeting caused by their deployment, the airbrakes were brought closer to the wing and lowered in relation to the fuselage datum line.

Once modified, the SM-2 became the SM-2A and later the SM-2B. The aircraft resumed its state trials in the summer of 1953. In fact, two SM-2s were built. In light of the test results both prototypes received the same modifications, especially those involving the stabilizer.

Specifications
Span, 9.04 m (29 ft 7.9 in); overall length, 13.9 m (45 ft 7.2 in); fuselage length, 10.285 m (33 ft 8.9 in); height, 3.95 m (12 ft 11.5 in); wheel

The stabilizer also had to be lowered to the base of the fin on SM-2 no. 02.

Among other modifications, the wing fences on SM-2 no. 02 were given a deeper chord.

Reengined with two AM-9Bs, the SM-2B became the SM-9/1—true prototype of the MiG-19.

track, 4.156 m (13 ft 7.6 in); wheel base, 4.398 m (14 ft 5.2 in), takeoff weight, 6,820 kg (15,030 lb).

Design Performance
Mach limit, 1.19.

MiG-19 / SM-9/1

After the failure of the SM-2, the only way for the project to move forward was to equip the AM-5F turbojet with a modified afterburner. Once modified, the engine was named the AM-9B (AM-9 being its preliminary design designation). Its dry thrust was now 2,550 daN (2,600 kg st), rising to 3,185 daN (3,250 kg st) when reheated and 6,370 daN (6,500 kg st) when paired—a figure that met the needs of the OKB engineers. Two test engines were installed in the SM-2B airframe. With these two AM-9Bs and a few modifications of the fuselage to accommodate the new afterburners and protect the structure from high temperatures, the SM-2B became the SM-9/1. The development of this aircraft was ordered by decree no. 2181-887 of the USSR council of ministers, dated 15 August 1953.

All flight controls were boosted, and the nonrotating tailplane was fitted with an elevator. The midwing sweepback C/4 was 55 degrees,

The air probe on the SM-9/1 could be hinged upward to avoid any damage that could be caused by ramp vehicles.

and the stabilizer sweep was 55 degrees at the leading edge. The cockpit was pressurized and air-conditioned with cold- and hot-air bleeds from the engine. The temperature inside the cockpit remained uniform thanks to a special temperature regulator with an automatic display. The ejection seat was of the curtain type, a device that protected the pilot's hands and face when ejecting at high speeds. The tail chute was housed in a canister under the rear fuselage. Armament consisted of three NR-23 cannons, two in each wing root and one on the lower right side of the fuselage.

The team in charge of the SM-9 tests was G. A. Sedov, chief pilot, V. A. Arkhipov, chief engineer, and V. A. Mikoyan, Arkhipov's assistant. The new engines were monitored by two specialists from the Mikulin OKB, I. I. Gneushev and V. P. Shavrikov. The SM-9's first flight took place on 5 January 1954 with Sedov at the controls. During that flight the engines ran smoothly but the afterburners were not used. The pilot found the aircraft easy to handle and capable of supersonic speed. During the second flight Sedov lit up the afterburners and broke the sound barrier, a procedure that he repeated many times in the course of the tests. On 12 September 1954 the factory tests ended; on 30 September the state acceptance trials commenced.

The SM-9 was clearly an aircraft with a future. The official test report made the point this way: "The SM-9/1's performance data are far better than those of the MiG-17F. The former is 380 km/h [205 kt]

The main features of the SM-9/1 are its Fowler-type flaps, fixed stabilizer with elevators, and deep-chord wing fences.

This photograph was taken before the wing-root cannons were installed. Their place was occupied by a provisional fairing.

214

faster at 10,000 m [32,800 feet], and its service ceiling is 900 m [2,950 feet] higher." This report, approved by Marshal Zhigarev, the air force commander-in-chief, recommended the SM-9 (designated the MiG-19 by the military) for its units.

Well before the end of the state trials, the council of ministers issued decree no. 286-133 on 17 February 1954 ordering the mass production of the MiG-19 in two factories, one in Gorki and the other in Novosibirsk. Initiating a rather uncommon procedure, the council of ministers ordered the ministry of aircraft production to build (and the ministry of defense to accept) the first fifty aircraft and the first hundred turbojets from the design office blueprints and not, as was customary, from the production sets of drawings, because the latter were not yet ready. The first MiG-19s were delivered to the air force in March 1955.

Performance
Max speed, 1,451 km/h at 10,000 m (784 kt at 32,800 ft); max speed at sea level, 1,150 km/h (620 kt); climb to 10,000 m (32,800 ft) in 1.1 min; to 15,000 m (49,200 ft) in 3.7 min; service ceiling, 17,500 m (57,400 ft).

MiG-19S / SM-9/2 and SM-9/3

Development of the next two prototypes, the SM-9/2 and SM-9/3, was intended to improve the handling of the MiG-19 with a stabilator or slab tailplane. While satisfactory on the whole, tests of the SM-9/1 uncovered some inadequacies, especially a decreasing linear acceleration at supersonic speeds. The answer was to design a linkage for the stabilator control that would generate acceptable control column forces and prevent the pilot from imparting a longitudinal swing to the aircraft through the whole range of speeds and altitudes. Test flights made by G. A. Sedov, K. K. Kokkinaki, and V. A. Nefyedov demonstrated the necessity of such a device. On several occasions the SM-9/2 reached very dangerous flight regimes, mainly when the aircraft started to swing and the pilot's use of the stabilator did nothing but increase the swing rate.

The SM-9/2 and SM-9/3 were built in 1954, one after the other. They differed from the SM-9/1 in their slab tailplane and other details:

—for the first time, ejection of the cockpit hood was controlled by pneumatic cylinders
—to increase the efficiency of the lateral control at high Mach numbers, spoilers mechanically linked to the ailerons were placed ahead of the flaps on the underwing

The SM-9/3 was the master aircraft for the mass-produced MiG-19.

—both pitch and roll channels were equipped with irreversible servo-controls driven by their own hydraulic circuit, the utility hydraulic system being used as a backup system; switching over the utility system was automatic when hydraulic pressure dropped below 65 kg/cm² (925 psi)
—the pitch control system (actuating rods) was, as a master control, equipped as well with an irreversible servo-control, the utility hydraulic system being also used as a standby system
—the slab tailplane had both third- and fourth-level emergency controls (an electromechanism actuated by the control column itself and a set switch on the column, respectively); the electromechanism cut in automatically when hydraulic pressure dropped below 50 kg/cm² (710 psi)
—the gear ratio between the control column and the slab tailplane changed according to the dynamic pressure and flight altitude—that is, according to the Mach number—thanks to the ARU-2A automatic feel control unit. The control column forces on the longitudinal axis were controlled by a Q-spring assembly in the ARU-2A mechanism. The aerodynamic hinge moment of the slab tailplane was not fed back to the column. This device allowed the pilot to master the aircraft's handling characteristics without having to think about the dynamic pressure or the Mach number. It was designed, tested, and built by a highly talented engineer and a historian of aviation, A .V. Minayev, who broke new ground in the field of flying control systems and was later appointed chief constructor and deputy minister of aircraft production.

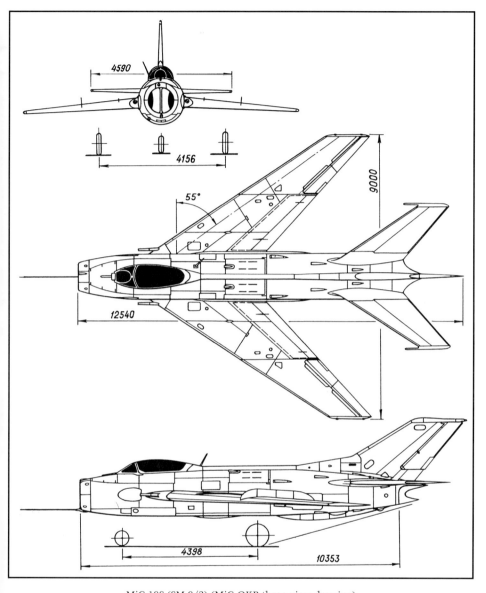

MiG-19S (SM-9/3) (MiG OKB three-view drawing)

—a 0.54-m^2 (5.8–square foot) ventral fin was added to improve directional stability

—both prototypes were equipped with three airbrakes, two flanking the rear fuselage and one under the middle part of the fuselage

The armament of the SM-9/2 consisted of three NR-23 cannons (two in the wing roots and one on the right side of the lower forward fuselage). It could also carry two or four rocket pods for 57-mm ARS-57 rockets. The main on-board systems included the RSIU-4 Dub ("oak") VHF, the SRO IFF transponder, a radar warning receiver, the SRD-3

The MiG-19S had a slab tailplane and a third airbrake under the fuselage developed on the SM-9/3.

Grad ("hail") or SRD-1M Konus ("cone") ranging radar linked to the ASP-5M gunsight, and the OSP-48 ILS. The SM-9/2 was moved to the flight test center in September 1954 and was taken up by G. A. Sedov on 16 September.

As of 4 May 1955 the OKB and GK NII VVS pilots had made fifty-eight flights and noted the outstanding qualities of the SM-9/2, especially its climb rate of 180 meters per second (35,400 feet per minute) at sea level. The appraisal of OKB pilots Sedov, Mosolov, Kokkinaki, and Nefyedov and NII VVS pilots Blagoveshchenskiy, Antipov, Ivanov, Molotkov, Beregovoy, and Korovin was very positive. The state trials proved that both prototypes were sufficiently long-legged for fighters with a range of 1,300 km (810 miles), and that the sound barrier was no longer a barrier at all. Because of the ARU-2, the slab tailplane, and many other technological advances, the shortcomings of the SM-2 were just a bad memory now.

The aircraft was continuously updated during the tests. For example, its balance range at takeoffs and landings was increased by shortening the displacement of the control column. Mosolov reached Mach 1.462 in the SM-9/2 by starting a dive at 9,300 m (30,500 feet). The SM-9/3 was rolled out on 26 August 1955 and went up for its first flight on 27 November with Kokkinaki at the controls.

The SM-9/3 differed slightly from the SM-9/2. The three NR-23 cannons of the latter were replaced by three NR-30s. A one-second salvo weighed 18 kg (40 pounds) as opposed to 9 kg (20 pounds) in the

SM-9/2. The SM-9/3 also reached Mach 1.46 and became the master aircraft for the MiG-19, which was mass-produced in two factories.

Specifications

Span, 9 m (29 ft 6.3 in); length (except nose probe), 12.54 m (41 ft 1.7 in); overall length, 14.64 m (48 ft 0.4 in); fuselage length, 10.353 m (33 ft 11.6 in); height with depressed shock struts, 3.885 m (12 ft 8.9 in); wheel track, 4.156 m (13 ft 7.6 in); wheel base, 4.398 m (14 ft 5.2 in); wing area, 25 m² (269 sq ft); takeoff weight, 7,560 kg (16,660 lb); max takeoff weight with two 760-l (201-US gal) drop tanks and two rocket pods, 8,832 kg (19,466 lb); fuel, 1,800 kg (3,970 lb); wing loading, 302.4–353.3 kg/m² (62–72.4 lb/sq ft).

Performance

Max speed, 1,452 km/h at 10,000 m (784 kt at 32,800 ft); with two 760-l (201-US gal) drop tanks, 1,150 km/h (620 kt); max operating limit Mach number, 1.44; climb to 5,000 m (16,400 ft) in 0.4 min; to 10,000 m (32,800 ft) in 1.1 min; to 15,000 m (49,200 ft) in 2.6 min; range, 1,390 km at 14,000 m (860 mi at 45,900 ft); with two 760-l (201-US gal) drop tanks, 2,200 km (1,365 mi); service ceiling, 17,500 m (57,400 ft); dynamic ceiling, 20,000 m (65,600 ft); takeoff roll with reheat, 515 m (1,690 ft); with dry thrust, 650 m (2,130 ft); with dry thrust and two 760-l (201-US gal) drop tanks, 900 m (2,950 ft); landing roll with main gear braking, 1,090 m (3,575 ft); with all-wheel braking, 890 m (2,920 ft); with all-wheel braking and tail chute, 610 m (2,000 ft).

MiG-19P / SM-7

Another derivative of the SM-2, this tactical interceptor was designed for day, night, and bad weather use. The SM-7 differed from the SM-9 in many points:

—it was fitted with the RP-1 radar linked to the ASP-5NM gunsight, an IFF interrogator, and a radar warning receiver
—the radio racks were rearranged, and the radar was housed separately in a more accessible bay
—armament was reduced to two NR-23 cannons in the wing roots, and the ammunition belts (up to 120 rpg) were placed in wing leading edge ducts
—from the start, it was expected that NR-30 cannons would replace the NR-23s

The SM-7/1, an interceptor developed from the SM-2, was equipped with the RP-1 radar.

The SM-7/1 as well as the SM-9/1 had a fixed stabilizer. One of the two radio-altimeters is visible under the right wing.

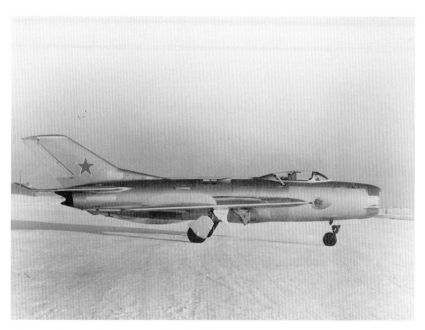

The SM-7/2 as well as the SM-9/3 had a slab tailplane.

The slab tailplane or all-flying tail of the SM-7/2 at its maximum pitch-up deflection.

The SM-7/2—the future MiG-19P—with all of its three airbrakes deployed.

—two RO-57-8 rocket pods with eight ARS-57 rockets apiece could be
    attached to wing pylons, as well as two 760-l (201-US gallon)
    underslung tanks
—the underslung tanks could be replaced by two additional rocket
    pods with sixteen ARS-57 rockets apiece
—because of the radar antennae, the forward section of the fuselage
    had to be modified up to the no. 9 frame and lengthened by 360
    millimeters (14.17 inches)
—the air intake ducts were reshaped accordingly, and their section
    was enlarged
—both wing tips were fitted with Pitot-static probes, the one on the
    right linked to the navigational instruments and the one on the
    left to all the others
—because of the changes made to the forward section of the fuse-
    lage, it was possible to widen the cockpit
—the pilot had a curtain-type ejection seat whose height could be
    adjusted to three positions over a range of 120 mm (4.7 inches)
—the seat back had a 16-mm-thick armor plate behind the rails of
    the ejection system, which were protected (as was the locking pin
    of the ejection pyrotechnic system) against shrapnel and even
    direct hits

—the control rods actuated by the column were moved to beneath the cockpit floor

—ventilation of the adjustable-area nozzle was improved by modifying the cooling air supplies

—the RSIU-3 VHF and the ARK-5 automatic direction finder were linked to a single antenna on the fuselage

—the DGMK-5 gyroscopic compass and the ARK-5 ADF shared a common indicator on the instrument panel

—the oxygen supply for the pilot was increased to five 2-1 (0.53-US gallon) spherical bottles

The first SM-7, powered by two Mikulin AM-9Bs (RD-9Bs) with rated thrust of 3,185 daN (3,250 kg st) each, rolled out of the factory in July 1954 and made its first flight under V. A. Nefyedov on 28 August. Factory tests lasted until 15 December; OKB pilots made forty-three flights for a total of twenty-five hours and thirty-six minutes. The aircraft was then moved to the NII VVS, where it was certified. It was mass-produced from 1955 and delivered to VVS, PVO, and VMF units under the military designation MiG-19P.

Following the production of the MiG-19S with a slab tailplane, the OKB built the SM-7/2 in 1955. This second prototype was also equipped with a slab tailplane and had the same equipment as the SM-7. It was tested, certified, and mass-produced as well.

Development of the SM-7 and SM-9/1 was basically the work of A. G. Brunov, deputy chief constructor; V. A. Arkhipov, Yu. A. Korolyev, and A. N. Soshin, main constructors; and G. Ye. Lozino-Lozinskiy, R. A. Belyakov, K. N. Rozanov, A. A. Nefyedov, Ya. I. Seletskiy, and N. I. Andrianov, brigade heads.

Specifications

Span, 9 m (29 ft 6.3 in); fuselage length, 10.48 m (34 ft 4.6 in); height, 4.02 m (13 ft 2.3 in); wing area, 25 m^2 (269 sq ft); empty weight, 5,468 kg (12,050 lb); takeoff weight, 7,384 kg (16,275 lb); takeoff weight with two 760-1 (201-US gal) drop tanks, 8,738 kg (19,260 lb); fuel, 1,700 kg (3,750 lb); wing loading, 295.4–349.5 kg/m^2 (60.6–71.6 lb/sq ft); max operating limit load factor, 7.95.

Performance

Max speed, 1,370 km/h at 10,000 m (740 kt at 32,800 ft); 1,255 km/h at 14,700 m (678 kt at 48,200 ft); max speed without reheat, 1,083 km/h at 10,000 m (585 kt at 32,800 ft); max Mach number, 1.384 at 9,800 m (32,150 ft), 1.296 at 8,700 m (28,540 ft); max speed at 1,200 m (3,940 ft), 1,180 km/h (637 kt); climb to 10,000 m (32,800 ft) with afterburner light-up at 5,000 m (16,400 ft) in 1.85 min; to 15,000 m (49,200 ft) with afterburner light-up at 5,000 m (16,400 ft) in 3.8 min; climb to 10,000 m (32,800 ft) without afterburner in 2.25 min; to 15,000 m (49,200 ft) in 6.1 min; service ceiling with afterburner, 17,600 m (57,700 ft); without afterburner, 16,150 m (52,970 ft); with afterburner and two 760-l (201-US gal) drop tanks, 16,000 m (52,500 ft); range, 1,474 km at 12,000 m (915 mi at 39,360 ft); with two 760-l (201-US gal) drop tanks, 2,318 km (1,440 mi).

MiG-19PM / SM-7/M   MiG-19PMU / SM-7/2M

The K-5 air-to-air missile was developed and mass-produced in the mid-1950s, and the Mikoyan OKB was ordered to design a version of the MiG-19 armed solely with these guided missiles. On January 1956 A. I. Mikoyan confirmed that the OKB was working on the preliminary design for the MiG-19PM armed with four K-5M missiles (M = *Modernizinrovannikh:* modernized). Once certified, the K-5M received the military designation of RS-2US. The future MiG-19PM was assigned the factory code SM-7/M.

The K-5M missiles were guided toward the target through a zone of equi-signals transmitted by the antenna of the RP-2U Izumrud-2 airborne radar. Due to the radar installation in the nose of the aircraft, the forward section of the fuselage had to be modified. The SM-7/M wing was identical to that of the MiG-19P except for the addition of the K-5M pylons and the removal of the wing cannons. The tailplane was also the same as that of the SM-7/1 and therefore had an elevator. The hydraulic system was identical as well. The SM-7/M was powered by two AM-9B (RD-9B) reheated turbojets that each generated 3,185 daN (3,250 kg st) of thrust. Its navigational instruments matched those of the MiG-19P with the exception of the DGMK-3 gyrocompass heading repeater, which was replaced by the GKI-1 earth inductor gyrocompass.

For the first time, the aircraft was fitted with an emergency right-left switchover from one wing tip probe to the other. The Izumrud-2 radar, an upgraded version of the RP-1, was linked with the ASP-5N sight for firing the K-5M missiles. This radar unit could spot a target ahead, plot its path in relation to the fighter's position (heading and distance) while it was still out of sight, bring the fighter toward the target to a suitable distance, and transmit coded pulses (together with the IFF interrogator) to establish the target's identity. It could detect targets in the forward sector at bearing angles of plus or minus 60 degrees and at elevation angles between plus-26 degrees and minus-14 degrees in relation to the aircraft's longitudinal axis. It was also capable of offering the pilot a choice of attack paths on the radar scope placed in the aircraft cockpit. Once the fighter had closed to within 3,500–4,000 m (11,480–13,120 feet) the Izumrud-2 automatically fed the ASP-5N sight the target's distance, bearing, and elevation coordinates, whatever the visibility conditions.

APU-4 launch rails were fastened to the wing pylons so as to fire either K-5M missiles or ARS-160 and ARS-212M unguided rockets. The missiles were electrically triggered by fire buttons located on the control column through a PUVS-52 active-inert control panel.

The SM-7/M made its debut in January 1957 with G. A. Sedov at the controls. After being certified, it was mass-produced with the military designation of MiG-19PM. Shortly thereafter the SM-7/2M was brought out for tests. It differed from the first prototype only in its slab tailplane. Most of the state trials and acceptance flights were made by S. A. Mikoyan, a military pilot. The SM-7/2M was flight-tested with K-5M missiles from 14 to 23 October 1957. It was certified and mass-produced as well under the military designation MiG-19PMU.

Specifications
Span, 9 m (29 ft 6.3 in); fuselage length, 10.48 m (34 ft 4.6 in); height, 4.02 m (13 ft 2.3 in); wing area, 25 m² (269 sq ft); takeoff weight, 7,730 kg (17,040 lb); takeoff weight with two 400-l (106-US gal) drop tanks, 8,464 kg (18,655 lb); wing loading, 309.2–338.6 kg/m² (63.4–69.4 lb/sq ft).

Performance
Max speed, 1,250 km/h at 10,000 m (675 kt at 32,800 ft); 1,130 km/h at 15,000 m (610 kt at 49,200 ft); without reheat, 1,100 km/h at 5,000 m (594 kt at 16,400 ft); 965 km/h at 14,000 m (520 kt at 45,900 ft); climb to 5,000 m (16,400 ft) with reheat in 4.8 min; climb to 5,000 m (16,400 ft) with dry thrust in 7.2 min; service ceiling with reheat, 16,700 m (54,800 ft); service ceiling with dry thrust, 15,000 m (49,200 ft); range, 1,000 km at 10,000 m (620 mi at 32,800 ft); with two 400-l (106-US gal) drop tanks, 1,415 km (880 mi).

MiG-19SV / SM-9V

In 1955 the cold war had turned up the heat on more than one political leader. Soviet airspace was being systematically violated by balloons carrying all sorts of detection equipment and by high-flying Canberra reconnaissance aircraft. It was at this time that the OKB first received information about the development in the United States of the Lockheed U-2, a reconnaissance aircraft that had a service ceiling of 25,000 m (82,000 feet). The situation was deemed serious because the USSR did not have a single aircraft capable of intercepting such high-altitude invaders.

A crash program was set up to counter the threat. The consensus was to build specialized high-altitude interceptors and in the meantime to modify the MiG-19S to improve its service ceiling—hence the "V" of the designation, which stands for *Visotniy* (altitude). The OKB quickly decided to make the following changes in the production aircraft:

—increase the wing area by 2 m² (21.5 square feet)
—remove the two NR-30 wing cannons; only the fuselage cannon was retained
—take the armor plate out of the pilot's seat back
—raise the turbine inlet temperature (TIT) of the AM-9B to 730° C (1,378° F); the modified engine was renamed the AM-9BF
—add a 12-degree flap setting to be used at 15,000 m (49,200 feet); the deployment of flaps during flight maneuvers marked a first in the USSR

G. K. Mosolov and V. A. Nefyedov dealt briskly with the SM-9V tests under the management of V. A. Arkhipov. The prototype was then handed to military test pilots. During high-altitude flights, the KKO-1 oxygen dispenser was tested. It secured a high oxygen pressure in the pilot's mask. Moreover, a research center designed and tested the VSS-04A pressure suit, a piece of equipment that was considered essential because the smallest pressure loss at high altitudes—whether caused by a direct hit or a tiny crack in the cockpit hood—could lead to the pilot's death. The pressure suit was also vital in case the pilot needed to eject at high altitudes and high speeds. The OKB brain trust, with Mikoyan in the lead, agreed to give the highest priority to the development of this pressure suit within the context of the SM-9V program.

The VSS-04A tests were carried out by two OKB pilots, K. K. Kokkinaki and V. A. Nefyedov, first in an altitude chamber and then in flight. In the altitude chamber, both pilots "climbed" to 25,000 m (82,000 feet), a first in the USSR. The suit was developed very quickly, and its use became normal practice. Soon afterward, the pressure suit was supplemented by the GSh pressure helmet. The combination allowed pilots to fly as high as 24,000 m (78,700 feet). OKB pilots G. A. Sedov, K. K. Kokkinaki, and G. K. Mosolov and military test pilots S. A. Mikoyan, V. P. Vasin, and V. S. Ilyushin quickly got used to the high-altitude equipment and to the SM-9V, which was mass-produced as the MiG-19SV. On 6 December 1956 N. I. Korovushkin, a GK NII VVS pilot, climbed to the record altitude of 20,740 m (68,030 feet) by using the zoom technique.

(A zoom is an optimized steep climb at high altitude, normally starting at the aircraft's maximum level Mach and trading speed for height to reach exceptional altitudes far above its sustainable level ceiling.)

The test report on the USSR's first high-altitude interceptor reads, "The MiG-19SV does not differ much from the MiG-19S as far as handling technique is concerned. On the other hand, at low speeds in the 350–380 km/h [189–205 kt] range the aircraft handles better and proved to be steadier in flight than the prototype." Several MiG-19SVs were powered by AM-9BF and BF-2 turbojets rated at 3,235 daN (3,300 kg st). One of them topped 1,572 km/h at 10,000 m (849 kt at 32,800 ft).

Specifications
Span, 10.3 m (33 ft 9.5 in); overall length without probe, 12.54 m (41 ft 1.7 in); with probe, 14.64 m (48 ft 0.4 in); wheel track, 4.156 m (13 ft 7.6 in); wheel base, 4.398 m (14 ft 5.2 in); wing area, 27 m^2 (290.6 sq ft); empty weight, 5,580 kg (12,300 lb); takeoff weight, 7,250 kg (15,980 lb); wing loading, 268.5 kg/m^2 (55 lb/sq ft).

Performance
Max speed, 1,420 km/h at 10,000 m (767 kt at 32,800 ft); service ceiling, 19,000 m (62,300 ft).

MiG-19P / SM-12/3 / SM-12/1 / SM-12/2

Plans for the SM-12/3 originated in the PVO's need for a fast, high-altitude interceptor. Compared with the MiG-19 prototype, the SM-12/3 had a longer forward fuselage and thinner rims around the engine air intake, which encircled a two-position nose dome housing the radar antenna. This arrangement was chosen to reduce the ram pressure losses in the intake. The AM-9B (RD-9B) turbojets were replaced by R3-26s rated at 3,725 daN (3,825 kg st) and built by a subsidiary of the Mikulin OKB managed by V. N. Sorokin. This change of power unit forced engineers to modify the nozzle throats and to install new heat shields in the engine bay.

Other modifications included: more reliable BU-13MSK and BU-13MK servo-controls for the slab tailplane and ailerons; the new APS-4MD electric stabilator trim actuator, cutting to a quarter the time required to set up the slab tailplane on the MiG-19S; and more unguided rockets to offset removal of the NR-30 cannon from the fuselage. These and other changes greatly enhanced the performance of the SM-12/3 over that of the MiG-19S. Maximum speed jumped from 1,430

The SM-12/3 was a reengined MiG-19S whose forward fuselage was lengthened noticeably. The two-position nose cone housed the radar.

km/h (772 kt) to 1,930 km/h (1,042 kt), and service ceiling improved from 17,500 m (57,400 feet) to 18,000 m (59,000 feet). The latter altitude could be reached in just 3.2 minutes. Thus a significant increase in speed and ceiling had been achieved without modifying the aircraft's structure noticeably and increasing its weight or the thrust of its power unit appreciably.

SM-12PM

This experimental interceptor, powered by two R3-26 turbojets, was also built in 1957. Like the SM-12/3, the air intake on the SM-12PM had a two-position nose dome, but this one was much more bulky because it had to house the TsD-30 radar antenna.

Instead of cannons the SM-12PM carried two RS-2U air-to-air missiles on wing pods. Its maximum speed reached 1,720 km/h (929 kt), and its service ceiling was 17,400 m (57,000 feet). It could climb to 10,000 m (32,800 feet) in four minutes. Its maximum range without drop tanks was 1,700 km (1,055 miles).

The SM-12/1 and SM-12/2, built at the same time as the SM-12/3, differed only in their equipment. The reshaping of the engine air intake foreshadowed already that of the MiG-21.

The SM-12PM had a much larger nose cone than the MiG-19. The two air-to-air missiles are of the RS-2U type.

The SM-12PMU had a new power plant, including a rocket engine, and was armed with two RS-2US (K-5M) air-to-air missiles.

SM-12PMU

The SM-12PMU, built in 1958, was an SM-12PM powered by two Sorokin R3M-26 experimental turbojets with 3,725 daN (3,800 kg st) of thrust and one U-19D booster container under the fuselage that featured a Sevruk RU-013 rocket engine with 2,940 daN (3,000 kg st) of thrust. The rocket engine could be relit several times in flight. The booster container also housed the necessary fuel and oxidizer tanks.

The SM-12PMU carried two semiactive homing K-5M (RS-2US) air-to-air missiles. Its maximum speed was identical to that of the PM at 1,720 km/h (929 kt) or Mach 1.69; the top speed attainable with the rocket engine was not recorded. Its navigational instruments were tested in late 1958 and early 1959. They were designed to receive and display guidance signals transmitted by ground stations. The SM-12PMU was also used to develop the SOD decimetric wave transponder and the RV-U, a new type of precision radio-altimeter.

MiG-19S / SM-30

In the mid-1950s the OKB masterminds had a bright idea: create a mobile launching ramp to enable a fighter to take off without an airfield. The council of ministers and the ministry of aircraft production signed two decrees in April 1955 approving the development of a powerful rocket booster tied to "a takeoff system with no takeoff roll" for the MiG-19S (a system referred to in the West as ZELL, for zero-length launch). The system was composed of a specially modified MiG-19S called the SM-30, a PU-30 launching ramp mounted on the chassis of a YaAZ-210 self-propelled vehicle, and a PRD-22 solid-propellant rocket booster. The management of the project was entrusted to M. I. Guryevich, with A. G. Agronik responsible for the test phase.

The big girder that carried the aircraft was also the launching ramp. This PU-30 ramp-trailer had a rotating lifting device to position the aircraft. The PRD-22 booster, whose impulsion reached 39,200 daN/sec (40,000 kg st/sec), was developed by an OKB team under the command of I. I. Kartukov. Its total operating time was limited to 2.5 seconds.

Once the ramp was positioned, the aircraft—with wheels up—was placed on special brackets and attached to its rocket booster and to the guide rails with shear bolts. The ramp was set at a 15-degree angle. The pistol was at the ready. The pilot climbed the cockpit access ladder, started the two RD-9B turbojets, went to full throttle, fired up the after-

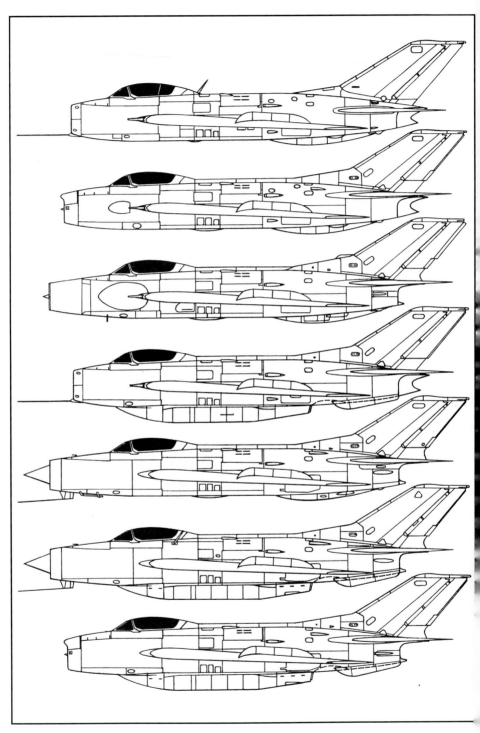

Top to bottom: MiG-19 (SM-9), MiG-19P (SM-7), SM-12/3, MiG-19SU (SM-50), SM-12PM, SM-12PMU, and MiG-19PU (SM-52) (MiG OKB drawings)

The SM-30 on its launch ramp. In early launches the aircraft's elevators and rudder were locked for the first three seconds.

burners, and depressed a knob to light up the booster rockets. The booster thrust added to the turbojet reheated thrust sheared the attachment bolts and imparted instantly to the aircraft a 4.5-g force. Locked at the start, the rudder and elevator were unlocked three seconds later as the aircraft left the guide rails.

Compared with the standard production aircraft, the SM-30 contained a number of modifications made necessary by the launch process:

—the air intake duct's upper skin panel was stiffened, as was the no. 15 frame, the lower hatches, the no. 2 fuel tank walls, and fuselage frame nos. 22, 24, 25, 26, and 30; two symmetric keels were placed under the rear fuselage to transmit the booster thrust to the aircraft
—the wing root attachment was reinforced by redesigning the mounting bolts
—the attachment fittings on fuel tank nos. 2 and 3 were strengthened
—a lock for the flying controls was fitted to the tail unit, as well as a pilot-operated emergency override device
—the ejection seat was equipped with a special helmet meant to immobilize the pilot's head

The first launch was made with an unmanned machine since the operation of the aircraft and all the systems had to be checked first without endangering anybody's life. The first launch confirmed the accuracy of the design data. Two highly experienced LII pilots, G. Shiyanov and S. Anokhin, were chosen for the manned tests, which were first conducted on a runway to make sure that the pilots could handle the very high g-loads to which they would be subjected on the PU-30 ramp. Other unmanned ramp launches served to measure those g-loads more precisely. At no time did they exceed 5 g. The first manned ramp launch took place on 13 April 1957 with Shiyanov at the controls. Before the rocket booster burned out, the aircraft had already exceeded the design safety speed that kept increasing. A slight bank attitude was easily countered by the pilot, who then flew the aircraft as usual and landed back at his base.

Shiyanov was launched five times. Anokhin took to the cockpit for the sixth manned attempt as well as the seventh, on 30 June 1957, in which the aircraft carried its full payload—two rocket pods and two 760-l (201-US gallon) drop tanks. After the eighth launch (made by Shiyanov) the ramp and the SM-30 were moved to the GK NII VVS test center, where V. G. Ivanov, a military test pilot, was launched six times. Five other military pilots, L. M. Kuvshinov, V. S. Kotlov, M. S. Tvelenyev, A. S. Blagoveshchenskiy, and G. T. Beregovoy (a future spaceman) made one launch each.

This extract from the final test report is especially noteworthy:

1. The launching phase does not present any particular difficulty and can be managed by any MiG-19 pilot
2. When launched with unlocked flying controls, the pilot feels much more secure because he can intervene at any time; besides, the locking system proved to be useless
3. To weigh the tactical pros and cons of such a system, it would be advisable to build a small batch of units (ramp + aircraft)
4. It is essential to develop a reliable landing system that does not require a runway

Expressed that way, the last point called on engineers to attempt the impossible. Nevertheless, they endeavored to reduce landing roll in two ways. The first involved the deployment of large drag chutes before touchdown, while the second entailed the use of arresting gear like that on aircraft carriers. Steel wires linked to hydraulic brakes by pulley blocks were set across the runway, and a MiG-19SV was equipped with an arrester hook whose control and position indicators were located in the cockpit. With this device the MiG-19 could be brought to a stop in 120 m (394 feet) after hooking the wires, with deceleration forces reaching –2 g. Once the tests were completed, a demonstration was given to Marshal G. K. Zhukov, minister of defense, who also endorsed the concept of landing without runways. The ramp

Because of its compound power plant, the MiG-19SU or SM-50 offered exceptional performance.

launching system was subsequently abandoned, but the idea of rocket-assisted takeoff (RATO) from runways continued to be explored.

MiG-19SU / SM-50/SM-51/SM-52

Still faster, still higher: those two imperatives summed up the development requests received from military authorities such as the VVS and the PVO. They also summed up the specifications of the SM-50 and SM-51, two prototypes of a high-speed interceptor with a lofty service ceiling.

The SM-50 was powered by two AM-9BMs whose reheated thrust was 3,135 daN (3,200 kg st) and by the U-19 booster container with two power ratings: 1,275 daN (1,300 kg st) and 2,940 daN (3,000 kg st). The rocket engine could not be relit in flight. The SM-51 was powered by two Sorokin R3M-26 experimental turbojets derived from the AM-9BM with 3,725 daN (3,800 kg st) of thrust and by the U-19D booster container. Its rocket engine had the same thrust as that of the U-19 but could be turned off and relit four times in flight. The SM-50 was developed from the MiG-19S, while the SM-51 was developed from the MiG-19P.

The single ventral fin of the MiG-19S had to be replaced by two well-spaced fins on the SM-50 because of the rocket engine exhaust.

The booster container, planned by D. D. Sevruk and built at the MiG OKB, was fastened under the fuselage. It was composed of:

—the RU-013 rocket engine
—three tanks: one for the TG-02 fuel, one for the AK-20 oxidizer, and one for the concentrated hydrogen peroxide
—the combustion chamber feed pumps
—the replenishment system for the three tanks
—the dump valves

The rocket engine weighed 338 kg (745 pounds); the fuel, 372 kg (820 pounds); the oxidizer, 112 kg (247 pounds); and the hydrogen per-oxide, 74.2 kg (163.5 pounds). The U-19 and U-19D booster containers operated almost autonomously; their only links to the cockpit were the electrical ignition control and the dump valve control.

Both the SM-50 and SM-51 were armed with two NR-30 cannons located in the wing roots. The SM-51 was equipped with an RP-5 Izum-rud radar. The takeoff weight of both aircraft—including the booster container—was 9,000 kg (19,835 pounds).

All factory test flights of the SM-50 and SM-51 were made by V. A. Nefyedov under the supervision of Yu. N. Korolyev, chief engineer.

The rocket engine contour gave the SM-50 a very strange silhouette.

Their maximum speed was 1,800 km/h (972 kt), their dynamic ceiling was 24,000 m (78,700 feet), and they could climb to 20,000 m (65,600 feet) in eight minutes. Their range—not an important factor for this type of aircraft—was limited to 800 km (497 miles). The state trials of the SM-50 were carried out by two LII pilots, M. M. Kotelnikov and A. A. Shcherbakov. Five SM-50s were built in factory no. 21.

The SM-52 was identical to the SM-51 with the exception of its radar, which was the Almaz ("diamond") model.

MiG-19S / SM-10

The purpose of this project was to give longer legs to the MiG-19 by refueling the aircraft in flight. Of course, the OKB had already conducted refueling experiments with the MiG-15. In May 1954 a decree of the council of ministers ordered the OKB to build a MiG-19 equipped with in-flight refueling devices and to convert a Tupolev Tu-16 into a tanker aircraft. The same decree secured the financing for the program. The OKB designed a flexible hose unreeled behind the Tu-16 whose tip was supposed to be seized by a kind of trap located near the left wing tip of the fighter.

The MiG-19S no. 415 (SM-10) had a special clutching device near the left wing tip for in-flight refueling trials.

The modified MiG-19S or SM-10 was built in mid-1955. On 29 September another governmental decree named the pilots in charge of the tests: V. A. Nefyedov, the OKB chief pilot, and V. N. Pronyakin, a military test pilot with the LII. The chief engineers in charge of supervising the tests were A. I. Komisarov for the OKB and I. I. Shelyest for the LII. To refuel, the fighter pilot first had to adapt his speed to that of the tanker and then lean the left wing tip of his fighter against the tip of the hose. As soon as contact was made the hose tip was securely connected, and with the help of its powerful pumps the Tu-16 transferred the fuel in a very short time because the flow rate was approximately 1,000 liters per minute (264 US gallons per minute). The refueling sequence took place at 450–500 km/h (243–270 kt) at 9,000–10,000 m (29,500–32,800 feet).

When the tanks of the SM-10 were full, the pumps stopped short and the two aircraft separated. The tanker operator could stop the refueling sequence at any time. (It is interesting to note that this tanker version or Tu-16N could use a unique wing tip–to–wing tip transfer technique to refuel Tu-16 bombers.) The refueling sequence could be repeated several times during a single flight and was possible in the daytime or on a clear night with the help of a wing-mounted floodlight. Because the oxygen reserve of the MiG-19S was found to be insufficient, that of the SM-10 was increased to 18 l (4.7 US gallons).

The tip of the flexible hose unreeled by the tanker aircraft was clutched by a kind of trap located near the SM-10's wing tip.

After its factory tests, the SM-10 passed its state acceptance trials in 1956. But this version was not produced for reasons that Western military authorities would not have thought possible: that year a large share of the Soviet defense budget was funneled to the design and production of surface-to-surface ballistic missiles, so the VVS could not afford to develop the tanker aircraft it needed. This policy was revised in the early 1980s, and today most of the operational MiG-25s and all of the MiG-31s are equipped with in-flight refueling systems.

MiG-19S / SM-20 / SM-20/P / SM-K/1 / SM-K/2

These four modified MiG-19S were used to develop and test air-to-surface winged homing missiles. During flight tests the SM-20 was carried under the fuselage of a Tu-95 bomber. The pilot was in the fighter cockpit, but the engines were off. When the preselected altitude was reached, the pilot started the engines and released his airplane from the bomber. He then had to check on the operation of the radio remote control enabled by the mother aircraft and to return to his base when the test was over. The first SM-20 was built in 1956, and tests started in October of that year.

The second "flying simulator" or SM-20/P was built in early 1957. The aircraft was modified to help the engines start at high altitudes—a problem that had plagued the SM-20. On this version, ignition of the combustion chamber was aided by a carburetor. In all other respects the SM-20/P was identical to the SM-20. Both aircraft were flight-tested by Sultan Amet-Khan of the LII and V. G. Pavlov of the ministry of radio equipment industry research center. The chief engineer responsible for both programs was A. I. Vyushkov.

Two more flying simulators were tested in 1957 and 1958, the SM-K/1 and SM-K/2. Both were derived from the MiG-19S and were also intended to help in the development of flight management and guidance systems for air-to-surface winged missiles.

MiG-19S / SM-9/3T

The K-13 air-to-air homing missile was developed from a U.S.–built Sidewinder recovered in China, and it was decided to arm all new Soviet fighters with this weapon. A modified version of the MiG-19, the SM-9/3T, was used to test the new missile. Two K-13s were attached under an APU-13 launch rail that was itself held by the launch mechanism of the all-purpose pylon. The missile's fire control system was neutralized as long as the gear was not retracted. The tests conducted with the SM-9/3T related to the separation from the APU-13 rail, the performance of the K-13 immediately after it was fired, and the effects of the missile's solid-propellant combustion gases on the performance of the aircraft engines.

The SM-9/3T was tested with two K-13s up to Mach 1.245 at 10,800 m (35,400 feet) and up to 910 km/h at 7,600 m (491 kt at 24,900 feet). Those tests showed that the K-13s under the wing had no influence on the aircraft's handling characteristics. The SM-9/3T was first piloted by A. V. Fedotov on 11 February 1959. Other test flights were made by another OKB pilot, P. M. Ostapyenko, ending on 3 March.

MiG-19 / Flying Test Beds for Various Systems

Various experiments have been carried out on MiG-19s to upgrade some of their components and systems. That is how the KT-37 wheels of the main gear were replaced by KT-61s and the standard drum brakes were replaced by disc brakes, reducing markedly the heat generated by the brakes after landing. New gear legs were also tested for

The SM-9/3T was a MiG-19S without cannons that was used for testing K-13 air-to-air missiles.

Close-up of the K-13 air-to-air missile. Its launch rail was held by the MiG-19S's all-purpose pylon.

The SM-6 was a MiG-19S used for testing K-6 air-to-air missiles, hence its designation.

Close-up of the K-6 air-to-air missile. The fairing of the target illuminator used during the firing tests is visible just above the missile homing head.

MiG-19S no. 420 was used for testing rocket pods firing five ARS-57s apiece.

The SM-21 was a MiG-19S used for testing 210-mm S-21 large-caliber rockets, hence its designation.

MiG-19S no. 420 was also used for testing large-caliber rockets fired from triple launch tubes.

KT-87 wheels. Other MiG-19s were used for firing tests of unguided rockets (by themselves or in pods) and homing missiles (infrared or radar-seeking). Below is a list of some modified prototypes and their missions.

SM-11: Tests of the Yastryeb infrared sensor for night fighters in 1956.

MiG-19R: Tests of several new cameras for the MiG-19R daytime reconnaissance aircraft. A small number of this version were built.

MiG-19P (SM-52P): (1) Tests of the Gorizont-1 navigational equipment, which transmitted coded radio signals giving speed, altitude, heading, and other information to ground stations in order to guide the aircraft to its target. (2) Tests of the Svod radio navigational aid, which in bad weather (zero visibility) could automatically calculate the aircraft's coordinates (bearing and distance) in relation to the nearest ground station and vector the aircraft to a beacon. This same SM-52P was used to test the Almaz ("diamond") fire control radar linked to the Baza-6 range finder.

MiG-19S: Tests of the unique BK-65 astronomical compass repeater, which could calculate the aircraft's true heading and keep it on a predetermined course in northern latitudes (from 40 to 90 degrees) and into conditions of sun visibility at altitudes above "aircraft horizon" up to 70 degrees. In 1956 the Ilim experimental automatic direction finder was tested. It was meant to guide the aircraft to radio stations or radio beacons. Several types of antennae were tested as well. In 1956 the RUP-4 instrument landing system was tested on a MiG-19. Its computer worked simultaneously with an automatic direction finder and an earth induction gyrocompass to plot the approach, using active guidance methods such as a "to/from" pointer. For the production MiG-19 a new NI-50IM dead reckoning position computer was tested and ordered. It could be operated at up to 20,000 m (65,600 feet) and 2,000 km/h (1,080 kt).

SM-21: MiG-19S no. 406, used for 210-mm S-21 unguided rocket tests, hence its designation.

MiG-19S (SM-2D, SM-6, SM-24): Tests of various navigational systems and weaponry.

I-370, I-380, I-410, and I-420 Series

I-370 / I-1 / I-2

During the first half of the 1950s the OKB carried out—concurrently with the difficult development of the MiG-19—research work in other directions in its quest to break the sound barrier. One result was the I-370 (prototypes I-1 and I-2), forefather of a new family of fighters that did not rely on Mikulin engines. The I-370 represented a synthesis of the MiG-17 and MiG-19 as far as structure and aerodynamic design were concerned. Powered by a Klimov VK-7 but retaining the cannon arrangement of the MiG-17 (one N-37 on the right and two NR-23s on the left), the I-1 borrowed the planform and 55-degree C/4 sweepback of the MiG-19 wing as well as the forward fuselage (except for a few modifications made necessary by the new power plant). The wing, attached to the fuselage just like the MiG-19, was of the monospar type with inside stiffeners.

The axis of rotation of the variable incidence stabilizer (with elevators) was just above the afterburner duct. Preliminary sketches showed the stabilizer atop the fin, but after the SM-2 tests it was put under the base. Because one VK-7 "inhaled" a lot less air than two AM-9Bs, the air intake section was reduced. To counterbalance the diminution in yaw stability, two ventral fins angled at 45 degrees were added under the tail. The ailerons and flaps were identical to those of the MiG-19.

As a synthesis of the MiG-17 and MiG-19, the I-370 and its engine were developed in case of a setback to the MiG-19 program.

The cockpit hood was molded in one piece with a frame member in *elektron,* a light magnesium alloy. In case of emergency it had to be jettisoned whatever the aircraft's attitude. The windshield was bullet-proofed and electrically heated. When the pilot activated the ejection seat, the cockpit hood was automatically jettisoned and the airbrakes were deployed.

The VK-7 no. K-733 had a nominal thrust of 3,455 daN (3,525 kg st) and 5,130 daN (5,235 kg st) with afterburner. In the midst of the flight tests it was replaced by a slightly more powerful VK-7 no. K-338 with 4,115 daN (4,200 kg st) of nominal thrust and 6,145 daN (6,270 kg st) with afterburner. At first the top speed of the I-370 (I-1) reached 1,452 km/h (784 kt)—precisely that of the MiG-19S. But once the turbine inlet gas temperature was raised to 800° C (1,472° F), the I-370 reached 1,510–1,520 km/h (815–821 kt).

Rolled out in December 1954, the I-1 prototype was first piloted by F. I. Burtsev of the NII VVS on 16 February 1955. Tests continued until 2 June. According to Burtsev—who flew the aircraft thirteen times—the I-1 reached Mach 1.334 and remained steady in flight. Moreover, it handled just like the MiG-19 while taking off, climbing, descending, and landing.

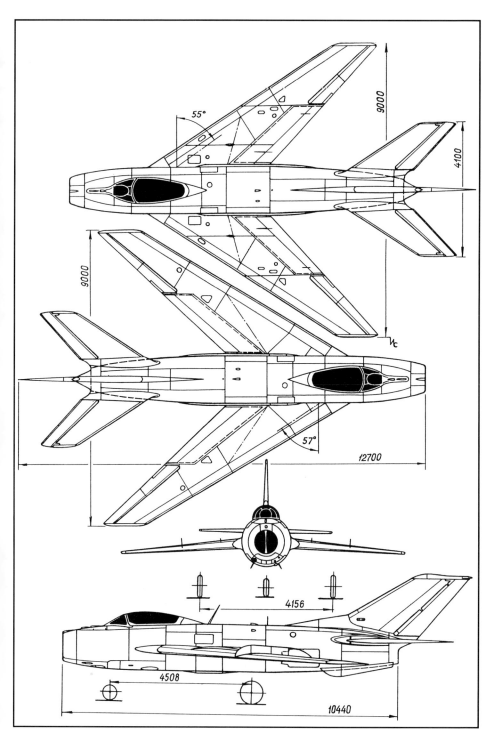

I-1 (I-370); *second from top,* the I-2 (MiG OKB four-view drawing)

The pickup of the VK-7 was not as good as that of the VK-1F. But so long as the engine speed did not exceed 12,100 rpm, the VK-7 running was virtually identical to the VK-1F at all ratings. The three fuselage tanks had a capacity of 2,025 l (535 US gallons). The first two tanks were of the bladder type, while the third was made of welded AMTsAM alloy in the rear fuselage. Two drop tanks could be attached under the wing. The weapon system included the ASP-4N gunsight tied to the Radal-M ranging radar, the RSIU-3 VHF, IFF interrogator, and the OSP-48 ILS receiver. A new wing that featured a sweepback C/4 of 57 degrees did not help the aircraft reach its design airspeed. With that new wing the aircraft was renamed the I-2.

Specifications
Span, 9 m (29 ft 6.3 in); overall length, 12.7 m (41 ft 8 in); fuselage length, 10.44 m (34 ft 3 in); wheel track, 4.156 m (13 ft 7.6 in); wheel base, 4.508 m (14 ft 9.5 in); wing area, 25 m² (269 sq ft); empty weight, 5,086 kg (11,210 lb); takeoff weight, 7,030 kg (15,495 lb); max takeoff weight, 8,300 kg (18,295 lb); fuel, 1,680 kg (3,700 lb); wing loading, 280.2–332 kg/m² (57.4–68 lb/sq ft).

Performance
Max speed, 1,452 km/h at 10,800 m (784 kt at 35,400 ft); climb to 5,000 m (16,400 ft) in 1.15 min; to 10,000 m (32,800 ft) in 3 min; service ceiling, 17,000 m (55,760 ft); approach speed, 180 km/h (97 kt); takeoff roll, 464 m (1,522 ft); landing roll, 730 m (2,395 ft); range with two drop tanks, 2,500 km (1,550 mi).

I-380 / I-3

The I-3 was a logical follow-on for the I-1/I-2 design philosophy among various single-engine fighters developed simultaneously with other types such as the MiG-19 that were already being mass-produced. Its design and structure were based on standard concepts shared by most other fighters of that time: all-metal structure, highly swept (over 50 degrees) and lift-augmented wing, airbrakes plus tail chute, powerful reheated turbojets, ejection seat, flying controls with artificial feel and gear ratio on the pitch channel, all-purpose on-board systems, and heavy armament (cannons).

The preliminary design of the I-3 (I-380) frontline fighter, mapped out around the new Klimov VK-3 turbojet, was completed in March 1954. The VK-3 possessed an axial flow compressor, an annular combustion chamber, and an afterburner. It was rated at 5,615 daN (5,730

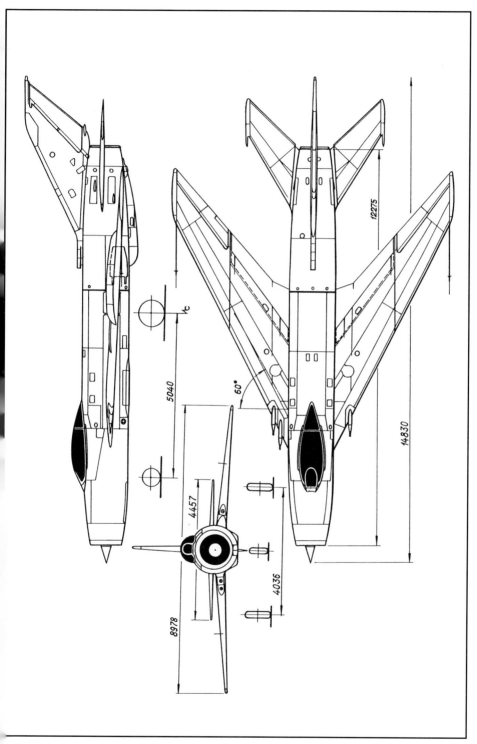

I-3 (I-380) (MiG OKB three-view drawing)

kg st) of design nominal thrust and 0,270 daN (0,440 kg st) with afterburner. The I-3's stabilator and aileron power units were of the irreversible type. On the other hand, the rudder servo-control units were
reversible. The wing had a sweepback of 60 degrees C/4. The airbrakes
covered an area of 1.2 m² (12.9 square feet) and flanked the fuselage
just behind the wing root. Their role was not only to shorten the landing roll and to improve the aircraft's handling in level flight but also to
reduce the aircraft's speed during a full-power vertical dive. The PT
2165-511 tail chute was 15 m² (161.5 square feet) in area. Armament
consisted of three NR-30 cannons, two on the right and one on the left
in the leading edge near the wing root. The pilot was protected by a 65-
mm-thick bulletproof windshield, a 10-mm armor plate in front of the
cockpit, and a 16-mm armor plate in the seat back and headrest. Development of the turbojet was delayed several times by technical problems; as it turned out, the I-3 was never powered by the VK-3 and was
later converted into an I-3U.

Specifications
Span, 8.978 m (29 ft 5.5 in); overall length, 14.83 m (48 ft 0.8 in); fuselage length (except cone), 12.275 m (40 ft 3.3 in); wheel track, 4.036 m
(13 ft 2.9 in); wheel base, 5.04 m (16 ft 6.4 in); wing area, 30 m² (322.9
sq ft); empty weight, 5,485 kg (12,090 lb); takeoff weight, 7,600 kg
(16,750 lb); max takeoff weight, 8,954 kg (19,735 lb); fuel, 1,800 kg
(3,970 lb); oil, 32 kg (70 lb); wing loading, 253.3–298.5 kg/m²
(51.9–61.2 lb/sq ft); max operating limit load factor, 9.

Performance
Max speed, 1,274 km/h (688 kt) at sea level; 1,311 km/h at 5,000 m
(708 kt at 16,400 ft); 1,775 km/h at 10,000 m (958 kt at 32,800 ft); climb
to 5,000 m (16,400 ft) in 0.81 min; to 10,000 m (32,800 ft) in 1.9 min;
service ceiling, 18,800 m (61,660 ft); landing speed, 190 km/h (103 kt);
endurance, 1 h 46 min; range, 1,365 km (848 mi); takeoff roll, 390 m
(1,280 ft); landing roll, 726 m (2,380 ft).

I-410 / I-3U / I-5

The I-3U was a revised and corrected edition of the I-3. Its role was to
intercept and destroy hostile aircraft at any speed and altitude and in
any weather conditions. It differed from the basic aircraft in its automatic flight management and fire control systems. The latter, called
the Uragan-1, included the Almaz ranging radar, the OKB-857 computer, and the AP-36-118 autopilot. The radar range was 17 km (10.5

The I-3U/I-5 was equipped with the Almaz ranging radar, housed in an off-center nose cone.

The cockpit canopy of the I-3U/I-5 was hinged to open upward and forward—an arrangement retained on the first-series MiG-21s.

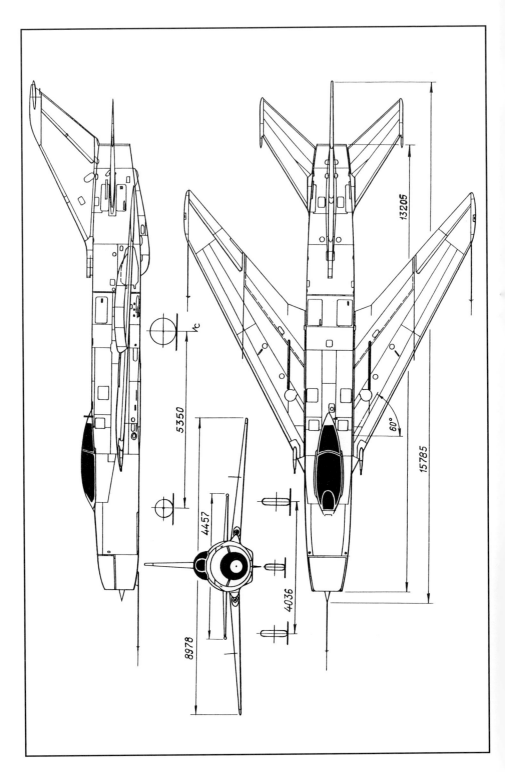

miles). The fighter was led to its target by ground stations. The main airborne systems included the ASP-5M gunsight, the ARK-5 automatic direction finder, the MRP-48 marker receiver, IFF interrogator, and radar warning receiver (RWR). Armament consisted of two NR-30 cannons with 65 rpg located in the leading edge near the wing root; the aircraft could also carry two rocket pods with a total of sixteen ARS-57 rockets.

Like the I-3, the I-3U was a victim of its disappointing engine. The ill-fated VK-3, which was supposed to deliver a dry thrust of 5,615 daN (5,730 kg st) and a reheated thrust of 8,270 daN (8,440 kg st), never did live up to expectations. Because no other turbojet in that category was available, the I-3U, rolled out in 1956, never left the ground. This explains why only design performance data are given.

Specifications
Span, 8.978 m (29 ft 5.5 in); overall length (except probe), 15.785 m (51 ft 9.5 in); fuselage length (except cone), 13.54 m (44 ft 5 in); wheel track, 4.036 m (13 ft 2.9 in); wheel base, 5.35 m (17 ft 6.6 in); wing area, 30 m^2 (322.9 sq ft); empty weight, 6,447 kg (14,210 lb); takeoff weight, 8,500 kg (18,735 lb); max takeoff weight, 10,028 kg (22,100 lb); fuel, 1,800 kg (3,970 lb); wing loading, 283.3–334.3 kg/m^2 (58–68.5 lb/sq ft); max operating limit load factor, 8.

Design Performance
Max speed, 1,610 km/h at 5,000 m (870 kt at 16,400 ft); 1,750 km/h at 10,000 m (945 kt at 32,800 ft); climb to 5,000 m (16,400 ft) in 0.47 min; to 10,000 m (32,800 ft) in 1.12 min; service ceiling, 18,300 m (60,000 ft); landing speed, 210 km/h (113 kt); endurance, 1 h 28 min; range, 1,290 km (800 mi); takeoff roll, 560 m (1,840 ft); landing roll, 580 m (1,900 ft).

I-420 / I-3P

This preliminary design did not come to fruition as planned. It was an interceptor version of the I-3, but while it was being built it was converted into an I-3U equipped with the Uragan-1 flight management and fire control system.

I-7U and I-75 Series

I-7U

The I-7U interceptor equipped with the Uragan-1 was developed once it became apparent that the I-3U would be grounded for lack of the right engine. The preliminary plans were completed in August 1956. The structure of the new aircraft was entirely reworked so that it could be powered by the Lyulka AL-7F turbojet, which delivered a dry thrust of 6,155 daN (6,240 kg st) and a reheated thrust of 9,035 daN (9,220 kg st). Except for a few standardized parts, the only piece of equipment common to both the I-3U and the I-7U was the Uragan-1 system. Everything else was completely modified.

All of the main airframe assemblies were redesigned after reconsideration of their basic principles. The fuselage diameter was increased, the wing sweepback C/4 was reduced to 55 degrees, and the gear kinematics were modified (the main gear retracted into the fuselage, their legs folding up inside the wing between the integral fuel tanks and the Fowler-type flaps). Many stamped panels were required for the wing and the tail unit. The ailerons and other movable surfaces contained no ribs but rather a solid core. Armament comprised two NR-30 cannons located on either side of the fuselage alongside the wing root ribs and four optional automatic rocket pods under the wing with a total of sixty-four ARS-57M rockets.

The aircraft was moved to the test center on 26 January 1957 and on 17 April performed its first taxiing tests, during which the aircraft was lifted a few feet. The I-7U made its first flight on 22 April with G. K. Mosolov at the controls. On the thirteenth flight the landing gear on the right side collapsed as the aircraft landed, damaging the right wing. The aircraft was returned to the workshop for repairs and later made six more flights, the last one on 24 January 1958. On 12 February tests were canceled by the general designer. The aircraft was once more returned to the workshop; fitted with the AL-7F-1 engine, it became the I-75F.

The tests had demonstrated the aircraft's quick acceleration as well as its outstanding climb rate on either dry or reheated thrust, a distinctive feature of the I-7U. On the other hand, the deflection travel of the stabilator proved to be sufficient at landing. When the aircraft reached Mach 1.6–1.65 it had a tendency to bank to the left, but its yaw stability remained satisfactory.

The resemblance between the 1-7U and the I-3U was quite superficial. The I-7U was in fact an entirely new machine.

The weapons system was the only common feature of the I-7U and I-3U. The cone housing the Almaz ranging radar is centered on the I-7U.

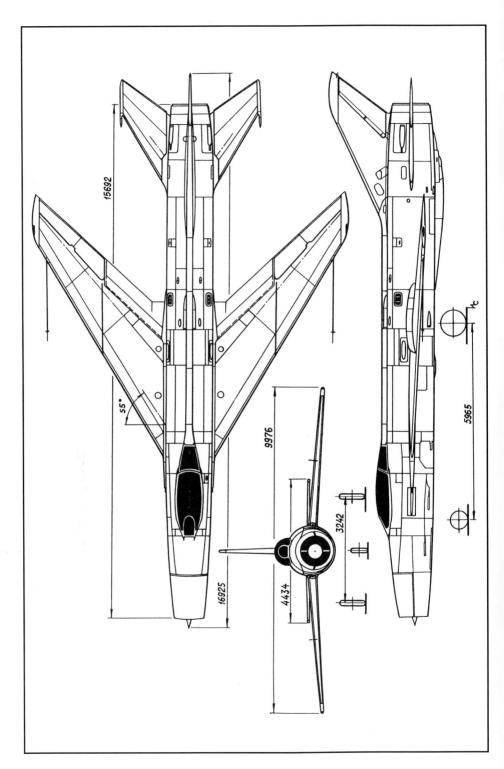

Specifications
Span, 9.976 m (32 ft 8.7 in); overall length, 16.925 m (55 ft 6.3 in); fuselage length (except cone), 15.692 m (51 ft 5.8 in); wheel track, 3.242 m (10 ft 7.6 in); wheel base, 5.965 m (19 ft 6.9 in); wing area, 31.9 m² (343.4 sq ft); empty weight, 7,952 kg (17,525 lb); takeoff weight, 10,200 kg (22,480 lb); max takeoff weight, 11,540 kg (25,435 lb); fuel, 2,000 kg (4,410 lb); wing loading, 319.7–361.7 kg/m² (65.5–74.1 lb/sq ft); max operating limit load factor, 9.

Performance
Recorded max speed with engine dry rating, 1,420 km/h (767 kt) (not recorded with reheated thrust because the Pitot-static probe readings were not corrected at high speeds; the max speeds that follow are design specifications); max speed with reheated thrust, 1,660 km/h at 5,000 m (896 kt at 16,400 ft); 2,200 km/h at 10,000 m (1,188 kt at 32,800 ft); 2,300 km/h at 11,000 m (1,242 kt at 36,080 ft); climb to 5,000 m (16,400 ft) in 0.6 min; to 10,000 m (32,800 ft) in 1.18 min; service ceiling, 19,100 m (62,650 ft); landing speed, 280–300 km/h (150–162 kt); endurance, 1 h 47 min; range, 1,505 km (935 mi); takeoff roll, 570 m (1,870 ft); landing roll, 990 m (3,250 ft).

I-75

The I-75 supersonic interceptor was a direct descendant of the I-7U equipped with the more sophisticated Uragan-5 interception system. The operation of the Uragan-5B modified radar was identical to that of the Izumrud or Almaz except for one thing: it had only one transmitter, one antenna, one display system, and two operating modes: search and target designation. The Uragan-5 system also comprised the AP-39 Uragan-5V autopilot, the Uragan-5D airborne computer, and the Uragan-5T-1 radar control unit, plus two K-8 air-to-air missiles that weighed 275 kg (605 pounds) apiece.

This aircraft was engineered to intercept automatically high-altitude supersonic bombers day or night or in bad weather and to destroy them as far away as possible from their intended targets. With two drop tanks its tactical radius of action was 720 km (447 miles). It could attack enemy aircraft at speeds between 800 and 1,500 km/h (432 and 810 kt) and altitudes between 10,000 and 20,000 m (32,800 and 65,600 feet). Its radar detection range was 30 km (18.6 miles), and it could lock onto targets 20 km (12.4 miles) away.

The I-75 differed from the I-7U in its forward fuselage, where the Uragan-5 accessory unit was housed. Moreover, in place of cannons it

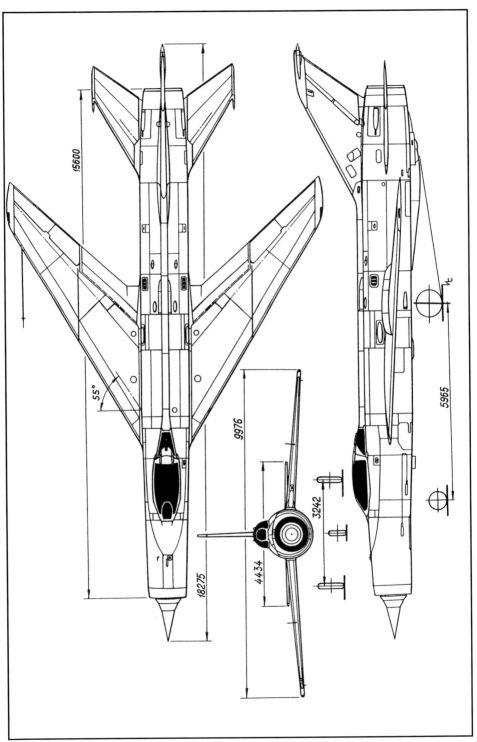

I-75 (MiG OKB three-view drawing)

The I-75 was in fact an I-7U minus the cannons, equipped with the new Uragan-5B radar, and armed with two K-8 air-to-air missiles.

The I-75's nose cone is much bulkier than that of the I-7U due to the larger radar antenna.

had wing pods for air-to-air missiles. To increase its efficiency, the fin height was increased and the sweep angle at the leading edge was reduced by 2.5 degrees. The cockpit was equipped with the only type of ejection seat available at MiG at the time. Tail chute canisters were located on either side of the ventral fin. The engine air intake cone was fixed. The air intake duct flow rate was controlled by a translating outer ring on the cylindrical end of the nose cone.

The I-75's flying controls were of the rigid type (tubular actuating rods in duralumin). The pitch channel was fitted with two BU-44B irreversible servo-controls and the ARU-3V artificial feel system. The BU-44 aileron servo-control and the BU-45 rudder servo-control were irreversible as well. Like the I-7U, the I-75 was powered by a Lyulka AL-7F turbojet that provided dry thrust of 6,155 daN (6,240 kg st) or 9,035 daN (9,220 kg st) with afterburner. The aircraft was moved to the test center on 1 March 1958 but did not fly until problems in the cockpit ejection system were ironed out.

Watched closely by A. N. Soshin, the chief engineer, G. K. Mosolov made the first flight on 28 April 1958 and four others thereafter. Between 15 May and 24 December the radar was installed and efforts were made to improve the engine and the Uragan-5 system. Flight tests resumed on 25 December and were terminated on 11 May after eighteen flights with the radar operating.

Specifications
Span, 9.976 m (32 ft 8.7 in); overall length, 18.275 m (59 ft 11.5 in); fuselage length (except cone), 15.6 m (51 ft 2.2 in); wheel track, 3.242 m (10 ft 7.6 in); wheel base, 5.965 m (19 ft 6.9 in); wing area, 31.9 m² (343.4 sq ft); empty weight, 8,274 kg (18,235 lb); takeoff weight, 10,950 kg (24,135 lb); max takeoff weight, 11,470 kg (25,280 lb); fuel, 2,000 kg (4,410 lb); oil, 17 kg (37 lb); wing loading, 343.3–359.6 kg/m² (70.4–73.7 lb/sq ft); max operating limit load factor, 9.

Performance
Max speed, 2,050 km/h at 11,400 m (1,107 kt at 37,400 ft); 1,870 km/h at 12,900 m (1,010 kt at 42,300 ft); 1,670 km/h at 12,400 m (902 kt at 40,670 ft) with two K-8 missiles; design climb to 6,000 m (19,680 ft) in 0.93 min; to 11,000 m (36,080 ft) in 3.05 min; service ceiling, 21,000 m (68,900 ft); landing speed, 240 km/h (130 kt); endurance with afterburner between 10,000 m and 15,000 m (32,800 and 49,200 ft), 25 min; range at 10,000 m (32,800 ft), 1,200 km (745 mi); at 12,000 m (39,360 ft), 1,470 km (910 mi); takeoff roll, 1,500 m (4,920 ft); landing roll, 2,000 m (6,560 ft).

The Ye-150 served as the test bed for the Mikulin-Tumanskiy R-15-300, which would later power the MiG-25.

Ye-150 and Ye-152 Series

Ye-150

The Ye-150 experimental prototype was designed as a test bed for the new Mikulin/Tumanskiy R-15-300 turbojet. The intent of the aircraft-plus-engine project was to lay the foundation for a new generation of interceptors. The aircraft was designed to fly at speeds of about 2,800 km/h (1,510 kt) and altitudes of 20,000 to 25,000 m (65,600 to 82,000 feet).

The initial plan called for the new engine to be tested on a remotely controlled aircraft. This turbojet had a very short lifetime, but in that brief period it was powered up on the test bench, examined in flight, and even used to power a missile. It had a dry thrust of 6,705 daN (6,840 kg st) and a reheated thrust of 9,945 daN (10,150 kg st); its afterburner also had a second-stage nozzle called an ejector that supplied 19,405 daN (19,800 kg st) of thrust at Mach 2.4–2.5 and helped to clean up the base drag. For components particularly sensitive to the thermal

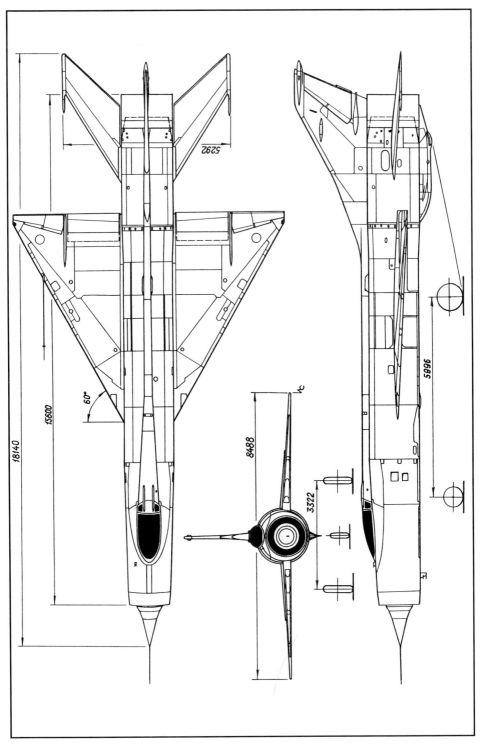

5292

18140

15600

60°

8488

3322

5896

Ye-150 (MiG OKB three-view drawing)

The boundary layer bleed in the "ejector" slot helped to clean up the base drag.

stresses (aerodynamic heating) that were the result of high speeds, the manufacturers decided to use heat-resistant materials such as stainless steel in place of duralumin.

The fuselage was shaped like a cylinder 1,600 mm in diameter except at the rear, where the diameter increased to 1,650 mm in the afterburner/ejector area. The shock cone in the engine air intake had a triple-angle profile and was made of dielectric material to house the antenna for the Uragan-5 interception system. The flow rate in the air inlet duct was controlled by a two-position translating ring. As soon as the aircraft reached Mach 1.65 the ring moved forward automatically; once the aircraft dropped back under that speed, the ring returned to its primary position.

The delta wing had a sweepback of 60 degrees at the leading edge, a thickness-chord ratio of 3.5 percent, Fowler-type flaps, and two-part ailerons with balance surfaces at the trailing edge. The wing could be fitted with two pylons for air-to-air missiles. The gear kinematics were standard: the nose gear strut retracted forward into the fuselage, and the main gear wheels also retracted into the fuselage while their struts folded into the wing. The cockpit was equipped with a curtain-type ejection seat. The fuel system included five fuselage tanks with a total capacity of 3,270 l (863 US gallons) plus two wet wing tanks that carried

245 l (65 US gallons) apiece. The stabilator controls were boosted by two BU-65 power units, and those of the ailerons and rudder by two BU-75 power units. There were two separate hydraulic systems, one primary circuit and one for the servo-controls. The main circuit served the gear, the flaps, the three airbrakes on the underside of the fuselage, the translating ring on the air intake, and the surge bleed valve (on the fuselage sides) while also acting as a backup for the servo-control units. The PT 5605-58 tail chute measured 18 m² (193.7 square feet). The cockpit hood was made of T2-55 glass, a 12-mm-thick material capable of withstanding 170° C (338° F) in aerodynamic heating.

The Ye-150 rolled out in December 1958 and was first piloted by A. V. Fedotov on 8 July 1960. During the fourth flight, on 26 July, aileron flutter was observed at Mach 0.925. The problem was quickly solved by fitting a damper on the aileron controls. After the fifth flight the tests had to be suspended because the casing of the engine gearbox had cracked. Tests resumed on 18 January 1961 with a brand-new R-15-300 turbojet. From 21 January to 30 March the aircraft made eight more flights and reached Mach 2.1 at 21,000 m (68,900 feet). After a second engine change, the Ye-150 made another twenty flights and hit a top speed of Mach 2.65 at 22,500 m (73,800 feet). At that point the ejector was replaced and the cockpit's thermal insulation improved; tests resumed on 14 November 1961 and ended on 25 January 1962. There were forty-two flights altogether. Tests of the Uragan-5 complex with two K-9 missiles were not carried out until the Ye-152A was ready a little later.

Specifications
Span, 8.488 m (27 ft 10.2 in); overall length (except probe), 18.14 m (59 ft 6.2 in); fuselage length (except cone), 15.6 m (51 ft 2.2 in); wheel track, 3.322 m (10 ft 10.8 in); wheel base, 5.996 m (19 ft 8 in); wing area, 34.615 m² (372.6 sq ft); empty weight, 8,276 kg (18,240 lb); take-off weight, 12,435 kg (27,405 lb); fuel, 3,410 kg (7,515 lb); wing loading, 359.2 kg/m² (73.6 lb/sq ft); max operating limit load factor, 5.1.

Performance
Max speed, 1,210 km/h (653 kt) at sea level; 2,890 km/h at 19,000 m (1,560 kt at 62,300 ft); climb to 5,000 m (16,400 ft) in 1 min 20 sec; to 20,000 m (65,600 ft) in 5 min 5 sec; service ceiling, 23,250 m (76,260 ft); landing speed, 275–295 km/h (148–160 kt); endurance, 1 h 50 min; range, 1,500 km (930 mi); takeoff roll, 935 m (3,065 ft); landing roll, 1,250 m (4,100 ft).

Ye-151/1 / Ye-151/2

Following close on the heels of the Ye-150, the OKB started work on the full-scale mock-up of a new prototype—the Ye-151—armed with a rotating twin-barrel cannon that was set on the forward fuselage structure and that revolved around the air intake case axis. With the cannon (a 23-mm TKB-495 whose axis of rotation was perpendicular to its annular support axis) at its widest angle, a noticeable torque occurred that disrupted the aircraft's three-axis stability and made it impossible to shoot accurately. For the Ye-151/2 the cannon and its support were moved behind the cockpit and thus closer to the aircraft's center of gravity.

The Ye-151's forward fuselage was longer than that of the Ye-150, but the dimensions of the air inlet duct did not need to be modified because the ammunition boxes and belts were transferred to behind the cockpit. Wind tunnel experiments proved that the internal aerodynamics of the extended duct improved the engine's operation. This arrangement was retained for future aircraft of this family, starting with the Ye-152A.

Ye-152A

The objective of this project was to develop a fighter capable of collision-course interception at 2,000 km/h (1,080 kt) between 1,000 and 23,000 m (3,280 and 75,440 feet). This tailed delta aircraft was powered by a pair of R-11F-300 turbojets rated at 3,800 daN (3,880 kg st) and 5,625 daN (5,740 kg st) with afterburner; and it was to be equipped with the Uragan-5B radar, which was still untested because of the eighteen-month delay of the Ye-150's engines.

The fixed cone was made of dielectric material to house the TsP radar antenna. Its triple-angle profile was selected to make the bow shock wave diverge. To control the flow in the air intake duct, the hydraulically controlled annular nose cowl moved on four tracks to three positions as dictated by the aircraft's speed and subsequently by ram air pressure. The Ye-152A's wing derived from the Ye-150. The fuselage was widened at the second spar level to accommodate two engines instead of one. The stabilator surfaces were identical to those of the Ye-150 except that their span was increased to 5.85 m (19 feet, 2.3 inches) from 5.292 m (17 feet, 4.3 inches) because of the wider fuselage. The Ye-152A had three airbrakes (one under the fuselage and two on its sides) and a double tail chute.

The twin-jet Ye-152A made its first flight before the Ye-150, which had to wait eighteen months for its engine.

The fuel tanks—six in the fuselage, one between the wheel wells, and two in the wing) had a total capacity of 4,400 l (1,162 US gallons). The aircraft could be armed with two K-9 air-to-air missiles developed by the MiG OKB (factory designation K-155). If the pilot had to eject he was protected by the cockpit hood, a precautionary measure that was used on other MiG fighters (including the MiG-21). Its main systems included the RSIU-4V VHF, the ARK-54N automatic direction finder, the SRO-2 IFF transponder, and the Meteorit radio-nav station. The SRP computer and the AP-39 autopilot were linked to the TsP radar.

The aircraft was rolled out in June 1959 and first piloted by G. K. Mosolov on 10 July. Tests opened on 10 June and ended on 6 August 1960 after fifty-five flights—fifty-one with clean wings, two with pylons, and two with pylons and K-9 missiles. The highest speed reached with wing pylons came at 13,000 m (42,640 feet). Ten in-flight engine relights were carried out at altitudes between 6,000 and 10,500 m (19,680 and 34,440 feet). Each time, the engine relit on the first try and built up full power in fifteen to twenty-five seconds.

Specifications
Span, 8.488 m (27 ft 10.2 in); overall length, 19 m (62 ft 4 in); fuselage length (except cone), 15.45 m (50 ft 8.3 in); wheel track, 3.322 m (10 ft

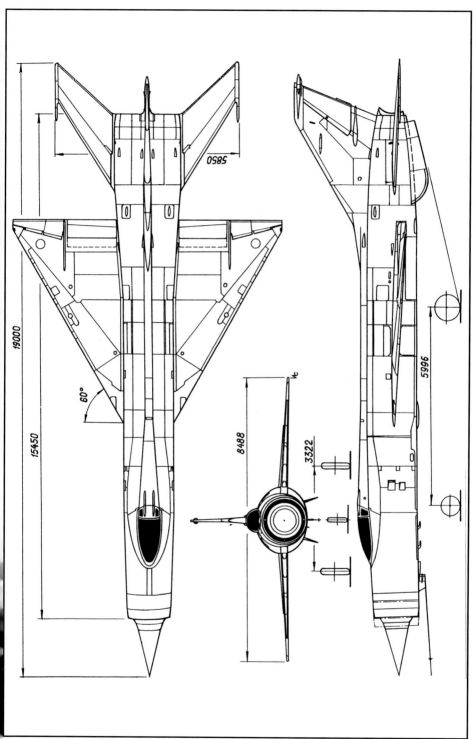

Ye-152A (MiG OKB three-view drawing)

The weapons system that combined the Uragan-5B radar and K-9 air-to-air missiles was tested on the Ye-152A.

10.8 in); wheel base, 5.995 m (19 ft 8 in); wing area, 34.02 m² (366.2 sq ft); takeoff weight, 12,500 kg (27,550 lb); max takeoff weight, 13,960 kg (30,770 lb), fuel, 3,560 kg (7,845 lb); wing loading, 367.4–410.3 kg/m² (75.3–84.1 lb/sq ft); operating limit load factor, 7.

Performance
Max speed, 2,135 km/h at 13,700 m (1,153 kt at 44,940 ft); 2,500 km/h at 20,000 m (1,350 kt at 65,600 ft); climb to 10,000 m (32,800 ft) in 1.45 min; to 20,000 m (65,600 ft) in 7.64 min; service ceiling, 19,800 m (64,950 ft); takeoff roll, 1,000 m (3,280 ft); landing roll, 1,600 m (5,250 ft).

Ye-152/1 / Ye-152/2

The Ye-152 represented the synthesis of two experimental prototypes: the Ye-150, test bed of the R-15-300 turbojet, and the Ye-152A, used to test the Uragan-5 automatic interception system and Mikoyan K-9 air-to-air missiles. The Ye-152 was actually rolled out after the Ye-152A. On the recommendation of two test pilots, G. K. Mosolov and A. V.

This photograph shows the Ye-152/1 with a drop tank under the fuselage and, at the wing tips, models of the K-9 air-to-air missile under development.

Fedotov, the new aircraft had a lower wing loading and better yaw stability; the wing tip shakes and the aileron flutter were eliminated, and its taxiing conditions were improved. The new wing was larger thanks to a deeper tip chord, a modification that allowed missiles to be installed there. Its enhanced handling in ground maneuvers was due to a wider wheel track made possible by the modification of the wing structure. The yaw stability was rectified by increasing both the fin chord and area of the ventral fin so that the tail section could play a more efficient part in the tail fin's work. The Ye-152 was designed to intercept and destroy on collision course any invader at 1,600 km/h at 10,000 m (864 kt at 32,800 feet) or 2,500 km/h at 20,000 m (1,350 kt at 65,600 feet) and beyond.

The diamond wing had a sweepback of 53 degrees, 47 minutes at the leading edge and a thin airfoil section (the thickness-chord ratio was 3.5 percent at the wing root and 5 percent at the wing tip). The triple-angle cone was fixed and made of dielectric material to house the radar antenna, like the Ye-150. Moreover, the Ye-152 had the same translating annular cowl as the Ye-152A. The cone's annular base plate was perforated to bleed the boundary layer in order to increase the total pressure recovery factor at the compressor inlet level. The sole airbrake was located under the fuselage, and the container for the PT-5605-58 tail chute was placed at the base of the ventral fin.

All flying controls were boosted by hydraulic servo-controls—two BU-65s for the slab tailplane, one BU-120M per aileron, and one BU-120M for the rudder. The hydraulic system used AMG-10 fluid and could handle pressures of 210 atmospheres. The autopilot was the AP-39. The total capacity of the fuel tanks in the fuselage and wing was 4,960 l (1,310 US gallons), and a PB-1500 drop tank attached under the fuselage could carry another 1,500 l (396 US gallons). In the event of ejection the pilot was protected by the cockpit hood. The R-15-300 turbojet of the Ye-152 was slightly more powerful than that of the Ye-150: 6,750 daN (6,800 kg st) of dry thrust, and 10,005 daN (10,210 kg st) with afterburner. The Ye-152, like the Ye-150, was equipped with an ejector.

The first prototype or Ye-152/1 was moved to the test center on 16 March 1961. For the first flight by G. K. Mosolov on 21 April, provisional ballast of 263 kg (580 pounds) was placed in the nose. Tests continued until 8 January 1962, started up once more on 2 March, and were finished by 11 September; sixty-seven flights were made—fifty-one with launching rails, five with missiles, and eleven in clean configuration.

The second prototype or Ye-152/2 took advantage of all the adjustments made during the Ye 152/1 tests and was significantly modified in two ways: (1) to expand the pitch stability margin, the fuel tanks' utilization sequence was altered; and (2) the boundary layer bleed device was improved by increasing the area of the perforated cone base plate. The flight envelope of the Ye-152/2 was tested up to a speed of 2,740 km/h (1,480 kt) at 22,500 m (73,800 feet) under clean conditions, and as far as Mach 2.28 at 18,000 m (59,000 feet) with K-9 missiles attached to the wing tips. Flying the Ye-152/2 proved to be very similar to flying the Ye-152/1. But cancellation of the Ye-152/1 tests as well as those of the K-9 missile sealed the fate of the Ye-152/2. Only 60 percent of its test schedule was completed.

Specifications
Span, 8.793 m (28 ft 10.2 in); overall length (except probe), 19.656 m (64 ft 5.9 in); fuselage length (except cone), 16.603 m (54 ft 5.7 in); wheel track, 4.2 m (13 ft 9.4 in); wheel base, 6.265 m (20 ft 6.7 in); wing area, 42.02 m^2 (452.3 sq ft); empty weight, 10,900 kg (24,025 lb); takeoff weight, 14,350 kg (31,630 lb); max takeoff weight, 14,900 kg (32,840 lb); fuel, 4,150 kg (9,145 lb); wing loading, 341.5–356.6 kg/m^2 (70–73.1 lb/sq ft).

Performance
Max speed, 2,510 km/h at 10,000 m (1,355 kt at 32,800 ft); 3,030 km/h at 15,400 m (1,636 kt at 50,500 ft); climb to 10,000 m (32,800 ft) in 3.67 min; to 20,000 m (65,600 ft) in 5.33 min; service ceiling, 22,670 m (74,360 ft); landing speed, 260–270 km/h (140–146 kt); range with one

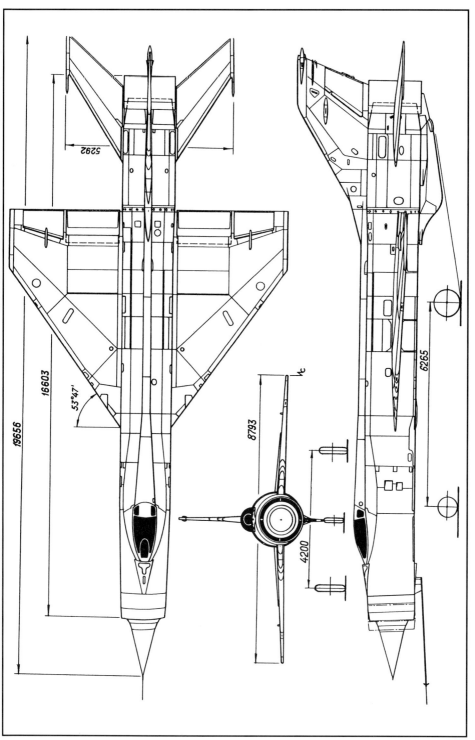

Ye-152 (MiG OKB three-view drawing)

19656

16603

53°47'

5292

8793

4200

6265

At the wing tips of this Ye-152P are models of the new R-4 air-to-air missile under development.

1,500-l (396-US gal) drop tank, 1,470 km (915 mi); takeoff roll with two K-9s, 1,185 m (3,885 ft); landing roll, 1,270–1,300 m (4,165–4,265 ft).

Ye-152P / Ye-152M (Ye-166)

The Ye-152M was designed as a basis for the development of a highly sophisticated interceptor equipped with the most modern navigation and interception systems. It differed from the Ye-152 in the arrangement of its fuel tanks: there were the usual six fuselage tanks (no. 1, 550 l [145 US gallons]; no. 2, 1,100 l [290 US gallons]; no. 3, 1,120 l [296 US gallons]; no. 4, 120 l [31.6 US gallons]; no. 5, 460 l [121 US gallons]; no. 6, 380 l [100 US gallons]; total capacity, 3,730 l [984 US gallons]) and four wing tanks (two in front of and two behind the main spar, each capable of holding 600 l [158 US gallons]), plus three tanks behind the cockpit in the dorsal spine of the fuselage (no. 1, 750 l [198 US gallons]; no. 2, 630 l [166 US gallons]; no. 3, 380 l [100 US gallons]; total capacity, 1,760 l [465 US gallons]). This overall capacity of 6,690 l (1,766 US gallons) could be augmented by the 1,500 l (396 US gallons) of the PB-

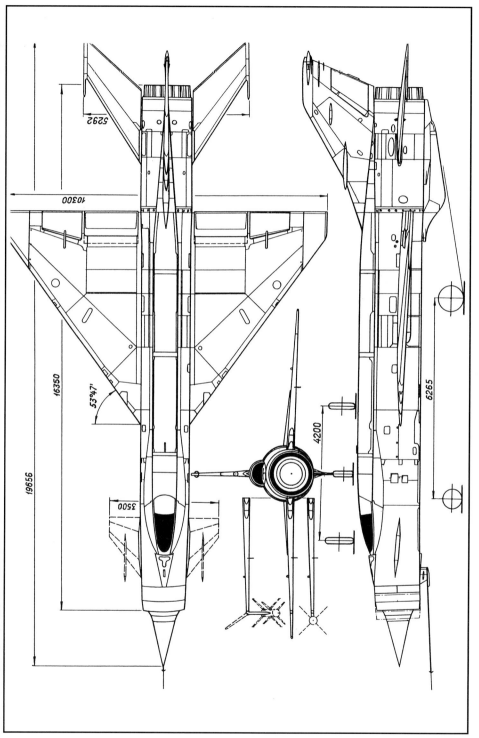

Ye-152M; the missile-equipped wing tips of the front view are those of the Ye-152P (MiG OKB three-view drawing)

273

Developed from the Ye-152M, the Ye-166 broke several world records in 1961 and 1962. The support attachments for the canard surface are visible just under the "Ye-166" marking.

1500 drop tank, bringing the maximum fuel weight to 6,800 kg (14,990 pounds).

Except for the fuel tanks, the Ye-152M's fuselage was identical to that of the Ye-152; but the ejector was replaced by a convergent-divergent exhaust nozzle that reduced the length of the fuselage by 253 millimeters. The tail units of the two aircraft were also identical.

The first version, called the Ye-152P, had a wing identical to that of the Ye-152 except for a small fence placed on the lower surface at midspan; also, the missiles were fired from the wing tips. Unfortunately, this arrangement proved to be a failure. Because the wing tips were not sufficiently stiff to keep the launch rails steady, the missiles followed an uncertain trajectory and usually missed their targets. Engineers tried to remedy the situation by fitting the wing tips with pylons that were also supposed to serve as winglets. This improved conditions somewhat but still could not match those of the Ye-152A with its midspan pylons. Missile tests were finally discontinued.

To reduce its load, the wing was equipped with large tips that increased its span by 1,507 mm (4 feet, 11.3 inches). Moreover, the fuselage nose section was fitted with an auxiliary structure to hold a canard surface having a span of 3.5 m (11 feet, 5.8 inches), intended to improve the aircraft's pitching stability above the sound barrier.

The Ye-152M did not fly with either the extended wing or the canard surface. But the aircraft became world-famous under the fancy designation Ye-166 when, with the short wing, it set the absolute world record for speed over a 100-km (62-mile) closed circuit at 2,401 km/h (1,297 kt) with A. V. Fedotov at the controls on 7 October 1961; the absolute world record for speed over a 15- to 25-km (9- to 16-mile) course at 2,681 km/h (1,448 kt) with G. K. Mosolov at the controls on 7 July 1962; and the world record for altitude at 22,670 m (74,360 feet) as well as for speed over a 15- to 25-km (9- to 16-mile) course with P. M. Ostapyenko at the controls on 11 September 1962. According to the documents submitted to the FAI to verify the record, the Ye-166 was powered by the R166 turbojet rated for 9,800 daN (10,000 kg st). This was not accurate: in fact, it used a reheated R-15B-300 capable of producing 10,975 daN (11,200 kg st). The Ye-152M test program was discontinued afterward, and the OKB's efforts were focused on an ambitious new design, the Ye-155—the future MiG-25.

Specifications
Span without enlarged wing tips, 8.793 m (28 ft 10.2 in); with enlarged wing tips, 10.3 m (33 ft 9.5 in); overall length (except probe), 19.656 m (64 ft 5.9 in); fuselage length (except cone), 16.35 m (53 ft 7.7 in); wheel track, 4.2 m (13 ft 9.4 in); wheel base, 6.265 m (20 ft 6.7 in); wing area (without enlarged wing tips), 42.89 m^2 (461.7 sq ft).

MiG-21 Series

Without a doubt, the MiG-21 is one of the most famous military aircraft in the world. Its name is known by specialists and the general public alike. Few of its competitors share the same level of name recognition: the Spitfire in Great Britain, the Mirage in France, and the Flying Fortress in the United States—that is about all. The MiG-21 owes its fame to many reasons. Even in World War II no other aircraft has had as many versions (more than thirty). No other aircraft has been operated by as many countries (forty-nine). And no other aircraft has found itself involved in as many armed conflicts.

Fifteen primary versions of the MiG-21 were mass-produced for twenty-eight years (from 1959 to 1987) in three factories in the USSR. The aircraft was also built under license in Czechoslovakia, China, and India. No fewer than seventeen world records were set by several special versions (the Ye-33, Ye-66, Ye-66A, Ye-66B, and Ye-76). As is the case for any aircraft whose family has developed over several decades, the combat effectiveness of the MiG-21 improved over time thanks to the technical progress made in three basic fields:

1. Improvement of the thrust-to-weight ratio for better performance
   - static thrust of the turbojet went up by over 40 percent: from 5,000 daN (5,100 kg st) to 6,960 daN (7,100 kg st)
   - maximum speed at sea level increased from 1,220 km/h (659 kt) to more than 1,300 km/h (700 kt)
   - initial rate of climb jumped from 130 meters per second (25,600 feet per minute) to 225 meters per second (44,300 feet per minute)
   - acceleration time from 600 km/h (324 kt) to 1,100 km/h (594 kt) at sea level decreased from 28 to 19.3 seconds
   - maximum operating limit load factor increased from 7 to 8.5
   - maximum operating indicated airspeed (IAS) was raised from 1,200 km/h (648 kt) to 1,300 km/h (702 kt)
   - maximum authorized hedge-hopping time at 1,000 km/h (540 kt) increased from 28 to 36 minutes
2. Reinforcement of the weapon system
   - the number of loading options expanded from twenty to sixty-eight because of the addition of multipurpose hard points
   - the minimum distance at which a flying target could be destroyed closed from 1,000 m (3,280 feet) to 200 m (655 feet) thanks to the installation of built-in cannons
3. Growth of the aircraft's safety of flight and operational availability
   - flight time per accident was stretched from 3,000 to 39,600 hours
   - the aircraft's lifetime was brought up to 2,100 hours
   - mission preparation time was reduced by 30 to 40 percent

When in 1954 all of the OKB's efforts were focused on the conception of a modern fighter capable of flying at twice the speed of sound or faster, its engineers had no preconceived ideas of which aerodynamic strategy to select. Sweepback wing? Delta wing? Both shapes had their proponents. Whatever the chosen approach, all of the specialists knew full well that their research would have to go off in hundreds of directions whether aerodynamics, power plants, or systems were concerned. The main problem was obviously to make the right choice for the aerodynamic design formula. This is why several experimental programs were launched simultaneously in two quite distinct directions: the sweepback-winged Ye-2 and Ye-2A and the delta-winged Ye-4, Ye-5, and Ye-6.

Everyone knows that the latter formula prevailed in the end. But it should be noted that the victor was in fact a tailed delta configuration.

The sweepback wing was tested on this Ye-2 airframe, but the MiG-21 silhouette was already taking shape.

MiG has always maintained that only this well-balanced scheme could (unlike the French Mirage III) secure a satisfactory degree of maneuverability at low speeds due to a high lift coefficient in this sector of the flight envelope. The OKB also decided to use the axial flow turbojet and the variable geometry air intake (with a multiposition cone) that helped to recover the engine inlet pressure over a wide range of angles of attack (AOA) and at supersonic speeds. Other criteria included simplicity of manufacture and ease of maintenance; in short, it was meant to be a trouble-free aircraft for maintenance personnel, the field support crew, and the pilots. Here begins the long story of the MiG-21.

Ye-1/Kh-1 / Ye-2/Kh-2

In the first half of the 1950s the military called for the development of a lightweight frontline fighter capable of speeds of Mach 2 and a service ceiling of 20,000 m (65,600 feet). Initial work was concerned with the test and aerodynamic refinement of a set of thin-airfoil wings characterized by a 55- to 57-degree sweepback at the leading edge, with the

The Ye-2 had half-span leading edge slats.

design of reliable flying controls for transonic speeds and Mach numbers between 1.5 and 1.7, and with the selective examination of all possible power plants—including several types of turbojets associated with supersonic air intakes.

Ye-2; *second from top, side view of the Ye-2A* (MiG OKB four-view drawing)

Such were the basic considerations that led in 1954 to a preliminary design named Ye-1 or Kh-1 and powered by the AM-5A turbojet. This first draft was quickly modified during that same year and became the Ye-2 powered by the AM-9B turbojet with afterburner, a Mikulin engine that had been retained to power the mass-produced MiG-19. (The AM-11 was the engineers' first choice, but it was not yet available.) The Ye-2 was planned as a single-engine fighter and thus had 2,550 daN (2,600 kg st) of thrust available, or 3,185 daN (3,250 kg st) with afterburner.

It was decided to keep the aircraft's master cross-section as little as possible because it was proportioned by the small volume of the cockpit (1 m³ [35.3 cubic feet]) and the dimensions of the engine. The two-spar stressed-skin wing had a sweepback of 55 degrees at the leading edge and a thickness ratio of 6 percent. The Fowler flaps were hydraulically controlled. The two-segment ailerons were mechanically linked to the spoilers. The outer halves of the leading edge were fitted with two-segment slats. The cockpit canopy structure and the ejection seat were identical to those aboard the I-3, and the canopy protected the pilot in case of ejection.

The windshield was heated by electrical wires set into the triplex glass interlayer. Flying controls were of the rigid type, the rudder being controlled by actuating rods that extended through the dorsal spine of the fuselage behind the cockpit. The slab tailplane was controlled by a servo-control unit located in the fin root. Each of the four fuselage tanks was cut off from the others by special nonreturn valves in order to improve the fuel system's fatigue resistance. The turbojet plus afterburner bay was cooled by ventilation between the combustion chamber shroud and the fuselage skin. As the aircraft accelerated, ventilation was relative to ram air pressure; at zero airspeed, the engine bay was cooled by the discharge of sucked-in air.

The landing gear took up much less space. The main gear wheels retracted into the fuselage, while their legs folded into the wing between the fuel tank/wing box and the rear spar on which the flaps were hinged. This mechanism was later used on several types of MiG fighters because it cut the structure's weight significantly. Armament consisted of two NR-30 cannons located in the lower part of the nose. The ammunition belts were housed in circular sleeves placed between fuselage bulkheads. Main equipment and accessories included the RSIU-4 VHF, the Uzel system's homing receiver, the ARK-5 automatic direction finder, the MRP-48 ILS, the Radal-M ranging radar linked to the ASP-5N gunsight, and a radar warning receiver.

The Ye-2 was first piloted by G. K. Mosolov on 14 February 1955 but finished only a part of its test schedule, which was completed by the reengined Ye-2A.

Specifications
Span, 8.109 m (26 ft 7.2 in); length (except probe), 13.23 m (43 ft 4.9 in); fuselage length (except cone), 11.737 m (38 ft 6.1 in); wheel track, 2.679 m (8 ft 9.5 in); wheel base, 4.41 m (14 ft 5.6 in); wing area, 21 m^2 (226 sq ft); empty weight, 3,687 kg (8,125 lb); takeoff weight, 5,334 kg (11,755 lb); fuel, 1,360 kg (3,000 lb); wing loading, 254 kg/m^2 (52 lb/sq ft).

Performance
Design max speed, 1,920 km/h (1,037 kt); landing speed, 250 km/h (135 kt); design service ceiling, 19,000 m (62,300 ft); takeoff roll, 700 m (2,295 ft); landing roll, 800 m (2,625 ft).

Ye-50

The Ye-50 was designed in 1954. Although it was a member of the Ye-2/Ye-2A family, it differed from them in many points. First, its power plant included the Mikulin AM-9Ye turbojet rated at 3,725 daN (3,800 kg st) and the Dushkin S-155 liquid propellant rocket engine capable of 1,275 daN (1,300 kg st). The AM-9Ye turbojet differed from the production AM-9B in several details that broadened its combat operating range (altitude and speed) and took into account the contribution of the rocket engine. Second, the fuselage had to be lengthened to make room for the three propellant tanks for the rocket engine (K-fuel, A-acid, T-hydrogen peroxide) as well as the combustion chamber above the turbojet's exhaust nozzle. The accessory drive and turbopumps of the rocket engine were located in the fairing of the fin base. The nozzle throats for the rocket engine and the turbojet lined up. Underneath the fuselage two lines for the combat emergency jettison device—treated on the inside against corrosive acids—ran across the skin, opening into the same plane as the nozzle throat. On the other hand, the forward fuselage, the wing, the slab tail, and the gear were not modified because those components were the subject of detailed engineering work with the Ye-2 and were at that time mastered by the manufacturing units. The Ye-50 was armed with two NR-30 cannons.

Three prototypes were built from 1955 to 1957. The Ye-50-1 rolled out in December 1955. It was first piloted by V. G. Mukhin of the LII MAP on 9 January 1956 (the same day as the Ye-5) and first lit up its rocket engine in the air on 8 June. The Ye-50-1 factory tests halted on 14 July after eighteen flights when the aircraft landed short of the runway and was destroyed.

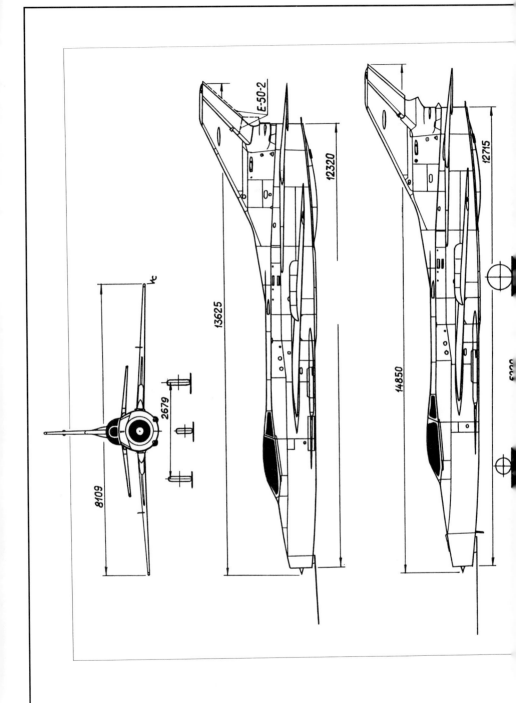

E-50-2

8109

2679

13625

12320

14850

12715

282

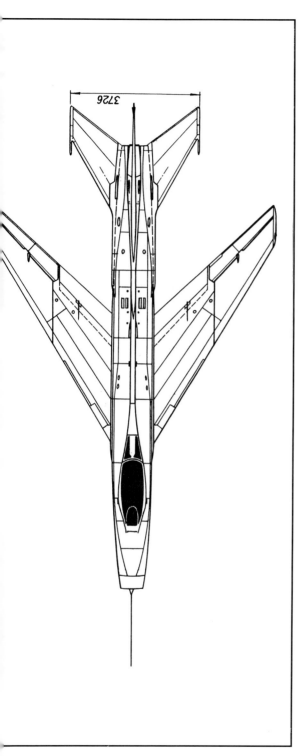

3726

Ye-50/1; *second from left,* the dotted line shows the Ye-50/2 modifications; *second from right,* side view of the Ye-50/3 (MiG OKB four-view drawing)

The Ye-50 differed from the Ye-2 only by its rear fuselage because of its dual power plant. This photograph is of the Ye-50/1.

The S-155 rocket engine was installed just above the jet nozzle of the Ye-50/1.

The Ye-50/3 differed from both the Ye-50/1 and the Ye-50/2 by a noticeable lengthening of the fuselage forward section.

The Ye-50/3 also differed from the two other prototypes by the new shape of its rear fuselage, which resulted in a shortened rudder.

The Ye-50-2 was rapidly completed, and V. P. Vasin (also of the LII MAP) was put in charge of the tests. The Ye-50-2 differed from the first prototype in the modified shape of the rear end and in the patch dubbed the "knife" placed along the trailing edge of the rudder to increase the vertical fin area. Several unofficial altitude and speed records were beaten with the rocket engine in use. On 17 June 1957 Vasin climbed to 25,000 m (82,000 feet) and a little later reached Mach 2.23 or 2,460 km/h (1,328 kt).

The Ye-50-3, built in 1957, had a larger kerosene tank and a smaller hydrogen peroxide tank than its predecessors. The former cell-type tank for kerosene was replaced by a sheet metal tank fitted with a transfer pump. Also, the fuselage nose was lengthened, and the air intake lips were sharpened. The air intake cone had a double angle, and the rear end of the fuselage was modified once more. During a high-altitude flight of this prototype with N. A. Korovin of the NII VVS at the controls, part of the tail fin caught fire. The aircraft became uncontrollable and went into a spin. The pilot ejected; unfortunately, the mechanism to separate the pilot's seat and the canopy did not work, and the pilot was killed.

After reviewing carefully the reasons for the many shortcomings of rocket engines and their systems, it was decided to discontinue the Ye-50 project. But the development of a new generation of fighters equipped with auxiliary power plants proceeded after a complete reappraisal of the basic data concerning this kind of power unit. The goals of this new effort were to increase the effectiveness of the rocket engine's control system and to safeguard the pilot and maintenance personnel (both aboard the aircraft and in the storage facilities) against any toxic components of the rocket engine propellants.

Specifications
Span, 8.109 m (26 ft 7.2 in); length (except probe), 13.625 m (44 ft 8.4 in) for the Ye-50-1; 14.85 m (48 ft 8.7 in) for the Ye-50-3; fuselage length (except cone), 12.32 m (40 ft 5 in) for the Ye-50-1; 12.715 m (41 ft 8.6 in) for the Ye-50-3; wheel track, 2.679 m (8 ft 9.5 in); wheel base, 5.22 m (17 ft 1.5 in); wing area, 21 m² (226 sq ft); takeoff weight, 8,500 kg (18,735 lb); wing loading, 404.8 kg/m² (83 lb/sq ft).

Performance
Max speed, 2,460 km/h (1,328 kt); climb to 10,000 m (32,800 ft) in 6.7 min; to 20,000 m (65,600 ft) in 9.4 min; static ceiling, 23,000 m (75,440 ft); zoom ceiling, 25,600 m (83,970 ft); landing speed, 280 km/h (151 kt); range, 450 km (280 mi); takeoff roll, 900 m (2,950 ft); landing roll, 860 m (2,820 ft).

Ye-4   Ye-5 / Kh-5 / I-500 / MiG-21

The Ye-4 and Ye-5 frontline single-seat fighters were built in 1955 and 1956. Unlike the Ye-2 and Ye-2A, built at the same time, the Ye-4 and Ye-5 had a delta wing with a sweepback of 57 degrees at the leading edge. But the fuselages were identical; so because the wheel wells for the main gear were at the same place, the legs had to be modified to retract into the delta wing. That modification gave the Ye-4 and Ye-5 a slightly larger wheel track. A high degree of commonality between the two competing aircraft was to have allowed engineers to choose the wing shape for the future frontline fighter with full knowledge of the implications of their choice.

The wing profile was a TsAGI-S9s, the ailerons were axially trimmed, and the single-slotted Fowler flaps were rectangular. The airbrakes were placed beneath the fuselage—two of them near the wing root, the third hinged farther back on frame no. 28. The three sets of fuselage fuel tanks had a total capacity of 1,570 l (414 US gallons), and the aircraft could store an additional 400 l (106 US gallons) under the fuselage in a "supersonic" drop tank. The two NR-30 cannons (at sixty rounds per gun) flanked the lower part of the forward fuselage. The shell cases were ejected, but the ammunition belts and links were collected in the empty part of the shell sleeves that ran across three fuselage sections. A pylon under the fuselage could carry either an automatic rocket pod (16 ARS-57) or one FAB 500 bomb. The ASP-5N gunsight was linked to the SRD-1M Konus ranging radar. The slab tailplane was actuated by a BU-44 servo-control unit and the aileron controls were boosted by a BU-45 servo-control unit, both of the irreversible variety. The slab tailplane was fitted with the ARU-3V feel computer to modify the gear ratio between the stick and the stabilator and to control the stick load according to flight speed and altitude. The ejection seat was identical to that of the MiG-19S (the curtain type). Main equipment included the RSIU-4V VHF, the ARK-5 automatic direction finder, and a radar warning receiver.

The last stage of the long development process that was to lead to the MiG-21, the Ye-4 was powered by the AM-9 turbojet rated at 3,185 daN (3,250 kg st with reheat). This was virtually the only thing differentiating this aircraft from the Ye-5. The Ye-4 arrived at the flight test center in June 1955. After the usual proceedings (weight determinations, ground roll tests, and a few "leapfrog" takeoffs) the Ye-4 was first piloted by G. A. Sedov on 16 June. Factory tests ended on 20 September 1956.

During these fifteen months, several methods were tested to straighten the airflow on the wing—either six fences over the upper surface (four of which protruded ahead of the leading edge) or two

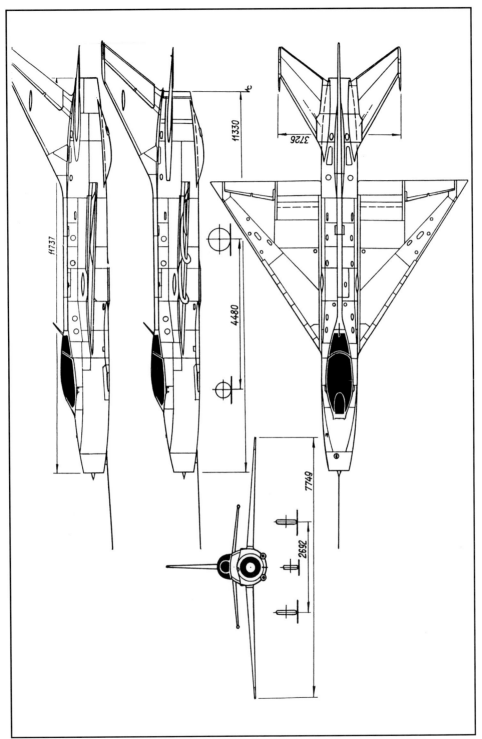

Ye-4; *second from top,* side view of the Ye-5 (MiG OKB four-view drawing)

The pros and cons of the delta wing approach were assessed on the Ye-4. The wing had two fences on the lower surface, but the Ye-4 has also flown with a clean wing.

On the Ye-4 with fences on the lower surface of the wing, the wing tips were markedly pointed.

On the same Ye-4, fitted this time with six fences on the wing's upper surface, the square wing tips reduced the span by 600 millimeters (23.6 inches).

The Ye-5 retained the delta wing and upper-surface fences of the Ye-4, but the wing tips are again markedly pointed.

fences over the lower surface and an increase in wingspan (sharp wing tips). Finally a third solution was found: a chord extension on about one-third of the leading edge that formed a vortex-generating "tooth." After this modification the Ye-4 was transferred to the LII, where it was tested a number of times at great angles of attack.

The Ye-5 was powered by the AM-11 turbojet with 3,725 daN (3,800 kg st) of dry thrust and 5,000 daN (5,100 kg st) of maximum thrust with afterburner. Its wing had six fences (four of which protruded ahead of the leading edge). It was moved to the flight test center on 10 December 1955 and was first piloted on 9 January 1956 by V. A. Nefyedov. During a ground run-up on 20 February the engine caught fire, and the turbine was destroyed. After being repaired at the factory the Ye-5 made eight flights between 26 March and 19 May. Tests were suspended once more when the turbine broke and the engine had to be removed. On 18 October the aircraft was grounded. On 27 October the AM-11 was returned to the factory to increase the volume of the afterburner, and on 24 November the nose of the fuselage was returned to the workshop to be lengthened by 400 millimeters in order to move the aircraft's center of mass forward and increase the fuel capacity to 1,810 l (478 US gallons). Those modifications had a strong influence on the final choice between the Ye-5's delta wing and the Ye-2A's sweepback wing.

The Ye-5 resumed its test schedule on 1 April 1957 and went up thirteen times before 26 May for reassessments of the aircraft's flight envelope after the modifications to its engine and fuselage. No fewer than ten AM-11 test models and one production engine were used to complete the aircraft's factory tests. According to the test pilot, the Ye-5 cockpit was roomier than that of the MiG-19, and the approach and landing procedures of both aircraft were quite similar. Extension of the gear had no effect on the aircraft's trim characteristics. There was no change in the longitudinal trim anywhere in its flight envelope, and it was easier to fly than the MiG-19 at supersonic speeds. The forward movement of the cone ahead of the air intake plane was automatic from Mach 1.4 upward.

The factory tests of the Ye-5 or MiG-21 came to a close without major problems. All of its design parameters were met except for the range, which fell short because of the excessive specific fuel consumption of the engine that was renamed the R-11. Altogether seven Ye-5s were built, two prototypes and five preproduction machines.

The following details refer to the Ye-4.

Specifications
Span, 7.749 m (25 ft 5 in); fuselage length (except cone), 11.737 m (38

ft 6.1 in); wheel track, 2.692 m (8 ft 10 in); wheel base, 4.48 m (14 ft 8.4 in); wing area, 23.13 m² (248.9 sq ft); takeoff weight, 5,200 kg (11,465 lb), fuel, 1,300 kg (2,865 lb); wing loading, 224.8 kg/m² (46.03 lb/sq ft).

Performance
Max speed, 1,970 km/h (1,064 kt); climb to 5,000 m (16,400 ft) in 1.6 min; service ceiling, 16,400 m (53,800 ft); range in clean configuration, 1,120 km (695 mi).

The following details refer to the Ye-5.

Specifications
Span, 7.749 m (25 ft 5 in); fuselage length (except cone), 11.93 m (37 ft 2 in); wheel track, 2.692 m (8 ft 10 in); wheel base, 4.48 m (14 ft 8.4 in); wing area, 23.13 m² (248.9 sq ft); takeoff weight, 5,700 kg (12,565 lb); fuel, 1,500 kg (3,305 lb); wing loading, 246.4 kg/m² (50.47 lb/sq ft).

Performance
Max speed, 1,970 km/h (1,064 kt); climb to 5,000 m (16,400 ft) in 0.6 min; service ceiling, 17,650 m (57,900 ft); range, 1,330 km (825 mi); takeoff roll, 730 m (2,395 ft); landing roll, 890 m (2,920 ft).

Ye-2A / MiG-23*

This lightweight fighter was a direct offspring of the Ye-5—and not, as would appear logical, the Ye-2—but differed from it in the wing and the main gear legs. The delta wing of the Ye-5 was replaced by a sweepback wing, minus the leading-edge slats of the Ye-2. The upper surface of this wing was fitted with two fences at midspan to play the role of vortex generators in order to increase the ailerons' effectiveness at great angles of attack. The automatic slats were also discarded because their asymmetrical deployment at times (while sideslipping, for example) triggered a severe buffet that endangered the aircraft's pitch and roll stability.

Like the Ye-5, the Ye-2A was powered by the AM-11 turbojet rated for 5,000 daN (5,100 kg st) and was first taken up on 22 March 1956 by G. K. Mosolov. In addition to the prototype, only five other machines were built; yet these were described as "production" aircraft and named

---

*For various reasons, the designation MiG-23 was granted to other aircraft; but its final holder was the 23-11, a mass-produced VG fighter.

The Ye-2A derived from the Ye-5 more than the Ye-2, but it had a sweepback wing without leading edge slats and was fitted with stall fences on the upper surfaces of the wing (see the side view in the Ye-2 drawing).

MiG-23s even though an assembly line was set up for true mass production in factory no. 21 as early as 1956. Oddly, the factory tests were carried out with the Ye-2A/6—the fifth "production" aircraft built in factory no. 21—by G. K. Mosolov, V. A. Nefyedov, G. A. Sedov, and others.

The Ye-2A/3 was transferred to the LII MAP for tests of such special procedures as power-off landings. Six flights made by test pilot A. P. Bogorodskiy proved that those landings did not present any specific problems. But in the meantime the Ye-5 (with the longer fuselage) proved to be the more promising aircraft, and all Ye-2A flight tests were canceled.

Specifications
Span, 8.109 m (26 ft 7.2 in); length (except probe), 13.23 m (43 ft 4.9 in); fuselage length (except cone), 11.33 m (37 ft 2 in); wheel track, 2.679 m (8 ft 9.5 in); wheel base, 4.41 m (14 ft 5.6 in); wing area, 21 m^2 (226 sq ft); empty weight, 4,340 kg (9,656 lb); takeoff weight, 6,250 kg (13,775 lb); fuel, 1,450 kg (3,195 lb); wing loading, 297.6 kg/m^2 (61 lb/sq ft).

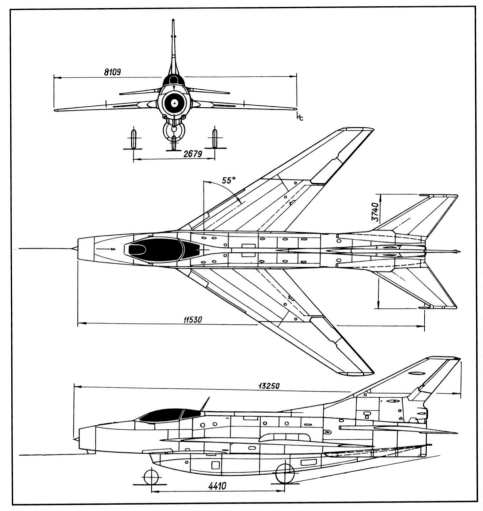

Ye-50A (MiG OKB three-view drawing)

Performance

Max speed, 1,900 km/h (1,026 kt); climb to 10,000 m (32,800 ft) in 7.3 min; service ceiling, 18,000 m (59,050 ft); landing speed, 280 km/h (151 kt); range, 2,000 km (1,240 mi).

Ye-50A

The Ye-50A high-altitude interceptor was designed in 1956 and built in the Gorki production factory after approval of a full-scale mock-up that

Artist's rendition of the Ye-50A, which was never completed. It was designed with a compound power unit, like the Ye-50.

retained the Ye-2A airframe structure. The power unit was composed of the AM-11 turbojet and the S-155 rocket engine. Like the Ye-50, the new prototype had its rocket engine, accessories, and hydrogen peroxide tank gathered in the fin base. A frame was added in the tail of the fuselage, and a jet air pump nozzle was installed in the engine bay.

The wing, the stabilator, the cockpit hood, and the gear were identical to those of the Ye-2A. The inside fuselage arrangement was also identical to that of the Ye-2A all the way to frame no. 20. Beyond that point and back to the fin attachment fittings, it was identical to that of the Ye-50. The fuel system was modified slightly, tank nos. 6 and 7 being removed. The rocket engine control and supply systems were gathered in a long, easy-to-remove fairing on the underside of the fuselage. The production factory was ordered to build a batch of twenty Ye-50As, but they were never finished because L. S. Dushkin OKB closed and therefore could not supply the needed rocket engines.

Specifications
Span, 8.109 m (26 ft 7.2 in); length (except probe), 13.25 m (43 ft 5.7 in); fuselage length (except cone), 11.53 m (37 ft 9.9 in); wheel track, 2.679 m (8 ft 9.5 in); wheel base, 4.41 m (14 ft 5.6 in).

The Ye-6/1, the first MiG-21 prototype, was ill fated. Engine failure led to the death of its test pilot, V. A. Nefyedov.

The Ye-6/2, the second MiG-21 prototype, was equipped experimentally with launch rails at the wing tips. In this photograph, they hold K-13 air-to-air missiles.

The first production MiG-21 was the MiG-21F, shown in this photograph with UB-16-57U rocket pods under the wing.

MiG-21 / Ye-6/1 / Ye-6/2 / Ye-6/3 (Ye-66)

The first three MiG-21 prototypes, Ye-6/1, Ye-6/2, and Ye-6/3, were built and flight-tested in 1957 and 1958. They were powered by a new version of the AM-11 turbojet, the R-11F-300 (developed from the experimental R-37F) rated at 3,800 daN (3,880 kg st) dry or 5,625 daN (5,740 kg st) with afterburner. Their stabilators were lower than that of the Ye-5, forcing designers to rearrange the airbrakes in these units; the two canted ventral fins on the fuselage under the tail were replaced by a single unit; the nozzle throat was lengthened; and the rear part of the cockpit hood was redesigned. Only the Ye-6/1 retained the six wing fences first seen on the Ye-5.

The MiG-21's airbrakes closely followed the shape of the NR-30 gun fairings.

In no time the Ye-6/1 reached Mach 2.05 at 12,050 m (39,520 feet). But the seventh flight, on 28 May 1958, ended in tragedy after the engine failed at about 18,000 m (59,040 feet). The test pilot, V. A. Nefyedov, struggled desperately to return to the airfield in order to save the aircraft and the recording of all its flight data. He made it to the runway, but as the plane touched down it overturned and caught fire. Severely burned, Nefyedov died in a hospital a few hours later. The official inquiry established that the pilot was betrayed by the pressure drop in the hydraulic system due to engine failure. Because the stabilator was hydraulically controlled the standby electrical control was automatically activated, but it took the backup unit far too long to set the stabilator at the proper angle. As a consequence the hydraulic system on the Ye-6/2 was duplicated and backed up by an emergency pump, and the electrical control unit was removed. K. K. Kokkinaki was given responsibility for the Ye-6/2 test program. This second prototype, numbered 22, was equipped experimentally with missile launching rails at the wing tips.

The Ye-6/3 made its first flight in December 1958 and became world-famous a few months later under the fanciful designation Ye-66 while beating two world records:

1. 31 October 1959. Speed over a 15- to 25-km (9- to 16-mile) course at unrestricted altitude, 2,388 km/h (1,289.52 kt). Pilot, G. K.

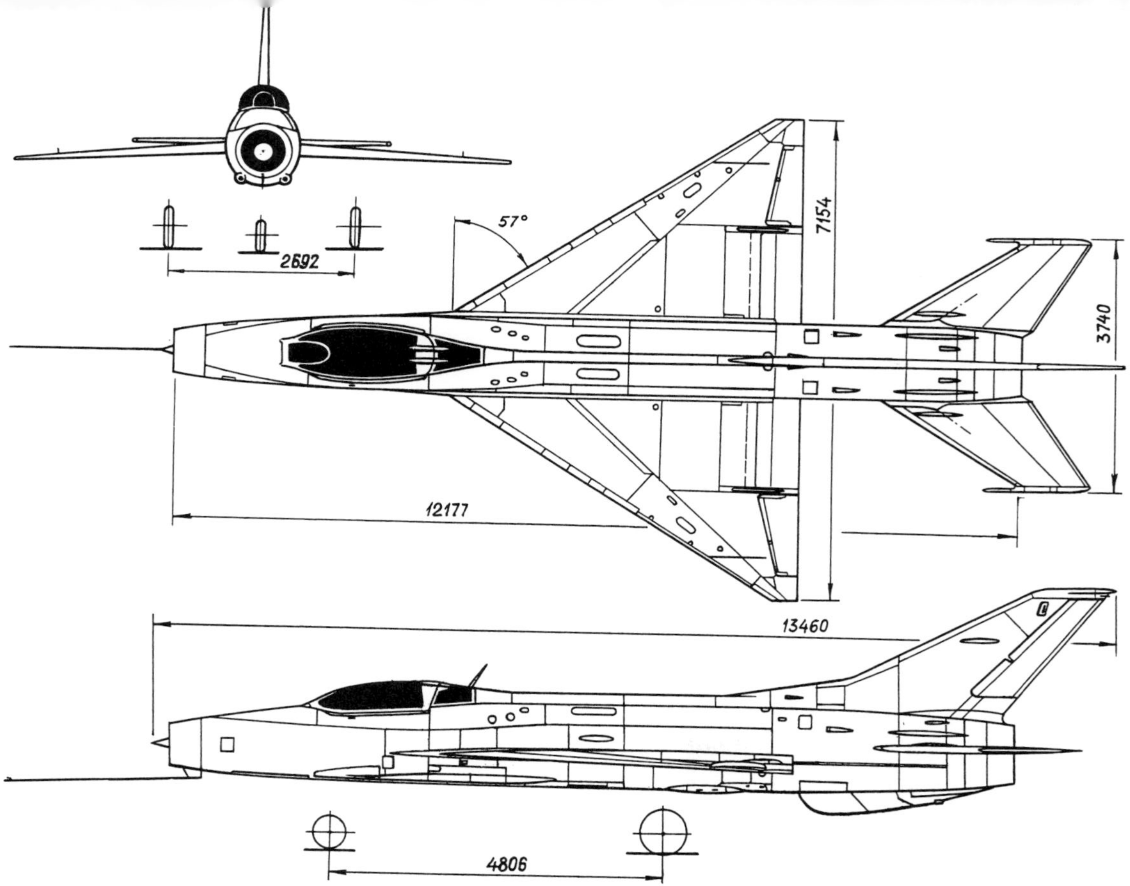

MiG-21F (MiG OKB three-view drawing)

3740

7154

57°

12177

2692

13460

4806

A MiG-21F equipped experimentally with K-13 air-to-air missiles under wing pylons. The cannons were removed.

Mosolov. Highest speed attained during this flight, 2,504 km/h (1,352.16 kt)
2. 16 September 1960. Speed over a closed circuit of 100 km (62 miles), 2,148.66 km/h (1,160.28 kt). Pilot, K. K. Kokkinaki. Highest speed attained during this flight, 2,499 km/h (1,349.46 kt) or Mach 2.35

MiG-21F / Ye-6T / *Tip 72*

Introduced in 1958, the MiG-21F (Ye-6T) was the first production model of the family. The delta wing had a 57-degree sweepback at the leading edge like the preceding delta-wing prototypes, with Fowler flaps designed by TsAGI. The power unit was the R-11F-300 turbojet, rated at 5,625 daN (5,740 kg st) of reheated thrust. Its control system could set the air intake shock cone in three different positions. It was thus possible to change the cross-section area of the air intake duct as well as the direction of the shock waves according to flight regime. Its military instrumentation was still relatively basic, limited to the ASP-SDN gunsight, the SRD-5 ranging radar, and the IFF transponder.

The Ye-6T/3, the third MiG-21F prototype, was used to test canard surfaces.

There was no automatic direction finder. The curtain-type ejection seat was identical to that of the MiG-19. The tail chute was housed in a small container under the rear of the fuselage. The ten fuel tanks—six in the fuselage and four in the wing—had a total capacity of 2,160 l (570 US gallons).

Armament included two NR-30 cannons with sixty rounds per gun and store stations under the wing for two UB-16-57U rocket pods with either sixteen 57-mm S-5M air-to-air rockets (ARS-57) apiece or sixteen 57-mm S-5K air-to-surface rockets (KARS-57); two 240-mm ARS-240 heavy air-to-surface rockets; or two 50- to 500-kg (110- to 1,100-pound) bombs. The third prototype, the Ye-6T/3, was tested with a small mobile canard surface set near the nose; this foreplane was to appear later on the Ye-8 experimental machine. The Ye-6T/3 was also used to develop the launching system of the air-to-air missiles that were to arm future versions of the MiG-21.

Tests of the MiG-21F ended in 1958. Forty machines were assembled in the Gorki factory in 1959 and 1960.

Specifications

Span, 7.154 m (23 ft 5.7 in); length (except probe), 13.46 m (44 ft 1.9 in); fuselage length (except cone), 12.177 m (39 ft 11.4 in); wheel track, 2.692 m (8 ft 10 in); wheel base, 4.806 m (15 ft 9.2 in); wing area, 23 m² (247.6 sq ft); takeoff weight, 6,850 kg (15,100 lb); fuel, 1,790 kg (3,945 lb); wing loading, 297.8 kg/m² (61 lb/sq ft); max operating limit load factor, 7.

Performance
Max speed, 2,175 km/h at 12,500 m (1,175 kt at 41,000 ft); max speed at sea level, 1,100 km/h (594 kt); climb rate at sea level in clean configuration, 175 m/sec (34,450 ft/min); climb to 18,500 m (60,700 ft) in 7.5 min; service ceiling, 19,000 m (62,300 ft); landing speed, 280 km/h (151 kt); range at 14,000 m (45,900 ft) in clean configuration, 1,520 km (945 mi); takeoff roll, 900 m (2,950 ft); landing roll with tail chute, 800 m (2,625 ft).

MiG-21F-13 / Tip 74

This was the first MiG-21 armed with air-to-air missiles, the first to be mass-produced, the first to be exported, and the first to be built outside the USSR (in Czechoslovakia, in India, and in China). It may be regarded as the basic model of the entire family. That is why we will devote special attention to this machine. But first it must be stated that the MiG-21F-13 did not come into its final silhouette until aircraft no. 115 left the assembly line. Starting with this aircraft, the fin height was reduced and its chord was increased. The following details refer to the final model, except where otherwise noted.

The MiG-21F-13 delta wing, like that of its predecessors, had a 57-degree sweepback at the leading edge and a 2.2 aspect ratio. The wing airfoil was a TsAGI S-12 with a thickness ratio of 4.2 percent at the wing root and 5 percent at the wing tip; incidence, 0 degrees; anhedral, 2 degrees; maximum chord, 5.97 m (19 feet, 7 inches); mean aerodynamic chord (MAC), 4.002 m (13 feet, 1.6 inches). The small fence that can be seen in front of each aileron had a height equal to 7 percent of the MAC. The trailing edge of each wing was fully occupied by a single-slotted flap of 0.935 m² (10.1 square feet) and an inset aileron of 0.51 m² (5.5 square feet). The flaps were hydraulically controlled, and the ailerons were hydraulically boosted by two BU-45 servo-control units. The wing structure was organized around three spars:

1. A front spar preceded by twenty-five ribs that were square to the leading edge, and a front false spar; the front wet wing tank was located ahead of that spar, close to the wing root
2. A center beam that was square to the fuselage datum line; the wheel was stored between the front spar and the center beam,

The MiG-21F-13 was the first production MiG-21 armed with K-13 air-to-air missiles, hence its designation. After it was accepted by the air force, the K-13 was renamed R-3.

and the bracket for the gear leg hinge was placed at the juncture of those two spars

3. A rear spar preceded by ten ribs that were set parallel to the fuselage datum line, and a rear false spar

The rear wet wing tanks were located between rib nos. 1 and 6. The upper and lower walls of the fuel tanks were made of machined plates of the V-95 alloy, the same one used for skin panels without stiffeners. The wing was attached to the fuselage by five fittings, three to transmit the moment (matched with the front, center, and rear spars) and two to convey the bending strain (matched with the front and rear false spars). The all-metal semimonocoque fuselage could be split into two parts between frame nos. 28 and 28A (twenty-eight frames in front, thirteen to the rear). Other basic components included body longerons in front, stringers at the rear, and relatively thick skin to strengthen the whole structure. The air intake, like that of the MiG-21F, housed a three-position cone that controlled the duct area according to the aircraft's speed: up to Mach 1.5 the cone did not move, between Mach 1.5 and 1.9 it moved forward partway, and beyond Mach 1.9 it moved farther forward.

This MiG-21F-13 is equipped with a finned "supersonic" drop tank having a capacity of 490 l (129 US gallons).

The side airbrakes, hinged to frame no. 11, each measured 0.38 m² (4.1 square feet) and had a maximum deflection of 25 degrees; the belly airbrake measured 0.47 m² (5.06 square feet) and had a maximum deflection of 40 degrees. Under the tail section of the fuselage was a ventral fin 0.352 m (13.85 inches) high with a canister for a 16-m² (172.2–square foot) tail chute on its left. When the chute was deployed the landing roll was cut by about 400 m (1,310 feet). The engine bay was located between frame nos. 29 and 34. The cockpit was pressurized and air-conditioned; a special regulator kept the temperature in the 15° C range give or take 5° C.

The fuselage had maximum diameter of 1.242 m (4 feet, 0.9 inches) and a maximum cross-section of 1.28 m² (13.8 square feet). The tail fin had a sweepback of 60 degrees at the leading edge and an area of 3.8 m² (40.9 square feet), versus 4.08 m² (43.9 square feet) for the MiG-21F and the first 114 MiG-21F-13s; the rudder had an area of 0.965 m² (10.4 square feet) and a maximum deflection of plus-or-minus 25 degrees; tail fin airfoil, TsAGI S-11; thickness/chord ratio, 6 percent. The stabilator had a 55-degree sweepback at the leading edge, an area of 3.94 m² (42.4 square feet), a span of 3.74 m (12 feet, 3.2 inches), no dihedral, an A6A symmetrical airfoil, and a thickness/chord ratio of 6 percent. This all-flying tail had a variable incidence ranging from +7.5 to –16.5 degrees and the ARU-3V feel computer.

The tricycle landing gear was composed of a forward nosewheel unit (tire size 500 x 180) that retracted forward between frame nos. 6

and 11 and a main wheel unit that retracted inward. The gear legs lodged in the wing, but the wheels (660 x 200) turned themselves 87 degrees to stow vertically inside the fuselage. The gear was hydraulically controlled and had a backup pneumatic system.

The cockpit was fitted with the SK ejection seat, a system that used the canopy to protect the pilot. The instrument panel included (besides the basic instrumentation) the KAP-2K autopilot with roll limitation of plus-or-minus 35 degrees, the R-802V (RSIU-5V) VHF, the MRP-56P marker receiver, the ARK-10 automatic direction finder, the RV-UM radio-altimeter for 0–600 m (0–1,970 feet), the SOD-57M decimetric transponder, the Sirena 2 radar warning receiver, and the SRO-2 IFF transponder.

The R-11F-300 turbojet was rated at 3,820 daN (3,900 kg st) dry and from 4,800 daN (4,900 kg st) to 5,625 daN (5,740 kg st) with a throttleable afterburner. The fuel tank capacity—2,280 l (602 US gallons) in the aircraft's first series—was increased to 2,550 l (673 US gallons). The MiG-21F-13 could also carry a drop tank under the fuselage with 490 or 800 l (129 or 211 US gallons) of fuel.

Armament consisted of a single NR-30 cannon with thirty rounds on the right side of the fuselage and either two R-3S air-to-air missiles (IR seeker, firing distance of 1 to 7 km [0.62 to 4.3 miles]), two UB-32-57U rocket pods (S-5 rockets), two 240-mm S-24 rockets, or two bombs (up to 500 kg [1,100 pounds] apiece) on two underwing pylons. The ASP-5ND gunsight was linked to the SRD-5M Kvant ("quantum") ranging radar. For limited reconnaissance missions the aircraft could be equipped with the AFA-39 camera.

The MiG-21F-13 was mass-produced in the Gorki factory between 1960 and 1962 for the VVS and in the MMZ Znamya Truda factory in Moscow between 1962 and 1965 for export.

Specifications
Span, 7.154 m (23 ft 5.7 in); length (except probe), 13.46 m (44 ft 1.9 in); overall length, 15.76 m (51 ft 8.5 in); height, 4.1 m (13 ft 5.4 in); wheel track, 2.692 m (8 ft 10 in); wheel base, 4.806 m (15 ft 9.2 in); wing area, 23 m^2 (247.6 sq ft); empty weight, 4,980 kg (10,975 lb); takeoff weight with two R-3S missiles, 7,370 kg (16,245 lb); max takeoff weight with 490-l (129-US gal) drop tank and two 500-kg (1,100-lb) bombs, 8,625 kg (19,010 lb); fuel, 2,115 kg (4,660 lb); wing loading, 320.4–375 kg/m^2 (65.7–76.9 lb/sq ft); max operating limit load factor, 7.

Performance
Max speed, 2,175 km/h at 13,000 m (1,175 kt at 42,640 ft); max speed at sea level, 1,150 km/h (621 kt); climb rate at sea level in clean configuration, 175 m/sec (34,450 ft/min); climb to 19,000 m (62,300 ft)

The Ye-6V/1 (as well as the Ye-6V/2) was used to test various devices for short takeoff and landing. The canister for the brake chute is located at the base of the fin.

This Ye-6V/2 is equipped with a finned drop tank, two K-13 air-to-air missiles, and two JATO boosters.

On 9 July 1961 Fedotov executed a jet-assisted takeoff in public in the Ye-6V/2.

with 490-l (129-US gal) drop tank and two R-3S missiles in 13.5 min; service ceiling, 19,000 m (62,300 ft); landing speed, 280 km/h (151 kt); range, 1,300 km at 11,000 m (810 mi at 36,080 ft); with 800-l (211-US gal) drop tank, 1,670 km (1,040 mi); takeoff roll, 800 m (2,625 ft); landing roll with tail chute, 800 m (2,625 ft).

Ye-6V

Two *Tip* 74 airframes, the Ye-6V/1 and Ye-6V/2, were modified to assess various positions for the tail chute container, to test several types of chutes, and to make jet-assisted takeoffs with solid propellant rockets. The Ye-6V had two tail chute canisters, one on the left side of the ventral fin and the other at the base of the tail fin. The latter position was retained on all MiG-21s starting with the PFM. Both Ye-6V prototypes were also tested with wheel-ski compound gear. Fedotov demonstrated jet-assisted takeoffs in the Ye-6V/2 at the Tushino air show on 9 July 1961.

Ye-66A

In early 1961 the OKB undertook a complete refit of the Ye-6T/1 to reengine this prototype with the R-11F2-300 rated at 6,000 daN (6,120 kg st) and the U-21 booster package, including the Dushkin S3-20M5A liquid propellant rocket engine. This engine and its tanks were housed in a long fairing under the fuselage. The total thrust of the turbojet and

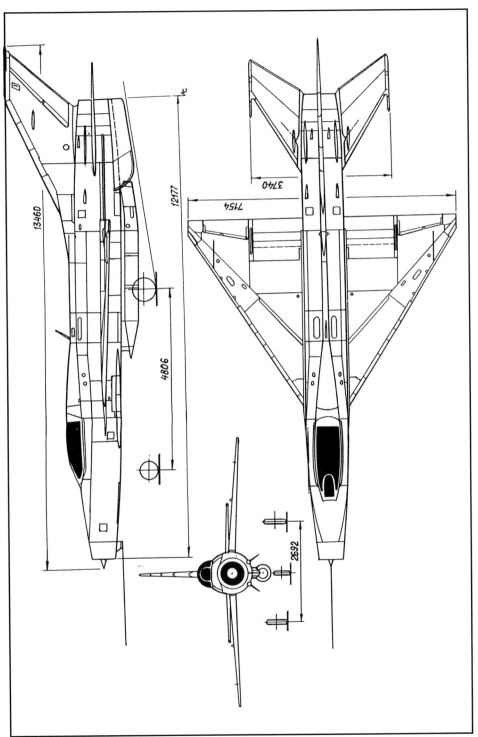

Ye-66A (MiG OKB three-view drawing)

The Ye-66A, a record breaker, was in fact the Ye-6T/1 prototype reengined with a more powerful turbojet and a liquid propellant rocket engine (Photo RR).

the rocket was 11,465–11,830 daN (11,700–12,070 kg st), the two figures corresponding to the afterburner throttleability range of the turbojet.

This reengining necessitated a few structural modifications. For instance, the lone ventral fin was replaced by two rather high fins whose total area amounted to 11.5 percent of the wing area; the tail fin was enlarged to 4.44 m² (47.8 square feet); the fin nose was made thinner; and an auxiliary fuel tank was installed behind the cockpit. With this aircraft G. K. Mosolov beat the absolute world record for altitude on 28 April 1961, ending an incredible zoom at 34,714 m (113,862 feet). In the documents sent to the FAI (Fédération Aéronautique Internationale) to verify the record, the aircraft's power unit was described as one R-37F TRD (turbojet) at 5,880 daN (6,000 kg st) and one U-21 ZhRD (rocket engine) at 2,940 daN (3,000 kg st).

Specifications
Span, 7.154 m (23 ft 5.7 in); length (except probe), 13.46 m (44 ft 1.9 in); fuselage length (except cone), 12.177 m (39 ft 11.4 in); wheel track, 2.692 m (8 ft 10 in); wheel base, 4.806 m (15 ft 9.2 in); wing area, 23 m² (247.6 sq ft).

Like the Ye-7/1, the Ye-7/2—prototype of the MiG-21P, pictured here with a finned drop tank—had a dorsal fuel tank with a capacity of 170 l (45 US gallons).

The Ye-7 emanated directly from the Ye-6T. This photograph shows the Ye-7/1 equipped with the new MiG-21P-13 interception system (TsD-30T radar) and two K-13 missiles.

MiG-21P / Ye-7/1 / Ye-7/2

The Ye-7/1 and Ye-7/2 were both direct descendants of the Ye-6T. The MiG-21P was therefore a direct descendant of the MiG-21F, with the same R-11F-300 turbojet but a new 170-l (45-US gallon) fuel tank behind the cockpit. To make the aircraft usable at rough strips the main gear was fitted with bigger wheels (type KT-50/2, tire size 800 x 200), and to shorten its takeoff roll two attachment points were added under rear fuselage for two solid propellant "accelerators" that could be dropped after ten seconds of burning time. The Ye-7s had the KAP-1 autopilot, but oscillations were damped on the roll axis only.

The MiG-21P was the first member of the family without cannons. The new air-battle concept prevailing at that time called for missiles to be the only armament of fighter aircraft. It was thought that the considerable increase in fighter speed had ended the era of close combat. Confined conflicts such as the Vietnam War would reveal the errors of that doctrine.

The MiG-21P was also the first member of the family to be equipped with a real interception system, the MiG-21P-13, which included TsD-30T radar (with surveillance, acquisition, tracking, and fire control modes), command receiver, SOD-57M decimetric transponder, Vozdukh-1-Lazur guidance system, KSI navigation system, IFF interrogator, and two K-13 IR homing air-to-air missiles. In place of missiles, the MiG-21P could carry unguided rocket pods, bombs, and even napalm containers. For ground-attack missions the pilot had the PKI-1 gunsight, which could also be used in the event of radar failure. The ejection seat was of the SK type.

The Ye-7/1 prototype made its first flight on 10 August 1958, the Ye-7/2 on 18 January 1960. The factory tests, conducted by P. M. Ostapyenko and I. N. Kravtsov, ended on 8 May 1960, and production was launched in June. The performance of the MiG-21P was identical to that of the MiG-21F except for the service ceiling, which increased to 19,100 m (62,650 feet); the climb rate at sea level in clean configuration, which was reduced to 150 m/sec (29,530 ft/min); and the landing roll with tail chute, which was reduced to 650 m (2,130 feet). Its maximum operating limit load factor was 7.8.

MiG-21PF / *Tip* 76 / Ye-76

If the MiG-21P descended from the MiG-21F, the MiG-21PF descended from the MiG-21F-13. However, it contained the more powerful R-11F-

The new shape of the spine allowed engineers to add two fuel tanks and bring the aircraft's total fuel capacity to 2,750 l (668 US gallons).

300, rated at 3,870 daN (3,950 kg st) or 6,000 daN (6,120 kg st) with afterburner. The greater volume of the nose cone (which housed a bigger radar unit) and the increased airflow essential for the new turbojet meant that the diameter of the air intake had to be increased from 690 mm (27.2 inches) to 870 mm (34.25 inches). It is noteworthy that this change had already taken place on the MiG-21P.

A new system, the UVD-2M, was developed to ensure steady control of the nose cone at all flight regimes. (On the ground, the cone extended 1,213 mm [47.8 inches] ahead of the air intake plane; in flight, according to speed, this could be reduced to 200 mm [7.9 inches].) This MiG-21PF inherited such modifications from the MiG-21P as larger wheels and attachment points for solid propellant boosters. However, its silhouette was somewhat modified by the removal of the rear window in the canopy and the new shape of the dorsal fairing immediately behind the cockpit; this housed two more fuel tanks, bringing the total fuel capacity to 2,750 l (726 US gallons). In addition, the PDV-5 air data probe—which on earlier versions was set axially under the air intake—was moved above and to the right of the intake.

The RP-21 Sapfir ("sapphire") radar made its debut on the MiG-21PF. The ASP-5ND gunsight was replaced by a PKI-1. Other equipment included the KAP-2 autopilot (still only for roll stabilization), a guidance command receiver, and an IFF interrogator. Like the MiG-21P, the PF had no cannon. Its armament was limited to two K-13 air-to-air missiles that could be replaced by the usual weaponry options

On late-series MiG-21PFs, the tail chute canister was moved to the base of the tail fin after this position was tried out on the Ye-6V.

(rocket pods, bombs, etc.). On PF late series the tail chute canister was set at the base of the tail fin, a position that was tried out on the Ye-6V.

In early 1962 a decree signed by the minister of defense accepted the MiG-21PF into the military inventory of the USSR. The aircraft was mass-produced in the Gorki factory between 1962 and 1964 for the VVS and in the MMZ Znamya Truda factory in Moscow between 1964 and 1968 for export. A MiG-21PF—renamed Ye-76 for this purpose—broke several female world records in 1966 and 1967:

1. 16 September 1966. Speed over a closed circuit of 500 km (310 miles), 2,062 km/h (1,113.5 kt). Pilot, M. Solovyeva
2. 11 October 1966. Speed over a closed circuit of 2,000 km (1,240 miles), 900.267 km/h (559.07 kt). Pilot, Ye. Martova
3. 18 February 1967. Speed over a closed circuit of 100 km (62 miles), 2,128.7 km/h (1,149.5 kt). Pilot, Ye. Martova
4. 28 March 1967. Speed over a closed circuit of 1,000 km (620 miles), 1,928.16 km/h (1,041.2 kt). Pilot, L. Zaitseva

The documents sent to the FAI mentioned that the Ye-76 was powered by an R-37F turbojet rated at 5,830 daN (5,950 kg st).

Specifications
Span, 7.154 m (23 ft 5.7 in); fuselage length (except cone), 12.285 m
(40 ft 3.7 in); wheel track, 2.692 m (8 ft 10 in); wheel base, 4.806 m (15
ft 9.2 in); wing area, 23 m² (247.6 sq ft); takeoff weight, 7,750 kg
(17,080 lb); max takeoff weight, 9,500 kg (20,940 lb); max takeoff
weight on rough strip or metal-plank strip, 8,800 kg (19,395 lb); fuel,
2,280 kg (5,025 lb); wing loading, 337-413-382.6 kg/m² (69.1-84.7-78.4
lb/sq ft); max operating limit load factor, 8.

Performance
Max speed, 2,175 km/h at 13,000 m (1,175 kt at 42,640 ft); max speed
at sea level, 1,300 km/h (702 kt); climb to 18,500 m (60,700 ft) in 8
min; service ceiling, 19,000 m (62,300 ft); climb rate at sea level (half
internal fuel, full thrust) with two R-3S missiles, 205 m/sec (40,350
ft/min); landing speed, 280 km/h (151 kt); range, 1,400 km (870 mi);
with 800-l (211-US gal) drop tank, 1,770 km (1,100 mi); takeoff roll, 850
m (2,790 ft); landing roll with tail chute, 850 m (2,790 ft).

MiG-21FL / *Tip 77*

This frontline fighter-interceptor was a special version of the MiG-21PF
developed to be built under license in India and for export. Externally
both aircraft were very similar, but the MiG-21FL had the same engine
as the MiG-21F—the R-11F-300—and the total capacity of the fuel tanks
was increased to 2,900 l (766 US gallons). Moreover, the RP-21 radar
was replaced by the R-2L, a less advanced export model.

The MiG-21FL was built in the MMZ Znamya Truda factory in
Moscow between 1965 and 1968 and by HAL in India from 1966
onward.

Specifications
Span, 7.154 m (23 ft 5.7 in); fuselage length (except cone), 12.285 m
(40 ft 3.7 in); wheel track, 2.692 m (8 ft 10 in); wheel base, 4.806 m (15
ft 9.2 in); wing area, 23 m² (247.6 sq ft); takeoff weight in clean configuration, 7,830 kg (17,255 lb); max takeoff weight, 9,400 kg (20,715 lb);
max takeoff weight on rough strip or metal-plank strip, 8,100 kg
(17,850 lb); wing loading, 340.4-408.7-352.2 kg/m² (69.8-83.8-72.2 lb/sq
ft); max operating limit load factor, 8.

Performance
Max speed, 2,175 km/h at 13,000 m (1,175 kt at 42,640 ft); max speed
at sea level, 1,130 km/h (610 kt); climb rate at sea level in clean confi-
uration, 175 m/sec (34,450 ft/min); climb to 18,500 m (60,700 ft) in 8

min; service ceiling, 19,000 m (62,300 ft); landing speed, 280 km/h (151 kt); range, 1,450 km (900 mi); with 800-l (211-US gal) drop tank, 1,800 km (1,120 mi); takeoff roll, 850 m (2,790 ft); landing roll with tail chute, 850 m (2,790 ft).

Ye-8

Work on the Ye-8 frontline fighter-interceptor got under way in 1961 by government decree. It was referred to as the MiG-23 (the second aircraft to go by that name). Basically it was a modified MiG-21PF airframe reengined with the more powerful R-21 turbojet and equipped with the new Sapfir-21 radar in the nose. The antenna diameter of this radar forced the manufacturer to move the air intake under the cockpit. To simplify the mass production of the future MiG-23, which was intended to replace the MiG-21PF on the assembly lines, the Ye-8 received the same on-board systems as the MiG-21. The Ye-8 requirement called for a fighter capable of intercepting and destroying intruders in the front and rear sectors twenty-four hours a day and in any weather conditions.

The production MiG-21 wing retained for both Ye-8 prototypes was not fitted with the SPS system of flap blowing, and the stabilator, also taken from the MiG-21 assembly line, was lowered by 135 millimeters (5.3 inches) below the fuselage datum line. Other noteworthy technical innovations were also made:

1. Under the tail of the fuselage there was a ventral fin that folded to starboard when the landing gear extended, the hinge control being slaved to the gear's follow-up linkage. Tested for the first time on the Ye-8, this type of ventral fin was later included on the production MiG-23 (the one with a VG wing).
2. A foreplane—or more precisely a rotating delta canard surface—with a span of 2.6 m (8 feet, 5.4 inches) was set immediately behind the radome. This "destabilizing" surface was not controlled by the pilot. In subsonic flight it behaved like a weathercock; at Mach 1 and beyond, it was mechanically set to a neutral position in relation to the aircraft's datum line, modifying the aerodynamic center and reducing the margin of pitch stability (unnecessary at supersonic speeds).

The R-21F that powered the Ye-8 was in fact an R-11F modified by N. Metskhvarishvili and rated at 4,605 daN (4,700 kg st) dry or 7,055 daN (7,200 kg st) with afterburner (an outstanding afterburning ratio of

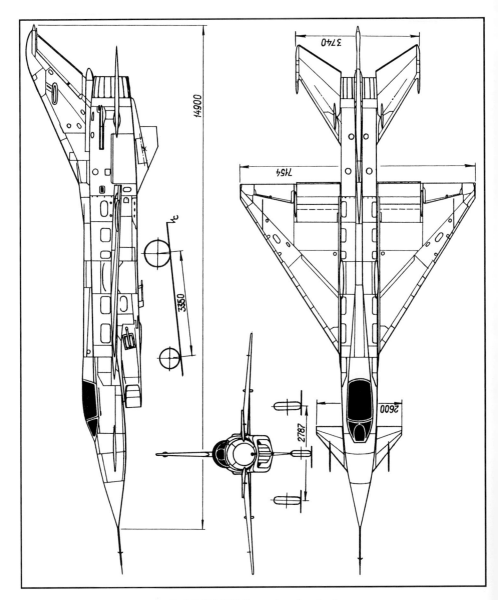

Ye-8 (MiG OKB three-view drawing)

53 percent). Compared to the R-11F, the R-21F had a diameter of 845 mm (33.26 inches) versus 772 mm (30.39 inches), a nozzle throat diameter of 987 mm (38.86 inches) versus 902 mm (35.51 inches), and a dry weight of 1,250 kg (2,755 pounds) versus 1,165 kg (2,568 pounds). The ventral air intake was divided into two ducts by a three-step splitter

The 81 or Ye-8/1 consisted of a MiG-21PF airframe drastically modified to make room for the Sapfir-21 radar antenna.

that could be adjusted electrohydraulically. The front gear leg retracted into the splitter. The main gear, taken from a production MiG-21 of the Ye-7 type, was strong enough to withstand takeoffs and landings on rough strips. The flying controls and the hydraulic system were those of the MiG-21, except for devices linked to the canard surfaces and the folding ventral fin. The armament specified for the Ye-8 was two K-13 air-to-air missiles, but neither the missiles nor the radar were ever installed.

The Ye-8/1, which bore the marking "81" on the sides of its fuselage just under the cockpit, was moved to the test center on 5 March 1962. Its R-21F no. 21-205 turbojet was intended only for ground tests and was later replaced by the flight-cleared R-21F no. 21-106. The first flight, on 17 April 1962 with G. K. Mosolov at the controls, went off without incident. For the first five flights all systems were tested, the engine was put through its paces (including relight in flight at up to 8,000 m [26,240 feet]), and the directional stability was controlled. The next six flights were dedicated to measuring accelerations at various Mach numbers and reaching the service ceiling. The operation of the canard surface was checked at the same time.

After one engine surge and one flameout on the twenty-first and twenty-fifth flights, the R-21F no. 21-106 was replaced by no. 21-108, an

The 82 or Ye-8/2, here armed with K-13 missiles, was flown only thirteen times before its tests were halted.

engine with a larger turbine nozzle. On 11 September 1962 the engine burst at Mach 1.7 at 10,000 m (32,800 feet). Mosolov ejected but was seriously wounded and had to be taken to a hospital.

The Ye-8/2 (or "82") was flight-tested thirteen times by A. V. Fedotov from 29 June to 4 September 1962. But all flights were canceled after the Ye-8/1 crash. The inquiry revealed that the accident was due to the breakup of a part of the sixth compressor stage rotor. Once it came loose it ripped through the engine casing and the aircraft's skin and hit the wing, demolishing the aileron. The plane entered a tailspin at 5,000 m (16,400 feet). The sudden loss of thrust led to a surge in the compressor and the air intake ducts. During the subsequent rapid deceleration the aircraft suffered severe lateral oscillations, a phenomenon observed in previous flights after the pilot had intentionally cut off the engine. At this point the aircraft was practically uncontrollable.

Specifications
Span, 7.154 m (23 ft 5.7 in); length (except probe), 14.9 m (48 ft 10.6 in); wheel track, 2.787 m (9 ft 1.7 in); wheel base, 3.35 m (10 ft 11.9 in); wing area, 23.13 m^2 (249 sq ft); takeoff weight, 6,800 kg (14,985 lb);

max takeoff weight, 8,200 kg (18,070 lb); wing loading, 294–354.5 m^2 (60.3–72.7 lb/sq ft).

Performance
Max speed, 2,230 km/h (1,204 kt); service ceiling, 20,000 m (65,600 ft).

MiG-21PFM / MiG-21PFS / Ye-7SPS / *Tip* 94

The MiG-21 silhouette was again retouched. The tail fin chord was increased to improve the yaw stability margin: total area rose to 5.35 m^2 (57.26 square feet) without increasing the size of the rudder. The tail chute container found itself back—for good—at the base of the tail fin, and the chute canopy was given a cruciform shape. But the most significant modification was the addition of the SPS system (*Sduv Pogranichnogo Sloya:* boundary layer blowing). Air bled from the turbojet HP compressor was blown over the flaps' upper surface, accelerating the boundary layer speed and thereby delaying its separation. The overall wing lift was thus increased when the aircraft landed (pilots used SPS by choice). The MiG-21s equipped with this system were easily recognizable by their bulky flap-actuator fairings located at flap midspan. The area of the flaps was slightly reduced to 0.92 m^2 (9.9 square feet) per unit from the 0.935 m^2 (10.06 square feet) of previous versions. Their deflection was 25 degrees at takeoff and a maximum of 45 degrees at landing. The engine was the same as that of the MiG-21PF, but the letter *s* (for *Sduv*) was added to its official name. It thus became the R-11F2S-300 (or 37F2S in some documents). It was rated at 6,050 daN (6,175 kg st) with afterburner.

The cockpit canopy was thoroughly modified. Instead of opening upward with hinges to the front, it was divided in two elements: a fixed windshield and a canopy hinged to starboard. This change was linked to the installation of the new KM-1 third-generation ejection seat (the canopy was no longer needed to protect the pilot during the initial ejection sequence). The KM-1 was not yet a true zero-zero ejection seat because its operating range was, with the exception of altitude (0–25,000 m [0–82,000 feet]), limited by speed (130–1,200 km/h [70–648 kt]). This cockpit rearrangement led to a slight reduction of the fuel tanks' capacity: 2,650 l (700 US gallons) versus 2,750 l (726 US gallons) in the PF.

So that it could be used from unprepared strips, the MiG-21PFM could be fitted with two SPRD-99 solid rocket boosters each rated at 2,450 daN (2,500 kg st). The aircraft thus possessed a complete package

The MiG-21PFM had five different systems to shorten its landing roll: flaps, flap blowing, airbrakes, tail chute, and wheel brakes.

Takeoff of a MiG-21PFM assisted by two SPRD-99 solid-fuel boosters.

of systems to improve takeoff performance (afterburner, boosters) as well as landing performance (flaps, SPS, airbrakes, PT-21UK tail chute, wheel brakes).

Armament included two air-to-air RS-2US (K-5M) semiactive radar homing missiles (leading to the installation of the RP-21M radar and a modification of the aircraft's wiring diagram) as well as Kh-66 air-to-surface missiles. This weaponry could be supplemented by the GP-9 gun pod (a twin-barrel 23-mm GSh-23 cannon) under the center part of the fuselage and the ASP-PF-21 gunsight in the cockpit. The radar warning receiver was a Sirena-3M, and the new IFF interrogator had the rather curious appellation of Khrom-Nikyel ("chromium-nickel"). Before the MiG-21PFM was ready, a small batch of MiG-21PFSs were allotted to a fighter regiment. This version differed from the PFM only in its engine, which featured an additional mode of throttleable afterburning to improve significantly the aircraft's acceleration time.

The MiG-21PFM was mass-produced in the Gorki factory between 1964 and 1965 for the VVS and in the MMZ Znamya Truda factory in Moscow between 1966 and 1968 for export.

Specifications
Span, 7.154 m (23 ft 5.7 in); fuselage length (except cone), 12.285 m (40 ft 3.7 in); height, 4.125 m (13 ft 6.4 in); wheel track, 2.787 m (9 ft 1.7 in); wheel base, 4.71 m (15 ft 5.4 in); wing area, 23 m² (247.6 sq ft); takeoff weight, 7,820 kg (17,235 lb); max takeoff weight, 9,080 kg (20,010 lb); max takeoff weight on rough strip or metal-plank strip, 8,800 kg (19,395 lb); fuel, 2,200 kg (4,850 lb); wing loading, 340–394.8 kg/m² (69.7–80.9 lb/sq ft); max operating limit load factor, 8.5.

Performance
Max speed, 2,230 km/h at 13,000 m (1,204 kt at 42,640 ft); max speed at sea level, 1,300 km/h (702 kt); climb rate at sea level (half internal fuel, full thrust) with two R-3S missiles, 125 m/sec (24,600 ft/min); climb to 18,500 m (60,680 ft) in 8 min; service ceiling, 19,000 m (62,320 ft); landing speed, 250 km/h (135 kt); range, 1,300 km (810 mi); with 800-l (211-US gal) drop tank, 1,670 km (1,035 mi); takeoff roll, 850 m (2,790 ft); landing roll with SPS and tail chute, 550 m (1,800 ft).

MiG-21R / *Tip* 94R

The MiG-21R was a tactical reconnaissance aircraft derived at first from the MiG-21PF, but the prototype did not have its broad-chord tail

The MiG-21R's first reconnaissance pods and the necessary wiring were tested on a MiG-21PF airframe.

With 340 l (90 US gallons) of fuel in the dorsal spine and broad-chord vertical tail surfaces, the MiG-21R was quite similar to the MiG-21S. In this photograph it carries the D-99 reconnaissance pod.

fin. All reconnaissance systems were gathered into a long streamlined pod under the center of the fuselage. The aircraft could also defend itself with two air-to-air missiles under its wing.

Powered by a R-11F2S-300 rated at 6,050 daN (6,175 kg st), the MiG-21R took advantage of the SPS system. The capacity of the fuel tanks in the dorsal fairing was raised to 340 l (90 US gallons) to bring the total fuel capacity to 2,800 l (740 US gallons). Because the aircraft could not carry a drop tank under the fuselage, the wing was equipped with the fuel pipes needed for two drop tanks holding 490 l (129 US gallons) apiece. The broad-chord tail fin was also retained. After all of these modifications, the aircraft looked very much like the MiG-21S.

Several types of pods were developed for this reconnaissance version: day and night reconnaissance photo equipment (forward-facing or oblique cameras), electronic intelligence (elint) sensors, as well as laser, infrared, and television detection systems. The aircraft's wiring had to be modified accordingly. Among other significant changes, it is noteworthy that the KAP-2 autopilot (which provided only roll stabilization) was replaced by the three-axis AP-155. Moreover, the aircraft was equipped with the SPO-3 radar warning receiver. The airborne radar was the TsD-30, and underwing armament comprised two K-13T (R-3S) IR homing air-to-air missiles, and/or UB-16 and UB-32 rocket pods, S-24 rockets, and bombs. The GP-9 gun pod could also be added.

The MiG-21R was mass-produced for the VVS and for export in the Gorki factory between 1965 and 1971.

Specifications
Span, 7.154 m (23 ft 5.7 in); fuselage length (except cone), 12.285 m (40 ft 3.7 in); height, 4.125 m (13 ft 6.4 in); wheel track, 2.787 m (9 ft 1.7 in); wheel base, 4.71 m (15 ft 5.4 in); wing area, 23 m² (247.6 sq ft); takeoff weight, 8,100 kg (17,850 lb); fuel, 2,320 kg (5,115 lb); wing loading, 321.2 kg/m² (65.8 lb/sq ft); max operating limit load factor, 6.

Performance
Max speed, 1,700 km/h at 13,000 m (918 kt at 42,640 ft); max speed at sea level, 1,150 km/h (621 kt); climb rate at sea level (half internal fuel, full thrust) with reconnaisance pod and two R-3S missiles, 105 m/sec (20,670 ft/min); climb to 14,600 m (47,890 ft) in 8.5 min; service ceiling, 15,100 m (49,530 ft); landing speed, 250 km/h (135 kt); range, 1,130 km (700 mi); with two 490-l (129-US gal) drop tanks, 1,600 km (995 mi); takeoff roll, 900 m (2,950 ft); landing roll with SPS and tail chute, 550 m (1,800 ft).

This MiG-21S is armed with four UB-16-57 rocket pods and one GP-9 gun pod for a 23-mm twin-barrel cannon.

MiG-21S / *Tip* 95 / Ye-7S / Ye-7N

This new interceptor inherited most of the MiG-21's features: four hard points under the wing (two for weaponry and two for 490-l [129-US gallon] drop tanks), the fuel tank in the dorsal fairing with a capacity of 340 l (90 US gallons), the R-11F2S-300 turbojet and SPS system, and the three-axis AP-155 autopilot. But the guidance system was the more sophisticated Lazur-M ("azure"). The MiG-21S came equipped with the new RP-22S radar, and the old PKI-1 gunsight was replaced by the ASP-PF. Armament included two R-3R air-to-air missiles and bombs or rocket pods under the wing as well as the GP-9 gun pod (a twin-barrel GSh-23 with 200 rounds) under the fuselage. A direct offspring of the MiG-21S, the Ye-7N was fitted with a pod under the fuselage to carry a small tactical nuclear bomb.

The MiG-21S was mass-produced for the VVS in the Gorki factory between 1965 and 1968.

Specifications
Span, 7.154 m (23 ft 5.7 in); fuselage length (except cone), 12.285 m (40 ft 3.7 in); wheel track, 2.787 m (9 ft 1.7 in); wheel base, 4.71 m (15 ft 5.4 in); wing area, 23 m² (247.6 sq ft); takeoff weight, 8,150 kg (17,960 lb); fuel, 2,320 kg (5,115 lb); wing loading, 354.4 kg/m² (72.7 lb/sq ft); max operating limit load factor, 8.5.

Performance

Max speed, 2,230 km/h at 13,000 m (1,204 kt at 42,640 ft); max speed at sea level, 1,300 km/h (702 kt); climb rate at sea level (half internal fuel, full thrust) with two R-3S missiles, 115 m/sec (22,640 ft/min); climb to 17,500 m (54,400 ft) in 8.5 min; service ceiling, 18,000 m (59,000 ft); landing speed, 250 km/h (135 kt); range, 1,240 km (770 mi); with 800-l (211-US gal) drop tank, 1,610 km (1,000 mi); takeoff roll, 900 m (2,950 ft); landing roll with SPS and tail chute, 550 m (1,800 ft).

MiG-21SM / *Tip* 15

The MiG-21SM, the continuation of the "Meccano" game started with the Ye-6, evolved from a MiG-21S airframe on which two significant modifications were made. First, the new aircraft was powered by a more powerful turbojet developed by S. Gavrilov: the R-13-300, rated at 3,990 daN (4,070 kg st) dry and 6,360 daN (6,490 kg st) with afterburner. Second, it was armed with a built-in twin-barrel GSh-23L cannon with 200 rounds (the same weapon in a gun pod caused too much drag). The lessons of air battles in the Middle East finally registered, and the concept of a fighter armed only with missiles was flatly abandoned. The new ASP-PFD gunsight was designed specifically for combat situations requiring tight, high-g maneuvers. The aircraft was equipped with the RP-22 Sapfir-21 and the SPO-10 radar warning receiver.

In addition to the built-in cannon, armament included two K-13T (R-3S) or two K-13R (R-3R) air-to-air missiles, and/or UB-16 and UB-32 rocket pods, S-24 large-caliber rockets, four 100-kg (220-pound) bombs, and napalm containers. The total fuel capacity of this variant was reduced to 2,650 l (700 US gallons). The MiG-21SM was mass-produced for the VVS in the Gorki factory between 1968 and 1974.

Specifications

Span, 7.154 m (23 ft 5.7 in); fuselage length (except cone), 12.286 m (40 ft 3.7 in); wheel track, 2.787 m (9 ft 1.7 in); wheel base, 4.71 m (15 ft 5.4 in); wing area, 23 m^2 (247.6 sq ft); takeoff weight, 8,300 kg (18,295 lb); max takeoff weight, 9,100 kg (20,055 lb); max takeoff weight on rough strip or metal-plank strip, 8,800 kg (19,395 lb); wing loading, 360.9-395.7-382.6 kg/m^2 (74-81.1-78.4 lb/sq ft); max operating limit load factor, 8.5.

Performance

Max speed, 2,230 km/h at 13,000 m (1,204 kt at 42,640 ft); max speed at sea level, 1,300 km/h (702 kt); climb rate at sea level in clean configuration, 160 m/sec (11,930 ft/min); climb to 17,500 m (57,400 ft) in 9 min; service ceiling, 18,000 m (59,000 ft); landing speed, 250 km/h (135 kt); range, 1,050 km (650 mi); with 800-l (211-US gal) drop tank, 1,420 km (880 mi); takeoff roll, 800 m (2,625 ft); landing roll with SPS and tail chute, 550 m (1,800 ft).

MiG-21M / *Tip* 96

The MiG-21M was the export version of the MiG-21SM. Soviet aircraft manufacturers never put their most recent engines or equipment into such models; consequently, the MiG-21M was powered by the R-11F2S-300 with 6,050 daN (6,175 kg st); the RP-22 radar was replaced by the RP-21MA (a modified RP-21M) linked to the ASP-PFD gunsight; and older RS-2US missiles were substituted for the R-3Rs. The maximum weapon load was 1,300 kg (2,865 pounds).

The MiG-21M was mass-produced only for export in the MMZ Znamya Truda factory in Moscow between 1968 and 1971. India was granted the manufacturing license in 1971, and the first license-built MiG-21M—referred to as Type 96—was delivered to the Indian air force on 14 February 1973. The aircraft continued to be built there until 1981.

Specifications

Span, 7.154 m (23 ft 5.7 in); fuselage length (except cone), 12.285 m (40 ft 3.7 in); wheel track, 2.787 m (9 ft 1.7 in); wheel base, 4.71 m (15 ft 5.4 in); wing area, 23 m² (247.6 sq ft); takeoff weight, 8,300 kg (18,295 lb); max takeoff weight, 9,100 kg (20,055 lb); maximum takeoff weight on rough strip or metal-plank strip, 8,800 kg (19,395 lb); fuel, 2,200 kg (4,850 lb); wing loading, 360.9-395.7-382.6 kg/m² (74-81.1-78.4 lb/sq ft); max operating limit load factor, 8.5.

Performance

Max speed, 2,230 km/h at 13,000 m (1,204 kt at 42,640 ft); max speed at sea level, 1,300 km/h (702 kt); climb rate at sea level (half internal fuel, full thrust) with two R-3S missiles, 115 m/sec (22,640 ft/min); climb to 16,800 m (55,100 ft) in 9 min; service ceiling, 17,300 m (56,740 ft); landing speed, 250 km/h (135 kt); range, 1,050 km (650 mi); with 800-l (211-US gal) drop tank, 1,420 km (880 mi); takeoff roll, 900 m (2,950 ft); landing roll with SPS and tail chute, 550 m (1,800 ft).

The MiG-21MF was a modified MiG-21M powered by the same turbojet as the MiG-21SM. This one is armed with two R-3S missiles and four bombs.

MiG-21MF / *Tip 96F*

The MiG-21MF was a modified MiG-21M reengined with the R-13-300 turbojet rated at 3,990–6,360 daN (4,070–6,490 kg st) and reequipped with the RP-22 Sapfir-21 radar of the MiG-21SM. The capacity of its fuel tanks was limited to 2,650 l (700 US gallons), but the aircraft could carry either one 490-l (129-US gallon) or one 800-l (211-US gallon) drop tank under the fuselage and two 490-l (129-US gallon) drop tanks under the wing. Like its predecessors, the MiG-21MF could be fitted with two SPRD-99 solid rocket boosters capable of 2,450 daN (2,500 kg st) apiece.

Armament included a built-in GSh-23L cannon with 200 rounds under the fuselage and, under the wing, four air-to-air missiles (two R-3S and two R-3R), or two UB-32 and two UB-16 rocket pods (a total of ninety-six 57-mm S-5 rockets), or four 250- or 500-kg (550- or 1,100-pound) bombs, or any combination of these weapons, the maximum weapon load being 1,300 kg (2,865 pounds). The MiG-21MF could also be armed with R-60 and R-60M air-to-air missiles for close combat.

The PVD-7 air data probe consisted of a ram air pressure inlet, three rows of static pressure pickups, and two pairs of weathercocks (one to measure the angle of attack [AOA] and the other to measure the sideslip angle). The pressure inlets fed the air data computer, and the weathercocks fed the fire control computer. It is noteworthy that these weathercocks had disappeared from the air data probe with the "old" MiG-21F. This probe did not normally send its data to the cockpit instrument panel—this task was left to a short Pitot head on the front starboard side of the fuselage—but it could serve as the primary Pitot-static probe in case of emergency. The pilot was informed via the instrument panel by the AOA indicator on the front port side of the fuselage. Other equipment included the AP-155 autopilot, the Sirena-3M radar warning receiver, the SRO-2/SRZO-2 IFF transponder-inter-rogator, the SOD-57M decimetric ATC transponder, the RV-UM radio-altimeter for 0–600 m (0–1,970 feet), the Lazur command receiver, and the new TS-27AMSh cockpit periscope.

The MiG-21MF was mass-produced in the MMZ Znamya Truda fac-tory in Moscow between 1970 and 1974 and in the Gorki factory in 1975.

Specifications
Span, 7.154 m (23 ft 5.7 in); fuselage length (except cone), 12.285 m (40 ft 3.7 in); height, 4.125 m (13 ft 6.4 in); wheel track, 2.787 m (9 ft 1.7 in); wheel base, 4.71 m (15 ft 5.4 in); wing area, 23 m² (247.6 sq ft); empty weight, 5,350 kg (11,790 lb); takeoff weight with four R-3S/R missiles and three 490-l (129-US gal) drop tanks, 8,150 kg (19,725 lb); max takeoff weight with two R-3S/R missiles and three 490-l (129-US gal) drop tanks, 9,400 kg (20,720 lb); fuel, 2,200 kg (4,850 lb); wing loading, 356.5-389.1-408.7 kg/m² (73.1-79.8-83.8 lb/sq ft); max operat-ing limit load factor, 8.5.

Performance
Max speed, 2,230 km/h at 13,000 m (1,204 kt at 42,640 ft); max speed at sea level, 1,300 km/h (702 kt); climb to 17,700 m (58,055 ft) in 9 min; service ceiling, 18,200 m (59,700 ft); landing speed with SPS, 270 km/h (146 kt); landing speed without SPS, 310 km/h (167 kt); range, 1,050 km (650 mi); with 800-l (211-US gal) drop tank, 1,420 km (880 mi); with three 490-l (129-US gal) drop tanks, 1,800 km (970 mi); hi-lo-hi radius of action, 370 km (230 mi) with four 250-kg (550-lb) bombs; 740 km (460 mi) with two 250-kg (550-lb) bombs; and two 490-l (129-US gal) drop tanks; takeoff roll, 800 m (2,625 ft); landing roll with SPS and tail chute, 550 m (1,800 ft).

MiG-21MT / Tip 96T

As its appellation suggests, the MiG-21MT was also a modified MiG-21M. But the letter t (for *toplivo:* fuel) indicated that the aircraft's fuel capacity had increased significantly, from 2,650 l (700 US gallons) to

This MiG-21MT has three drop tanks and two infrared-guided R-3S air-to-air missiles.

3,250 l (858 US gallons). All the fuel for the uplift was held in the long dorsal fairing from the back of the cockpit to the base of the tail fin; this space contained a remarkable 900 l (238 US gallons) of fuel. The capacity of this segment was later reduced to 600 l (159 US gallons), cutting the aircraft's total fuel capacity to 2,950 l (779 US gallons).

The aircraft was powered by the R-13F-300 turbojet capable of 6,360 daN (6,490 kg st). Its armament and radar were identical to those of the MiG-21MF. The MiG-21MT was only a transitional aircraft, and only fifteen copies were built in 1971 in the MMZ Znamya Truda factory in Moscow. Five were allocated to the pilots of the VVS service evaluation department.

Specifications
Span, 7.154 m (23 ft 5.7 in); fuselage length (except cone), 12.285 m (40 ft 3.7 in); wheel track, 2.787 m (9 ft 1.7 in); wheel base, 4.71 m (15 ft 5.4 in); wing area, 23 m² (247.6 sq ft); takeoff weight, 8,800 kg (19,395 lb); fuel, 2,700 kg (5,950 lb); wing loading, 382.6 kg/m² (78.4 lb/sq ft); max operating limit load factor, 8.5.

Performance
Max speed, 2,175 km/h at 13,000 m (1,175 kt at 42,640 ft); max speed at sea level, 1,300 km/h (702 kt); climb rate at sea level in clean configuration, 150 m/sec (29,530 ft/min); climb to 16,800 m (55,100 ft) in 9 min; service ceiling, 17,300 m (56,745 ft); range, 1,300 km (810 mi); with 800-l (211-US gal) drop tank, 1,670 km (1,035 mi); takeoff roll, 900 m (2,950 ft); landing roll with SPS and tail chute, 550 m (1,800 ft).

MiG-21SMT / *Tip* 50

The MiG-21SMT frontline fighter-interceptor was a cross between the MiG-21M (same airframe, same armament) and the MiG-21MT (same R-13F-300 turbojet, same fuel capacity). The new engine led to modifications of the fuel system and a new setting of the cone control program. The VHF/UHF communications equipment was improved. The huge dorsal tank proved to be the cause of a major and unfortunate drawback: the aircraft's yaw stability margin had deteriorated. The capacity of the tank had to be reduced from 900 l (238 US gallons) to 600 l (159 US gallons), cutting the aircraft's total fuel capacity to 2,950 l (779 US gallons). The MiG-21SMT's radar was the RP-22 Sapfir-21, and its armament included a built-in GSh-23L, R-3S/R-3R air-to-air missiles (or R-60/R-60Ms for close combat), and/or UB-16 and UB-32 rocket pods, 240-mm S-24 rockets, and bombs.

The MiG-21SMT was mass-produced in the Gorki factory between 1971 and 1972.

Specifications
Span, 7.154 m (23 ft 5.7 in); fuselage length (except cone), 12.285 m (40 ft 3.7 in); wheel track, 2.787 m (9 ft 1.7 in); wheel base, 4.71 m (15 ft 5.4 in); wing area, 23 m² (247.6 sq ft); takeoff weight, 8,900 kg (19,615 lb); max takeoff weight, 9,100 kg (20,055 lb); max takeoff weight on rough strip or metal-plank strip, 8,800 kg (19,385 lb); fuel, 2,450 kg (5,400 lb); wing loading, 387-395.7-382.6 kg/m² (79.3-81.1-78.4 lb/sq ft); max operating limit load factor, 8.5.

Performance
Max speed, 2,175 km/h at 13,000 m (1,175 kt at 42,640 ft); max speed at sea level, 1,300 km/h (702 kt); climb rate at sea level (half internal fuel, full thrust) with two R-3S missiles, 200 m/sec (39,370 ft/min); climb to 16,800 m (55,100 ft) in 9 min; service ceiling, 17,300 m (56,745 ft); landing speed, 250 km/h (135 kt); range, 1,300 km (810 mi); with 800-l (211-US gal) drop tank, 1,670 km (1,035 mi); takeoff roll, 950 m (3,115 ft); landing roll with SPS and tail chute, 550 m (1,800 ft).

MiG-21 bis / Ye-7 bis / *Tip* 75

One of the lessons learned in combat over the Middle East and Vietnam was that in order to defeat a turboprop-powered fighter one had

The MiG-21bis was developed especially for low- and medium-altitude combat. Here it is armed with two K-60M and two K-13M air-to-air missiles.

to involve it in close combat at low altitudes. To meet that challenge, the jet fighter had to be armed to the teeth and supplied with a good amount of fuel. It also needed to be a stable and maneuverable machine. The dominant feature of all of the MiG-21 variants reviewed thus far was their good performance at medium and high altitudes. But it had never performed up to par at low altitudes, because of some of the distinctive characteristics of such power plants as the R-11F2-300 and the R-13-300.

This is why it was decided in February 1971 to construct a new MiG-21 that would be especially efficient at low altitudes and high indicated airspeeds. The new OKB offspring was named the MiG-21 bis. It was powered by a new engine, the R-25-300. It was rated at 4,020 daN (4,100 kg st) dry—roughly equivalent to the R-13-300—but had a much higher afterburning ratio in that its reheated thrust was rated at 6,960 daN (7,100 kg st). Moreover, at Mach 1 and beyond it could utilize a special afterburning regime called ChR (*Chrezvichayniy Rezhim:* exceptional rating) that permitted it to obtain a peak thrust of 9,700 daN (9,900 kg st) for as long as three minutes between 0 and 4,000 m (0 and 13,120 feet).

The MiG-21 bis could also be fitted with two SPRD-99 solid rocket boosters rated at 2,450 daN (2,500 kg st) apiece. Their kinetic energy topped 46,000 kgm/sec (451,000 W) for between 10 and 17.8 seconds according to temperature. The total capacity of the fuel tanks was 2,880

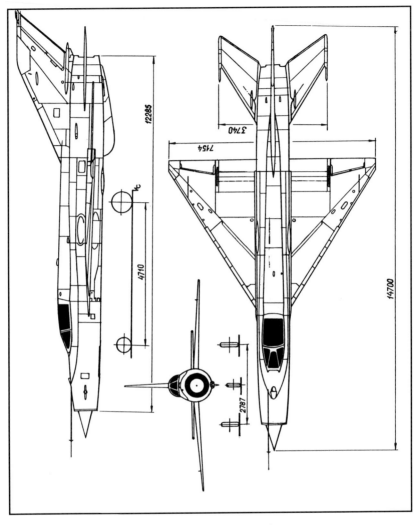

MiG-21bis (MiG OKB three-view drawing)

l (760 US gallons). Installation of the new engine forced engineers to modify the fuel system with a booster pump for the afterburner. The aircraft was fitted with the new RSBN-2S short-range navigation system, similar to the Shoran and replaced in late-series machines by the RSBN-5S; it also received new instrument approach equipment. Together, those new systems enabled the pilot to navigate accurately in even the worst weather. Automatic monitors for the airframe and engine significantly reduced maintenance downtime.

In addition to its built-in GSh-23 cannon, the MiG-21 bis was armed for close combat with R-55 and R-60M (K-60M) air-to-air missiles and with the R-13M (K-13M), which had twice the range of the R-3S and

The R-25-300, rated at 6,160 daN (7,100 kg st) with afterburner, gave the MiG-21bis excellent takeoff and climb performance.

could be fired under a much higher load factor. The airborne radar was the RP-22 Sapfir-21. The MiG-21 bis was accepted into VVS fighter air regiments in February 1972 and was mass-produced in the Gorki factory. India was granted the manufacturing license in 1974.

Specifications
Span, 7.154 m (23 ft 5.7 in); fuselage length (except cone), 12.285 m (40 ft 3.7 in); overall length (except probe), 14.7 m (48 ft 2.3 in); wheel track, 2.787 m (9 ft 1.7 in); wheel base, 4.71 m (15 ft 5.4 in); wing area, 23 m² (247.6 sq ft); takeoff weight with two R-3S missiles, 8,725 kg (19,230 lb); max takeoff weight, 9,800 kg (21,600 lb); max takeoff weight on rough strip or metal-plank strip, 8,800 kg (19,395 lb); max takeoff weight with KT-92D wheels and 42A or 058 (800 x 200) tires, 10,400 kg (22,920 lb); fuel, 2,390 kg (5,270 lb); wing loading, 379.3-426.1-382.6-452.2 kg/m² (77.8-87.4-78.4-92.7 lb/sq ft); max operating limit load factor, 8.5.

Performance
Max speed, 2,175 km/h at 13,000 m (1,175 kt at 42,640 ft); max Mach number, 2.05; max speed at sea level, 1,300 km/h (702 kt); climb rate at sea level (half internal fuel, full thrust) with two R-3S missiles, 230 m/sec (45,275 ft/min); climb to 17,000 m (55,760 ft) in 8.5 min; service ceiling, 17,500 m (57,400 ft); landing speed, 250 km/h (135 kt);

Four Typical Mission Profiles for the MiG-21 bis

Types of missions	Fuel	Weapons	Radius of action	Time
No. 1 bombing mission[a]	800-l (211-US gal) drop tank	two 500-kg (1,100-lb) bombs	290 km (170 mi)	1 min over target
No. 2 bombing mission[a]	800-l (211-US gal) drop tank	two 250-kg (550-lb) bombs	330 km (205 mi)	1 min over target
No. 1 interception mission[b]	no drop tank	two R-3S and two R-60 missiles	330 km (205 mi)	2 min in intercept area
No. 2 interception mission[b]	800-l (211-US gal) drop tank	two R-3S and two R-60 missiles	450 km (280 mi)	2 min in intercept area

Notes:
[a]These lo-lo-lo missions are flown at an average altitude of 200 m (650 ft), outbound and inbound.
[b]These missions are flown at high altitude: first 8,000 m (26,250 ft), then 5,000 m (16,400 ft) and 10,000 m (32,800 ft) outbound; 10,000 m (32,800 ft) for the return flight.

range in clean configuration, 1,225 km at 11,000 m (760 mi at 36,080 ft); 1,110 km at 14,000 m (690 mi at 45,920 ft); with 800-l (211-US gal) drop tank, 1,430 km at 14,000 m (890 mi at 45,920 ft); with two R-3S missiles and 490-l (129-US gal) drop tank, 1,355 km at 11,000 m (840 mi at 36,080 ft); with two R-3S missiles and 800-l (211-US gal) drop tank, 1,470 km at 10,000 m (910 mi at 32,800 ft); takeoff roll, 830 m (2,720 ft); landing roll with SPS and tail chute, 550 m (1,800 ft).

MiG-21U / *Tip* 66 / Ye-6U

The MiG-21 was the first mass-produced Mach 2 fighter in the USSR. It differed so extensively from the fighters of the preceding generation that pilots urgently needed a dedicated trainer. The decision to work on the preliminary design was made as early as November 1959 at a time when a few MiG-21Fs were taking shape on the assembly line, but the MiG-21F-13 airframe was finally selected to serve as the basis for the training aircraft. The layout retained was that of the tandem two-seater, with the student pilot in the front seat and the flight instructor in the rear seat. The cockpit hood was composed of a windshield and two side-hinged canopies that opened to starboard and thus could not be used to protect the crew on their SK ejection seats. The first MiG-21U employed the tail fin of the late-series MiG-21F-13, but the air data probe on the trainer was set above the air intake.

The Ye-6U, prototype of the MiG-21U two-seater, was developed from a MiG-21F-13 airframe.

In the MiG-21U the front seat is for the student pilot, the rear seat for the instructor. Both canopies open to starboard.

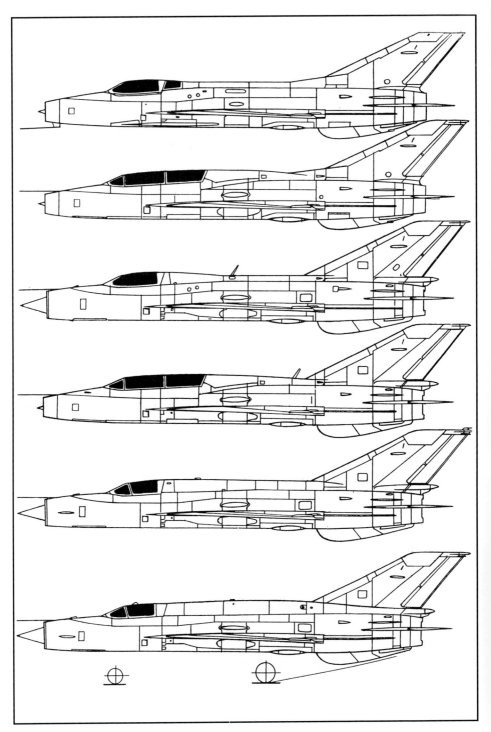

Top to bottom: MiG-21F-13, MiG-21U, MiG-21PFM, MiG-21UM, MiG-21bis, and MiG-21SMT (MiG OKB drawing)

The aircraft was powered by the R-11F-300, rated at 5,620 daN (5,740 kg st) with afterburner. The total capacity of the fuel tanks was 2,350 l (620 US gallons). The MiG-21U had neither radar nor built-in armament but could be equipped with a ventral pod containing an A-12.7 machine gun. It was first piloted on 17 October 1960 by P. M. Ostapyenko and was mass-produced for the VVS in the Tbilisi factory between 1962 and 1966 and for export in the MMZ Znamya Truda factory in Moscow between 1964 and 1968. A modified MiG-21U renamed Ye-33 broke two female world records in 1965:

1. 22 May 1965. Altitude, 24,336 m (79,822 feet). Pilot, N. Prokhanova
2. 23 June 1965. Altitude in horizontal flight, 19,020 m (62,386 feet). Pilot, L. Zaytseva

Specifications
Span, 7.154 m (23 ft 5.7 in); fuselage length (except cone and probe), 12.18 m (39 ft 11.5 in); wheel track, 2.692 m (8 ft 10 in); wheel base, 4.806 m (15 ft 9.2 in); wing area, 23 m² (247.6 sq ft); takeoff weight, 7,800 kg (17,190 lb); fuel, 1,950 kg (4,300 lb); wing loading, 339.1 kg/m² (69.5 lb/sq ft); max operating limit load factor, 7.

Performance
Max speed, 2,175 km/h at 13,000 m (1,175 kt at 42,640 ft); max speed at sea level, 1,150 km/h (621 kt); climb rate at sea level (half internal fuel, full thrust) with two R-3S missiles, 120 m/sec (23,620 ft/min); climb to 17,800 m (58,380 ft) in 8 min; service ceiling, 18,300 m (60,025 ft); landing speed, 280 km/h (151 kt); range, 1,210 km at 14,000 m (750 mi at 45,920 ft); with 800-l (211-US gal) drop tank, 1,460 km (905 mi); takeoff roll, 950 m (3,115 ft); landing roll, 800 m (2,625 ft).

MiG-21US / Tip 68 / Ye-33 / Ye-66B

The MiG-21US two-seat training aircraft was derived from the MiG-21U and differed externally in two points: the chord of the tail fin was broader, and the parachute container was located at the base of that fin. However, the most significant modifications were inside the aircraft. It was powered by the R-11F2S-300 turbojet rated at 6,050 daN (6,175 kg st) and consequently had the SPS system. Because the cone had no radar to house, there was no need to increase the diameter of the air intake, so the diameter remained at 690 mm (27.2 inches). The total capacity of the fuel tanks increased to 2,450 l (647 US gallons) and

The MiG-21US differed from the MiG-21U in two particulars: it had a broader-chord tail fin, and the tail chute canister was moved to the base of the vertical tail surfaces.

the SK ejection seats were replaced by the KM-1M (SK-3) model. The MiG-21US was mass-produced for the VVS and for export in the Tbilisi factory between 1966 and 1970.

A MiG-21US also renamed Ye-33 and piloted by S. Ye. Savitskaya broke four female world records on 6 June 1974:

1. Time to climb to 3,000 m (9,840 feet), 59.1 seconds
2. Time to climb to 6,000 m (19,680 feet), 1 minute, 20.4 seconds
3. Time to climb to 9,000 m (29,520 feet), 1 minute, 46.7 seconds
4. Time to climb to 12,000 m (39,360 feet), 2 minute, 35.1 seconds

Another MiG-21US renamed Ye-66B and piloted once more by S. Ye. Savitskaya topped those four records comfortably on 15 November 1974:

1. Time to climb to 3,000 m (9,840 feet), 41.2 seconds
2. Time to climb to 6,000 m (19,680 feet), 1 minute, 0.1 seconds
3. Time to climb to 9,000 m (29,520 feet), 1 minute, 21 seconds
4. Time to climb to 12,000 m (39,360 feet), 1 minute, 59.3 seconds

In the documents sent to the FAI to verify those records, it was mentioned that the Ye-66B was powered by one RDM at 6,860 daN (7,000 kg st) and two TTRDs at 2,250 daN (2,300 kg st). Those mysterious acronyms had to be deciphered; it was surmised that the RDM was in fact the R-11F2S-300 turbojet (somewhat revved up by toying with the engine combustion temperature and rotation speed) and that the TTRDs were two SPRD-99 solid rocket boosters for help at takeoff.

Specifications
Span, 7.154 m (23 ft 5.7 in); fuselage length (except cone and probe), 12.18 m (39 ft 11.5 in); wheel track, 2.692 m (8 ft 10 in); wheel base, 4.806 m (15 ft 9.2 in); wing area, 23 m² (247.6 sq ft); takeoff weight, 8,000 kg (17,630 lb); fuel, 2,030 kg (4,475 lb); wing loading, 347.8 kg/m² (71.3 lb/sq ft); max operating limit load factor, 7.

Performance
Max speed, 2,175 km/h at 13,000 m (1,175 kt at 42,640 ft); max speed at sea level, 1,150 km/h (621 kt); climb rate at sea level (half internal fuel, full thrust) with two R-3S missiles, 115 m/sec (22,640 ft/min); climb to 17,200 m (56,415 ft) in 8 min; service ceiling, 17,700 m (58,055 ft); landing speed, 250–260 km/h (135–140 kt); range, 1,210 km at 14,000 m (750 mi at 45,920 ft), with 800-l (211-US gal) drop tank, 1,460 km (905 mi); takeoff roll, 900 m (2,950 ft); landing roll with SPS and tail chute, 550 m (1,800 ft).

MiG-21UM / *Tip* 69

The last refinement of the two-seat version, the MiG-21UM differed from its predecessor chiefly in its upgraded instrumentation. The KAP-2 autopilot was replaced by the three-axis AP-155 (the MiG-21R and all subsequent versions of the aircraft were fitted with the AP-155). The aircraft was also equipped with the ASP-PDF computerized optical fire control. The forward equipment bay received a plug-in rack to reduce maintenance downtime. The MiG-21UM was also powered by the R-11F2S-300, rated at 6,050 daN (6,175 kg st). The total capacity of the fuel tanks was 2,450 l (647 US gallons).

The MiG-21UM succeeded the US on the assembly line in the Tbilisi factory in 1971 (for the VVS and export).

The MiG-21UM and MiG-21US differed mainly in their instrumentation.

Specifications

Span, 7.154 m (23 ft 5.7 in); fuselage length (except cone and probe), 12.18 m (39 ft 11.5 in); wheel track, 2.692 m (8 ft 10 in); wheel base, 4.806 m (15 ft 9.2 in); wing area, 23 m² (247.6 sq ft); takeoff weight, 8,000 kg (17,630 lb); fuel, 2,030 kg (4,475 lb); wing loading, 347.8 kg/m² (71.3 lb/sq ft); max operating limit load factor, 7.

Performance

Max speed, 2,175 km/h at 13,000 m (1,175 kt at 42,640 ft); max speed at sea level, 1,150 km/h (621 kt); climb rate at sea level in clean configuration, 150 m/sec (29,530 ft/min); climb to 16,800 m (55,100 ft) in 8 min; service ceiling, 17,300 m (56,745 ft); landing speed, 250–260 km/h (135–140 kt); range, 1,210 km at 14,000 m (750 mi at 47,920 ft); with 800-l (211-US gal) drop tank, 1,460 km (905 mi); takeoff roll, 900 m (2,950 ft); landing roll with SPS and tail chute, 550 m (1,800 ft).

MiG-21I / 21-11 / Analog

The MiG-21I was the test bed for the wing of the Tupolev Tu-144 supersonic airliner, allowing engineers to work out its airflow characteristics and test the whole flight control system. For the latter the OKB engineers had to start from scratch since this was their first pure delta-wing aircraft.

Two MiG-21Is were built to test the wing scheme of the Tu-144 supersonic airliner.

The airframe of the test bed was that of a MiG-21S fitted with a compound sweepback delta wing (78 degrees at the wing root over one-third of the leading edge, then 55 degrees). Its thickness-chord ratio tapered from 2.3 percent at the wing root to 2.5 percent at the wing tip. The trailing edge was shared evenly by the flaps and elevons. The plan view of the wing was almost identical to that of the Tu-144—hence the nickname "Analog."

Two MiG-21Is were needed for research purposes. No. 1 was first piloted by O. V. Gudkov on 18 April 1968, and the flight tests continued for about one year. Unfortunately, just after completion of the basic tests the prototype was destroyed. No. 2 was kept airworthy for several years. Both MiG-21Is were powered by the R-13F-300 capable of 6,360 daN (6,490 kg st). The total capacity of the fuel tanks was 3,270 l (864 US gallons). Both aircraft were used to train the first two pilots of the Tu-144, E. Yelyan and M. Kozlov.

Specifications
Span, 8.15 m (26 ft 8.9 in); length (except probe), 14.7 m (48 ft 2.8 in); fuselage length (except cone), 12.287 m (40 ft 3.7 in); wheel track, 2.787 m (9 ft 1.7 in); wheel base, 4.71 m (15 ft 5.4 in); wing area, 43 m^2 (462.85 sq ft); takeoff weight, 8,750 kg (19,285 lb); fuel, 2,715 kg (5,985 lb); wing loading, 203.5 kg/m^2 (41.7 lb/sq ft).

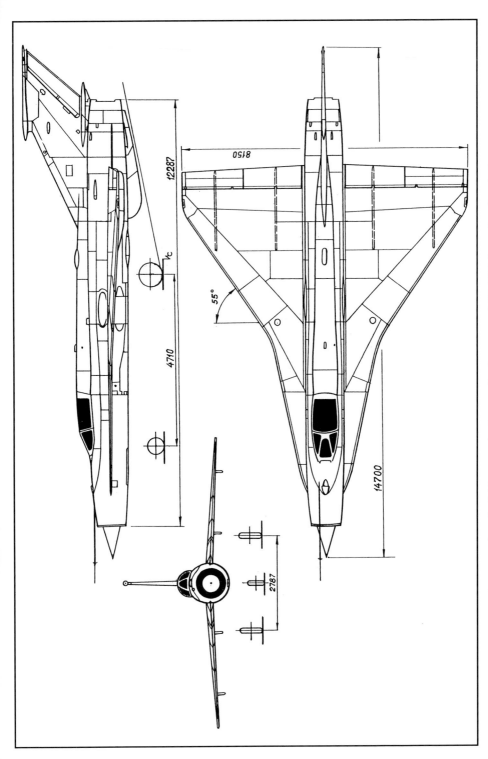

MiG-21I Analog; the fairing atop the fin contained a camera (MiG OKB three-view drawing)

8150

12287

4710

55°

14700

2787

The MiG-21PD was an experimental STOL aircraft developed from a MiG-21PFM airframe that was extended by 900 millimeters (35.4 inches).

Performance

Max speed, Mach 2 at 13,000 m (42,640 ft); max speed at sea level, 1,200 km/h (648 kt); landing speed, 225 km/h (122 kt).

MiG-21PD / Ye-7PD / 23-31 / *Tip* 92

This experimental STOL aircraft was designed to assess the MiG-21's operational capabilities on short strips. A MiG-21PFM airframe was equipped with one R-13F-300 turbojet rated at 6,360 daN (6,490 kg st) and two Kolyesov RD-36-35 lift jets that were set slightly forward near the aircraft's center of mass. This forced the engineers to add a fuselage "slice" 900 mm (35.43-inches) thick just behind the cockpit; the master cross-section became appreciably larger where the lift jets were installed. They were fed in air by a rearward-hinged door with louvers that was opened by an actuator at takeoff and landing. The action of the lift jets was limited to reducing the aircraft's ground roll. They were never operated in flight, and the inlet door was kept shut. The fixed tricycle gear had a wider wheel track.

The MiG-21PD (*Podyomnye Dvigatyeli*: lift jet) was first piloted on 16 June 1966 by P. M. Ostapyenko. B. A. Orlov took over for the factory tests, which ended in 1967. The aircraft made its public debut at

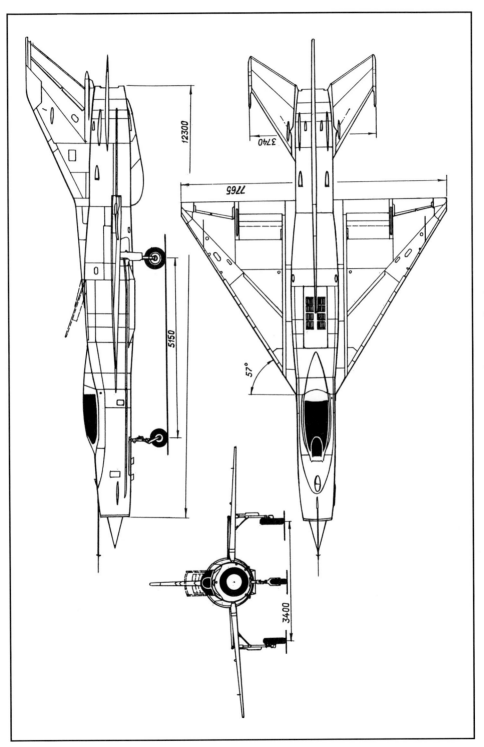

MiG-21PD (MiG OKB three-view drawing)

12300

3740

7765

5150

57°

3400

The MiG-21PD lift jets operated only to shorten the takeoff and landing roll.

the Domodyedovo air show in July 1967. There proved to be fewer pros to this STOL prototype than cons, so the aircraft's development was discontinued.

Specifications
Span, 7.765 m (25 ft 5.7 in); fuselage length (except cone and probe), 12.3 m (40 ft 4.3 in); wheel track, 3.4 m (11 ft 1.9 in); wheel base, 5.15 m (16 ft 10.7 in).

MiG-21Ye

In the mid-1960s the MiG OKB, in cooperation with the Kazan Aviation Institute (KAI), developed versions of the MiG-21PF and MiG-21PFM to be operated as remotely controlled target drones for VVS and PVO pilots as well as AAA gunners. For this purpose, fighters that had out-lived their operational parameters were used.

The radar in these aircraft was replaced by ballast to restore the aircraft's trimming. The ejection seats were removed to make room for remote control equipment and the drive mechanism for the control surfaces. The target drone was controlled by radio signals from the ground or from another aircraft specially equipped to steer the drone with preset routines. Those modifications were carried out in the VVS ARZs (air force overhaul workshops). The remotely controlled MiG-21Ye could take off and make maneuvers, but only within the subsonic flight envelope.

Because the MiG-21PD was an experimental aircraft, the landing gear was not retractable.

MiG-21K

This experimental version of the MiG-21 bis was designed to develop new on-board systems to be installed in cruise missiles and was, like the MiG-21 bis, powered by an R-25 turbojet.

MiG-23 Series

The MiG OKB's first approach to the variable geometry (VG) wing concept dates back to the early 1960s. The countless computations made by the design office showed that VG aircraft could offer an appreciable number of advantages. Many models were built and tested in TsAGI wind tunnels under different flight conditions: takeoffs, landings, and transonic/supersonic speeds. The tests confirmed most of the computations.

One of the basic problems the OKB had to face in the field of structure as well as aerodynamics was finding just the right place for the wing pivot, and thereby determining the chord and span of the wing. That problem was linked to the optimum pitching stability margin necessary according to the chosen sweep angle, since the shape of the wing as it pivoted and the mean aerodynamic chord were basically dependent on the position of the wing pivot. Another important item involved choosing the proper shape for the fixed wing panels (or gloves) and their wing-to-fuselage junctions. The difficulty there was linked to the distinctive features of the airflow around both the wing and the whole aircraft at great angles of attack in subsonic flight. The shape of the wing's fixed panels and the blending of their leading edge into the fuselage act extensively upon the vortex flow in these flight conditions; and obviously the vortex flow influences the lift capability and the static pitching stability.

The third hurdle was developing a flight control system capable of changing the wing's sweep angle and actuating an all-moving stabilator that operated differentially (taileron) plus all the moving surfaces hinged on the wing's main panels (spoilers and full-span trailing edge flaps). The spoilers were highly efficient at minimum sweep angles, but this efficiency dropped abruptly once the wing was set for a high sweep angle in subsonic flight regime. In transonic flight conditions, due to the airflow downwash onto the stabilator caused by their deployment, the spoilers experienced reverse aerodynamic feedback. This is why the roll control had two functions. When the pilot pushed the control column sideways, the spoiler on that side was extended and the opposite half part of the stabilator was deflected.

The spoilers' extension angle was greatest for the smallest wing sweep angle; as the sweep angle increased, the angle of the spoilers decreased all the way to zero. The slab stabilator operating differentially thus functioned in place of the aileron. To save weight and provide the yaw stability needed over its whole range of speeds, altitudes, and load factors, the aircraft was fitted with a large folding ventral fin (the first of its kind in the world).

Development of the MiG-23 was completed in record time by a group of highly motivated engineers who were never short of ideas, to judge from the number of patents registered as the prototype took shape. The MiG-23 silhouette emerged gradually. The OKB first built an aircraft of a totally different concept. It had a fixed delta wing, and its power pack included two lift jets to shorten takeoffs and landings and a primary power plant fed by two lateral air intakes (the first of their kind for a supersonic mixed-power aircraft), clearing space in the nose for the radar. That aircraft was the 23-01. In the course of development, which started in 1964, OKB engineers quickly realized that the lift jets became dead loads after takeoff and that the 23-01 was an uneconomical proposition. When the aircraft was almost completed, Mikoyan grew doubtful about the rationality of the project. Those doubts served as food for thought and were based on several arguments:

—even if the 23-01 could make short landings of 300–350 m
(985–1,150 feet), that is, two times less than average, there was
always a chance that one or both of the lift jets could fail on final
approach

—the space occupied by the lift jets could be better used to house
fuel tanks to increase the aircraft's range

The development of the 23-01 experimental machine was to some
extent tied to the customs of the day. At about this time France flight-
tested the Dassault Balzac experimental prototype powered by one
cruising turbojet and six smaller lift jets. Other countries such as Great
Britain and West Germany had also started to design similar machines.
The fourteen flights of the 23-01 and the sad end of the Balzac confirmed the pointlessness of the formula.

So another approach was tried: an aircraft powered by a turbojet
whose thrust could be vectored at takeoff, in flight, and at landing by
swiveling nozzles. The best-known examples are the British Harrier
VTOL aircraft and, in the USSR, the experimental Yak-36 and Yak-38
carrier-based combat aircraft, which features both vectored-thrust
engines and lift jets—but let us return to variable geometry. The final
parameters selected for the wing were minimum sweep angle of 16
degrees, maximum sweep angle of 72 degrees, and leading edge flaps.
The advantages of those choices are twofold.

1. Airflow characteristics: high lift-to-drag ratio in supersonic flight
   conditions due to a high sweep angle and a low thickness-chord
   ratio and in subsonic flight conditions due to a low sweep angle
   and a high wing aspect ratio; excellent lift coefficient at takeoff
   and landing because of a high aspect ratio and the full-span leading edge and trailing edge flaps; good lift-to-drag ratio and lift
   coefficient at transonic speeds with a midrange sweep angle
2. Flight data: better performance due to peak application of the
   sweep angle. On that subject it should be noted that the MiG-23
   pilot could choose any sweep angle between 16 and 72 degrees;
   each one presented a distinct advantage for a particular flight
   regime. Practical experience showed that the three most popular
   sweep angles were 16, 45, and 72 degrees. Because of its wide-
   ranging flight envelope, the MiG-23 was undoubtedly one of the
   best frontline fighters of the 1970s.

As soon as development was halted on the 23-01 VTOL, the highest
priority was assigned to the 23-11 VG project. This was further boosted
in 1965 by a decree of the ministry of aircraft production that detailed
the main specifications: "The MiG OKB is commissioned to design and

build a second prototype of the MiG-23 [the first was the 23-01] fitted with a high-lift variable geometry wing. The Rodina MKB [headed by general designer Selivanov] is in charge of designing the wing pivot." The preliminary design was drawn up in a very short time, from January to March 1966. A. A. Andreyev, a very capable designer, was put in charge of the project's technical management.

The R-27 turbojet was developed especially for the MiG-23 at a time when it was unclear whether the 23-01 or the 23-11 would win out. This is why it was developed and tested concurrently with the two aircraft. It was designed by K. R. Khachaturov as a modification of the R-11F2S-300 twin-spool turbojet, a reliable engine that had powered many of the MiG-21 variants and the whole Yak-28 line.

In the MiG-23 development process care was exercised to automate as many of the pilot's tasks as possible, especially while intercepting. A. V. Fedotov, newly appointed as the OKB's chief test pilot, played a dominant part in developing those systems. The 23-11 went for its first flight on 10 April 1967 with this experienced pilot at the controls and the wing at 16 degrees. As early as the second flight two days later, he tested the whole range of sweep angles. The aircraft proved to be easy to control whatever the sweep angle, a quality that triggered Fedotov's enthusiasm. His log entry for that day reads: "Flight with 16 to 72° sweep angle. It's a first! Terrific!"

That kind of emotional report seldom appears in a test pilot's logbook, but admittedly this was a rather unusual case. As early as the third flight, Fedotov broke the sound barrier and continued to accelerate until he reached Mach 1.2 with a 72-degree sweep angle. A few weeks later, on 9 July 1967, the MiG-23 made its public debut with Fedotov at the controls. It was clear after this brilliant display that the 23-11 would be the originator of a great aircraft family—and that was the case, even though the entry into service of such a new aircraft caused some problems of familiarization for pilots (before the delivery of a two-seat trainer to the fighter regiments) and field support crews.

The variable geometry concept was at the heart of some structural innovations. The fuselage structure was organized so that fuel tank no. 2 and the wing center section were as one. It was constructed of welded thin panels made out of VNS-2 alloy. This fuel tank was in fact the aircraft's primary structure. The stressed box that upheld the wing pivots was attached to that structure, and the air intake duct passed through it. This "wing box-tank" sustained high stress loads at all times and especially during high-g maneuvers. Considering the peculiarity of the aircraft's missions, the breaking strength of this structure was computed to withstand limit load factors up to 8.

During the factory experiments, state acceptance trials, and military tests, fuel tank no. 2 never caused trouble. And yet. . . . On 14

March 1972 test pilot A. G. Fastovets had to check the strength of a new type of wing that had a larger area (called the type 2 wing); to do that, he had to reach the limit load factor in pulling out of a long dive. Just as he hit 7.3 g on the accelerometer at 1,000 m (3,280 feet) the tank gave way, and the aircraft totally disintegrated. The pilot was lucky to eject in time.

The subsequent investigation blamed the failure of this primary structural element on cracks that had formed in the panels due to some sort of soot by hydrogen molecules that had found its way onto some of the rough castings. The production factory had to revise the whole of its welding process for the components of fuel tank no. 2 and to inspect all tanks already built. Several cases were reported of wing pivot failure due to the infiltration of hydrogen molecules in welded parts and rotating shafts as well. That problem was overcome by increasing the number of quality checks at every stage of manufacture and by strengthening the structure of the no. 2 fuel tank for all aircraft on the assembly line. For the aircraft already completed, heat carefully applied to the tank structure prevented the hydrogen molecules from spreading and the stresses from accumulating. Moreover, the pivot rotating shafts were made out of a better steel alloy called *khromansil.*

The area of the type 2 wing was augmented by a chord increase on the leading edge, but it had no leading edge flaps and had been dubbed the "dog-toothed" wing because of the typical shape of the end of its inner leading edge. This enlargement—5.25 m² (56.51 square feet) at 16 degrees, 4.27 m² (45.96 square feet) at 72 degrees—resulted in a sweep angle increase at the leading edge. The three most popular angles— 16°, 45°, and 72°—thus became 18°40', 47°40', and 74°40', a constant difference of 2°40'. But for convenience's sake it was decided not to modify the figures in the flight manual or on the instrument panel's sweep angle indicator, which therefore provided erroneous readings.

This enlarged wing would later be fitted with leading edge flaps and named the type 3 wing. The first MiG-23s equipped with that wing appeared in 1973, and from that date all MiG-23s and MiG-27s used it until assembly lines were closed in the early 1980s. The hydraulically driven flaps were added to raise the lift coefficient at great angles of attack. After the basic causes of flow breakaway (resulting in a severe buffeting) were suppressed, it became possible to fly at even greater AOAs. After a great deal of research, engineers developed an automated contrivance to protect against engine surges and flameouts while missiles and cannons were fired.

The more the aircraft was developed, the more the OKB and its client—the air force of the Soviet Army (VVS SA)—realized that it had to be upgraded. Its stability, handling characteristics, and maneuverability were significantly improved. It was possible to raise the maximum operating limit load factor not only by making the airframe sturdier but also by using sweep angle variations intelligently during high-g maneuvers.

The aircraft's handling characteristics at great AOAs were improved, the pilot helped by new visual and tactile warnings of critical AOAs that could prompt spins. Moreover, the sighting system was improved and the radar was modified so that it could operate in the close-combat mode; simultaneously, the aircraft received a target illuminator to guide semiactive radar homing missiles. New air-to-air missiles optimized for close combat were tested and certified.

In the 1970s a prolific family of attack airplanes based on the MiG-23 airframe developed. They could carry either bombs or rocket pods, air-to-surface missiles, six-barrel 30-mm guns, and many other front-line air support weapons. With every modification the MiG-23 became lighter. For instance, the takeoff weight of the MiG-23M (1971) was 15,750 kg (34,715 pounds), while for the MiG-23ML (1976) the comparable figure was 14,500 kg (31,960 pounds).

The rapid pace of advances in electronics and optoelectronics made it possible to produce new types of sensors related to outward sight, detection, IFF capabilities, computation of target coordinates, and the like. The power and capacity of the Sapfir radar improved significantly, and ground clutter was cleaned up. The radar was given new operating modes: separation of mobile targets in the lower sector, automatic and simultaneous tracking of several targets, and detection of small ground targets. The MiG-23P's automatic flight control system (SAU) featured a digital computer unit to control the aircraft's flight path.

MiG-23s were mass-produced in many versions until the early 1980s and are still operated in many countries, including Russia and the other republics. Today it is widely recognized that the MiG-23 represented an important step in the development of fighter and tactical air command in the USSR.

MiG-23PD / 23-01

This aircraft represented one-half of a dichotomous attack to a single objective. The specifications for the 23-01—as well as for the 23-11 or MiG-23PD (as for the MiG-21PD, *Podyomnye Dvigatyeli* = lift jet), built simultaneously—called for the aircraft to be capable of speeds of Mach 2–2.3 and also offer STOL performance.

The preliminary designs were completed in 1964, and assembly of the prototype started in 1965. V. A. Mikoyan (son of the president of

The STOL variant with lift jets was one of the innovations explored with the 23-01.

the Supreme Soviet and nephew of A. I. Mikoyan) was commissioned to take care of both projects. For the 23-01 the design bureau chose the tailed delta configuration of the MiG-21. The midwing was an enlarged replica of that of the MiG-21. The horizontal tail surfaces were of the all-flying type (slab tailplane). Because the concept selected for this prototype was based on the employment of lift jets, the primary turbojet could be fed only by semicircular lateral air intakes with shock cones identical to those of the French Mirage III. Both air ducts were separated slightly from the fuselage near the air intakes to create boundary layer bleeds, and both had blow-in doors above the wing's leading edge.

The two Kolyesov RD-36-35 lift jets rated at 2,300 daN (2,350 kg st) apiece were set in the middle of the fuselage with a slight forward inclination. They operated only during takeoffs and landings. For those short periods, a rearward-hinged, louvered door was opened by an actuator to supply air to the lift jets. Under the fuselage, both nozzle throats were fitted with a rotating grid that allowed the pilot to alter the direction of the thrust vector. In a way they operated as thrust reversers at landing; on the other hand, at takeoff the thrust of the lift jets was added to that of the whole power unit. The primary power plant was the R-27-300 rated at 5,095 daN (5,200 kg st) dry or 7,645 daN

Moving the air intakes to the side made room for a radar unit aboard the 23-01, here armed with R-23R and R-23T air-to-air missiles.

The 23-01 on final approach, with flaps fully extended. The louvered door that feeds the lift jets is open.

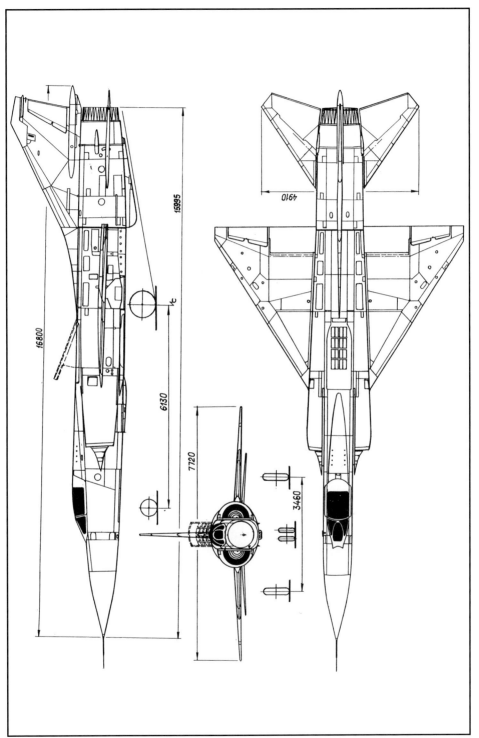

15995

16800

6130

7720

3460

4910

MiG-23PD (23-01) (MiG OKB three-view drawing)

The tail chute was one of several devices employed to shorten the landing roll.

(7,800 kg st) with afterburner. Like most of the MiG-21 variants the 23-01 had the SPS system (flap blowing by air bleed downstream from the last compressor stage of the R-27-300). The canister for the cruciform tail chute was located at the base of the tail fin's trailing edge.

Armament consisted of one twin-barrel GSh-23 under the fuselage and two air-to-air K-23 missiles under the wing (one K-23R and one K-23T). The 23-01 was first piloted on 3 April 1967 by P. M. Ostapyenko, who then took part in a series of complicated flight tests administered by V. M. Timofeyev. In three months Ostapyenko acquired sufficient experience with this machine to fly it in the air display planned for 9 July 1967 at Domodyedovo to celebrate the fiftieth anniversary of the October Revolution. Another OKB pilot, A. V. Fedotov, also flight-tested the 23-01; but this prototype had a very short life. Once the bureau's focus shifted to the variable geometry wing, the 23-01 tests were terminated—immediately after Ostapyenko's flyover at Domodyedovo—even though the aircraft's flight envelope was left practically unexplored with the exception of the takeoff and landing performance.

Specifications
Span, 7.72 m (25 ft 3.9 in); length (except probe), 16.8 m (55 ft 1.4 in); fuselage length (except probe), 15.995 m (52 ft 5.7 in); height, 5.15 m (16 ft 10.7 in); wheel track, 3.46 m (11 ft 4.2 in); wheel base, 6.13 m (20 ft 1.3 in); wing area, 40 m^2 (430.56 sq ft); takeoff weight, 16,000 kg (35,265 lb); war load, 2,500 kg (5,510 lb); max takeoff weight, 18,500 kg (40,775 lb).

In this photograph the 23-11/1 or 231 has its variable geometry wing set at the minimum sweep angle of 16 degrees.

The 23-11/1 with K-23 air-to-air missiles attached to pylons under the wing glove.

Performance

Takeoff roll with SPS and lift jets in clean configuration, 180–200 m (590–655 ft); landing roll with SPS, lift jets, and tail chute, 250 m (820 ft).

MiG-23 / 23-11/1

While one OKB team was at work on the 23-01, another tried to show that the variable geometry wing concept was well founded. Both projects were aimed at one objective: aircraft capable of speeds of Mach 2–2.3 and STOL performance. The 23-11 fuselage was shaped like a pointed cigar developing into a rounded-off angled square between frame nos. 18 and 20. The structure located in the midst of these two bulkheads was essential: it was the wing's center section, plus a fuel tank into which the air intake duct passed. The attachment fittings for the actuating cylinders of the main gear as well as the front ends of the gimbal joints were secured to its rear face (frame no. 20). The body then tapered to frame no. 28, where the whole rear fuselage could be detached to ease field maintenance and engine removal.

Bulkhead no. 31 at the rear of the fuselage supported the hinges for the four airbrakes, the support bearing for the stabilator, and the rear attachment fitting for the vertical fin. The skin was fabricated out of panels connected by fusion welds and then riveted. The wing box was made of the two main spars. The minimum sweep angle at the leading edge was 16 degrees, increasing steadily to a maximum of 72 degrees. Wing sweep was controlled by an SPK-1 hydraulic system whose two ball-screw actuators transformed spin to linear motion. Those actuators were linked directly to each wing's pivot arm. Pins were located on the center section 1,500 mm (59.06 inches) on either side of the fuselage datum line and lengthwise 128.5 mm (5.05 inches) ahead of bulkhead no. 20.

The wing's sweep angle could be modified by a control lever on the left console of the cockpit, and the pilot could follow the movement via the wing position indicator on the instrument panel. Each wing had leading edge (LE) flaps; single-slotted trailing edge (TE) flaps, in four sections; and two-section upper surface spoilers/lift dumpers forward of midflap sections. Extension of the LE and TE flaps was linked, but the LE controls featured a nonlinear mechanism that kept the angles from being identical. If the TE flaps were at 25 degrees at takeoff, the LE flaps were at 17 degrees; and when the TE flaps were at 50 degrees at landing, the LE flaps were at 19 degrees (their maximum). LE and

In this photograph the 23-11/1 has its wing set at the maximum sweep angle of 72 degrees.

TE flaps remained linked only if the wing was set at 16 degrees. Above that, the linkage rods automatically disengaged.

On the wing's upper surface, each spoiler was hinged on the rear main spar and acted like an aileron when operating differentially in conjunction with the horizontal tail surfaces. With a 16-degree wing sweep angle, its maximum deflection travel was 45 degrees. With a 72-degree wing sweep angle, the spoilers locked in the retracted mode and roll control was provided only by the horizontal tail surfaces operating differentially (tailerons). Between 16 and 72 degrees, the spoiler angle changed according to the sweep angle chosen by the pilot. Rudder control was provided by an irreversible servo-control unit supplemented by spring mechanisms to transmit the "feel." Operating spoilers instead of ailerons avoided the risk of wing twist when displacing ailerons at high speeds.

The 23-11 prototype was powered by the Khachaturov R-27F-300 (product 41) rated at 5,095 daN (5,200 kg st) dry or 7,645 daN (7,800 kg st) with afterburner. The nozzle area could be adjusted by means of a double ring of small flaps. Engine power was regulated at all ratings by a single linear throttle (the first of its kind at MiG), the variable geometry air intakes, and the blow-in doors (two for each intake duct). The specific fuel consumption of the 23-11/1 in level flight was 25 percent less than that of the MiG-21S with the much less powerful R-11F2-300.

The UVD-23 control system of the boundary layer splitter plates offered full thrust at any and all times and ensured that the engine functioned reliably at all ratings in the aircraft's flight envelope. The leading edges of the splitter plates at the air intakes stood 55 millimeters (2.16 inches) away from the fuselage wall, forming a boundary layer bleed duct. The UVD-23 apparatus was useful for setting the splitter plates to the most suitable position as the engine compressor pressure ratios ranged between 4 and 11. Automatic control took over when the aircraft reached Mach 1.15 and was governed by the deflection of the stabilator.

In the 23-11/1, 4,250 l (1,122 US gallons) of fuel were distributed among three fuselage integral tanks of 1,920, 820, and 710 l (507, 217, and 188 US gallons) and six wing structural tanks: two each of 62.5, 137.5, and 200 l (16.5, 36.3, and 52.8 US gallons). The first production machines also carried a drop tank under the fuselage. Because wing sweep varied, the fuselage-to-wing fuel and air lines passed through telescopic swivel joints.

The lower segment of the large ventral fin was hinged to fold to starboard when the landing gear was extended. The three gear legs were fitted with levered suspension. The main gear featured KT-133 trailed wheels with 830 x 225 tires (later increased to 830 x 300). The front leg had twin wheels with 520 x 125 tires and was fitted with the MRK-30 nosewheel steering mechanism, a shimmy damper, and a wheel centering device. All wheels were equipped with hydraulically controlled disc brakes. As the aircraft was being planned, conceiving the gear was like trying to square the circle. Because of the variable geometry concept the gear had to be housed entirely into the fuselage, but at the same time the wheel track had to remain fairly broad. This explains the seeming complexity of its kinematics.

The PT-10370-65 tail chute with an area of 21 m² (226 square feet) was housed in a cylinder at the base of the rudder with split cone-shaped doors. Armament of the 23-11/1 included four air-to-air K-23 missiles (two under the wing glove and two under the fuselage); during the tests K-13 missiles were also fired.

The 23-11/1 was moved to the test center on 26 May 1967, and after the usual ground and runway exercises the aircraft made its first flight on 10 June 1967 under the guidance of OKB chief pilot A. V. Fedotov. The first thirteen flights were devoted to preparing for the Domodyedovo air show. During that event on 9 July Fedotov gave a brilliant demonstration of all the capabilities of the VG concept. Subsequent flights explored the flight envelope and assessed the efficiency of the air intakes.

On 9 July 1967 at the Domodyedovo air show, Fedotov put on a remarkable demonstration of the 23-11 variable geometry aircraft. It was the prototype's fourteenth flight.

The R-27F-300 reached its twenty-five hour life limit on the prototype's forty-fifth flight. Tests resumed in January 1968 after the engine was replaced and the aircraft was equipped with the three-axis AP-155 autopilot. In early April the 23-11 moved to an airfield not far from a firing range to examine the operation of the air intakes and turbojets as well as the aircraft's handling characteristics when firing K-13 and K-23 missiles. Those tests took place between 8 April and 24 April. P. M. Ostapyenko and M. M. Komarov fired sixteen unguided missiles (the aircraft did not yet have radar). No surges or flameouts occurred between 5,000 m (16,400 feet) and 17,000 m (55,760 feet) and speeds of

Mach 0.7 to Mach 1.8 during the firing tests. The basic test schedule
cndcd in July after ninety-seven flights. The factory report concluded:

The MiG-23 variable geometry wing offers many advantages,
such as

— a significant reduction of the takeoff and landing rolls
(compared with those of all other existing aircraft in the
same category)
— a great ease of handling in the entire flight envelope and
especially at takeoff and landing
— a high indicated airspeed (IAS) at low altitude and, at maximum sweep, low g-forces in rough air
— a long range and a high flight endurance at cruise rating

Design performance should be met with the more powerful
R-27F2-300 turbojet (product 47) that the aircraft needs.

On 6 November 1968 A. I. Mikoyan confirmed the 23-11/1 factory
test report. This prototype, with its original markings, can be seen
today in the VVS museum on the Monino airfield near Moscow.

Specifications
Span (72° sweep), 7.779 m (25 ft 6.3 in); span (16° sweep), 13.965 m
(45 ft 9.8 in); fuselage length (except probe), 15.795 m (51 ft 9.8 in);
wheel track, 2.658 m (8 ft 8.7 in); wheel base, 5.772 m (18 ft 11.3 in);
wing area (72° sweep), 29.89 m² (321.74 sq ft); wing area (16° sweep),
32.1 m² (345.52 sq ft); takeoff weight in clean configuration, 12,860 kg
(28,345 lb); takeoff weight with four K-23 missiles, 13,300 kg (29,315
lb); wing loading (72° sweep), 424.2–445 kg/m² (86.9–91.2 lb/sq ft);
wing loading (16° sweep), 400.6–414.3 kg/m² (82.1–84.9 lb/sq ft); max
operating limit load factor, 3.1.

Performance
Max speed in clean configuration (72° sweep), 2,240 km/h or Mach
2.12 at 13,600 m (1,208 kt at 44,610 ft); max speed with two K-23 missiles (72° sweep), 2,255 km/h or Mach 2.13 at 13,400 m (1,217 kt at
43,950 ft); max speed with four K-23 missiles (72° sweep), 2,025 km/h
or Mach 1.905 at 12,800 m (1,214 kt at 41,985 ft); service ceiling, 17,200
m (56,415 ft); landing speed, 230 km/h (124 kt); takeoff speed, 270
km/h (146 kt); ferry range with two underbelly K-23 missiles (16°
sweep), 2,045 km (1,270 mi); takeoff roll, 320 m (1,050 ft); landing roll
with tail chute, 440 m (1,445 ft); landing roll without tail chute, 750 m
(2,460 ft).

MiG-23S / 23-11

The initial production model of the 23-11 was to be equipped with the new Sapfir-23 radar and the more powerful R-27F2M-300 engine rated at 6,760 daN (6,900 kg st) dry or 9,800 daN (10,000 kg st) with afterburner. The turbojet was ready in time; but unfortunately the radar was not, so the first aircraft had to make do with the Sapfir-21. With this older equipment the aircraft could carry at most four R-3S or R-3R missiles. Besides the radar and engine, the MiG-23S differed from the 23-11/1, 23-11/2, and 23-11/3 prototypes in its equipment: the ASP-PF computing fire control system, the TP-23 IR sensor, and the ARK-10 automatic direction finder. The radome was made of a new dielectric material.

The first MiG-23S was conveyed to the test center on 21 May 1969. The next eight days were spent determining the aircraft's balance, testing the systems, and running up the engine. The aircraft made its first flight on 28 May with A. V. Fedotov in the cockpit. On 10 July the aircraft was moved to the firing range to assess the performance of the engine during armament trials. By 20 August the MiG-23 had made thirty-two flights. Its weapon system (same as that of the MiG-21S) and the Sapfir-21 radar were tested on the fifth production aircraft, and no serious difficulties were uncovered. The built-in GSh-23L twin-barrel cannon was also fired. The SAU-23 automatic flight control system was checked, but only in the stabilization mode; it would be developed more fully as the test schedule proceeded. The SARP-12G emergency fault recorder was also developed. The MiG-23S was really just a transition model, and only fifty copies were built between mid-1969 and the end of 1970.

Specifications
Span (72° sweep), 7.779 m (25 ft 6.3 in); span (16° sweep), 13.965 m (45 ft 9.8 in); fuselage length (except probe), 15.65 m (51 ft 4.1 in); wheel track, 2.658 m (8 ft 8.7 in); wheel base, 5.772 m (18 ft 11.3 in); wing area (72° sweep), 29.89 m² (321.74 sq ft); wing area (16° sweep), 32.1 m² (345.52 sq ft).

Performance
Max speed in clean configuration (72° sweep), 2,405 km/h or Mach 2.27 at 12,800 m (1,298 kt at 42,000 ft); max speed with four R-3S missiles (72° sweep), 2,100 km/h or Mach 1.98 (1,133 kt); max operating Mach number, 2.27; service ceiling in clean configuration, 18,000 m (59,040 ft); service ceiling with four R-3S missiles, 16,500 m (54,120 ft); ferry range in clean configuration, 2,090 km (1,300 mi); ferry range

The MiG-23UB two-seater was both a trainer and a combat aircraft. In this photograph it is armed with four infrared-guided R-3S air-to-air missiles.

with four R-3S missiles, 1,800 km (970 mi); ferry range with 800-l (211-US gal) drop tank, 2,500 km (1,350 mi); takeoff roll, 550–700 m (1,800–2,295 ft); landing roll, 450–600 m (1,475–1,970 ft).

MiG-23UB / 23-51

The decision to build a two-seat trainer for the MiG-23 was made quickly: it was announced in a decree of the council of ministers dated 17 November 1967, fewer than six months after the prototype rollout. However, the ministry's directive went beyond a straightforward trainer and called for some sort of combat capacity—hence the designation UB (*Uchebniy Boyevoi:* training-combat). Derived directly from the MiG-23S, the MiG-23UB was powered by the same engine, the R-27F2M-300 rated at 6,760 daN (6,900 kg st) dry and 9,800 daN (10,000 kg st) with afterburner. The only structural modifications resulted from the rearrangement of the forward fuselage, the second cockpit taking the place of the equipment bay.

The ministry's decree allocated the following missions to the new aircraft:

1. Day and night training in clear and adverse weather conditions to teach pilots how to take off, handle the full flight envelope with different types of weapons or dummy missiles, and land
2. Combat within the limits of the aircraft's weaponry: the GSh-23L cannon, rockets, bombs, air-to-surface missiles (to attack ground or naval targets in visual mode), R-3S infrared-guided air-to-air missiles, or Kh-23 air-to-surface beam-rider guided missiles (since the two-seater had no radar, the latter's guidance equipment was housed in small pods under the wing glove)

All of this weaponry (except the cannon) was carried under four store points: two under the fuselage and two under the wing glove. In the front cockpit, the student pilot's equipment included the ASP-PFD fire control system (without the ranging device) and the weapon selection panel. All other controls were duplicated, and the instructor's set took priority. The nose was weighted to compensate for the lack of radar.

The MiG-23UB differed from the MiG-23S in many points:

1. Structurally, the nose section was modified up to the no. 18 bulkhead to make room for the second cockpit; the equipment bay and the standby hydraulic generator with its windmill were consequently moved back by reducing the capacity of fuel tank no. 1—normally 700 l (185 US gallons)—and, to compensate, adding a tank in the rear fuselage to carry 470 l (124 US gallons)
2. On-board equipment included the SOUA active angle-of-attack limiter (a few planes that were not so equipped used the SUA-1 critical AOA warning device and the RIS stick shaker), the UUA-1 attitude indicator, the Polyot-11-23 flight management system (including the RSBN-6S landing and short-range navigation device, the SKV-2N2 heading and vertical reference unit, and DV-30/DV-10 signal transmitters) linked to the SAU-23UB automatic flight control system, three-axis artificial feel units and trims, radio-altimeter, automatic direction finder, marker receiver, the SORTs warning light display panel, IFF interrogator and transponder, radar warning receiver, the SPU-9 intercom, and the MS-61 tape recorder

The MiG-23UB wing, like that of the single-seater, "jumped" from type 1 to type 3. With the type 1 wing the MiG-23UB could carry only a single drop tank under the fuselage; but with the type 3 wing it could carry one drop tank under the fuselage and two drop tanks on non-swiveling pylons under the outer wings for ferry flights. The gear wheels all had brakes, and the two cockpits were equipped with KM-1

A MiG-23UB takes off for a ferry flight. It has two drop tanks under the outer wing panels. The 16-degree sweep setting will be maintained for the entire flight.

ejection seats and a centralized emergency abandonment system. A periscope was installed on the jettisonable part of the rear canopy so that the instructor could see more clearly while taking off, landing, and taxiing.

The MiG-23UB was rolled out in March 1969 and was first piloted in May by M. M. Komarov. The factory tests (carried out by Komarov and P. M. Ostapyenko) and the state trials lasted until 1970. That year the aircraft was approved for duty in VVS and PVO fighter regiments, and it was produced in the Irkutsk factory until 1978.

Specifications
Span (72° sweep), 7.779 m (25 ft 6.3 in); span (45° sweep), 11.928 m (39 ft 1.6 in); span (16° sweep), 13.965 m (45 ft 9.8 in); fuselage length (except probe), 15.66 m (51 ft 4.5 in); wheel track, 2.658 m (8 ft 8.7 in); wheel base, 5.772 m (18 ft 11.3 in); wing area (72° sweep), 34.16 m² (367.7 sq ft); wing area (45° sweep), 35.5 m² (382.1 sq ft); wing area (16° sweep), 37.35 m² (402 sq ft); takeoff weight, 15,740 kg (34,690 lb); max takeoff weight, 18,000 kg (39,670 lb); landing weight, 12,400 kg (27,330 lb); fuel, 4,000 kg (8,815 lb); with three 800-l (211-US gal) drop tanks, 6,350 kg (13,995 lb); wing loading (72° sweep), 460.8–526.9 kg/m² (94.5–108 lb/sq ft); wing loading (45° sweep), 443.4–507 kg/m²

The MiG-23M was built in the greatest numbers. This one carries two R-60R and four R-60T air-to-air missiles.

(90.9–103.9 lb sq ft); wing loading (16° sweep), 421.4–481.9 kg/m² (86.4–98.8 lb/sq ft); max operating limit load factor, 7.

Performance
Max speed in clean configuration (72° sweep), 2,490 km/h or Mach 2.35 at 12,500 m (1,344 kt at 41,000 ft); max speed in clean configuration at sea level (72° sweep), 1,200 km/h (648 kt); max operating Mach number, 2.35; max operating Mach number with four R-3S missiles, 2; max operating Mach number with four R-3S missiles and 800-1 (211-US gal) drop tank, 0.8; service ceiling, 15,800 m (51,825 ft).

MiG-23M / 23-11 / MiG-23MF / MiG-23MS

The MiG-23M took the place of the MiG-23S on the assembly lines. It was the long-awaited production aircraft whose arrival had been hindered by delays in the development of its systems (especially the radar) and its engine (whose thrust was inadequate to achieve the design parameters). Its wing chord was broader, leading to that distinctive dogtooth at the inner end of the leading edge and also a larger

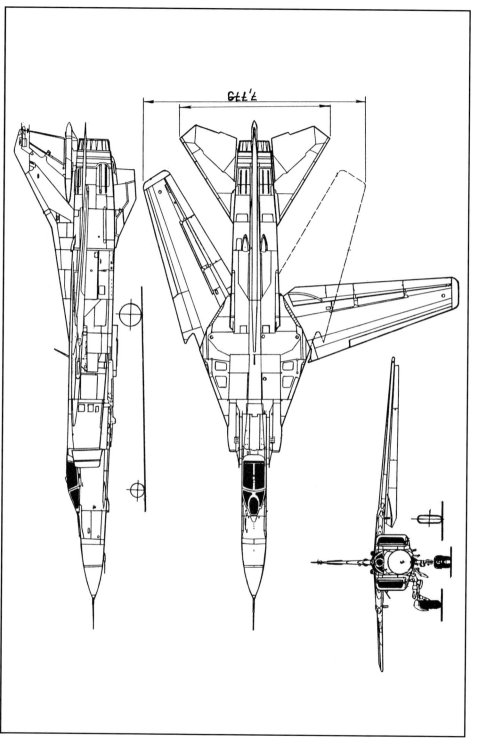

7.779

MiG-23MS (MiG OKB three-view drawing)

This MiG-23M carries R-23T and R-23R air-to-air missiles under the wing gloves. The four store stations under the fuselage are fitted with R-60Ts. The TP-23 infrared sensor is visible under the radome.

This MiG-23M has a type 2 wing, with its deeper chord and dog-toothed edge but no leading edge flaps. The four airbrakes are deployed.

368

The MiG-23MS was intended for export and equipped with less sophisticated radar and other systems.

wing area. This was the type 2 wing, without leading edge flaps; but the MiG-23M was later retrofitted with type 3 wing, its four-part leading edge flaps linked to those on the trailing edge as they retracted or extended.

It also had a different engine, the new Khachaturov R-23-300 rated at 8,135 daN (8,300 kg st) dry or 12,250 daN (12,500 kg st) with afterburner. And its systems were different: the S-23D-Sh forward-sector scanning and fire control system; the Sapfir-23-Sh radar; the TP-23 infrared sensor; the ASP-23D fire control device; and SAU-23A second-series automatic flight control system with the ARZ-1A feel computer on the pitch channel capable of taking the aircraft's speed, altitude, and sweep angle into account.

Because of the missions allotted to the MiG-23M—interception, air combat, and attack of ground and naval targets—its weapon system included the GSh-23L twin-barrel cannon embedded under the fuselage on its easy-access hoisting tray and, at four store stations (two under the fuselage, two under the wing glove), radar-guided R-23R, IR-guided R-23T, R-13M air-to-air missiles, R-3A training missiles, B-8 rocket pods (firing S-8 rockets), UB-32 rocket pods (firing S-5 rockets), bombs of various types and weights, submunitions dispensers, R-60 close-range air-to-air missiles, S-24 unguided air-to-surface rockets, and a pod housing the guidance system for air-to-surface missiles. The aircraft was decked out with an armament control panel.

Maximum internal fuel capacity was raised to 4,700 l (1,241 US gallons) thanks to a fourth fuel tank in the rear fuselage. The MiG-23M could also carry three drop tanks holding 790 l (209 US gallons) apiece, one under the fuselage and two under the wing glove). First piloted in June 1972 by A. V. Fedotov, this model was the most popular MiG-23 and originated two export versions, the MiG-23MF and MiG-23MS; they carried less-advanced systems, armament (R-3S/R-3R missiles), and engines (the MiG-23MS used the R-27F2M-300), and their camouflage paint varied according to where they operated. MiG-23M, MF, and MS aircraft have taken part in several local conflicts in the Middle East and Afghanistan.

Specifications
Span (72° sweep), 7.779 m (25 ft 6.3 in); span (16° sweep), 13.965 m (45 ft 9.8 in); fuselage length (except probe), 15.73 m (51 ft 7.3 in); wheel track, 2.658 m (8 ft 8.7 in); wheel base, 5.772 m (18 ft 11.3 in); wing area (72° sweep), 34.16 m^2 (367.7 sq ft); wing area (16° sweep), 37.35 m^2 (402 sq ft); takeoff weight, 15,750 kg (34,715 lb); max takeoff weight, 18,400 kg (40,555 lb); max takeoff weight with 790-l (209-US gal) drop tank, 19,130 kg (42,160 lb); with two 790-l (209-US gal) drop tanks, 19,940 kg (43,950 lb); with three 790-l (209-US gal) drop tanks, 20,670 kg (45,555 lb); internal fuel, 3,800 kg (8,375 lb); wing loading (72° sweep), 461–605 kg/m^2 (94.5–124 lb/sq ft); wing loading (16° sweep), 421.7–553.4 kg/m^2 (86.4–113.4 lb/sq ft); max operating limit load factor (45° sweep), 8 at ≤ Mach 0.85, 7 at > Mach 0.85.

Performance
Max speed in clean configuration (72° sweep), 2,490 km/h or Mach 2.35 at 12,500 m (1,344 kt at 41,000 ft); max speed in clean configuration (16° sweep), 935 km/h or Mach 0.8 at 3,500 m (505 kt at 11,500 ft).

MiG-23ML / 23-12

In the process of developing and upgrading the MiG-23 family, the MiG-23ML (rolled out in 1976) marked an important milestone that involved a complete refurbishing of the MiG-23M: a new engine, new systems, new missiles, and new radar capabilities. The R-29-300 turbojet was replaced by a first-series Khachaturov R-35 (R-35-300) rated at 8,380 daN (8,550 kg st) dry and 12,450–12,740 daN (12,700–13,000 kg st) with throttleable afterburner. Total fuel weight with three 800-l (211-US gallon) drop tanks reached 5,500 kg (12,120 pounds).

With its more powerful engine, the MiG-23ML marked a watershed in the aircraft's development. Armament includes two R-23 and four R-60 air-to-air missiles.

The MiG-23ML differed from the MiG-23M in many other aspects as well:

—the trailing edge flaps were divided into three sections
—the automatic flight control system was upgraded and renamed SAU-23AM
—the aircraft's weight was reduced by removing the fourth fuse-lage tank
—the new Polyot ("flight") flight management system was installed, including landing and short-range navigation systems, heading and vertical reference unit, and altitude and speed sensor; the Polyot was linked to the SAU-23M and could operate simultaneously
—the forward-sector scanning and fire control system, upgraded and renamed S-23ML, featured the Sapfir-23ML radar, the TP-23M infrared sensor, and the ASP-17ML sighting system
—six store points (four under the fuselage and two under the wing glove) carried the usual weaponry, including R-23R/R-23T air-to-air missiles to supplement the MiG-23M cannons and rockets
—two UPK-23-250 gun pods housing GSh-23L cannons could be mounted beneath the wing
—the shape of the tail fin was modified by shortening the dorsal fin

This photograph clearly shows the shape of the type 3 wing, with leading edge flaps. The GSh-23L twin-barrel cannon is visible between the air intakes. (Photo RR)

A MiG-23ML takes off with full reheat. The undercarriage is retracting, and the ventral fin is already fully unfolded. (Photo RR)

Close-up of the R-35 afterburner's flame holder rings. Foreground, the flap-type nozzle. (Photo RR)

A MiG-23ML lands in Finland with open tail chute. The new fin shape is a distinctive feature of this model. (Photo RR)

The MiG-23ML entered production in 1976 and was built until 1981. All were later upgraded to the MLD standard as quickly as they were returned to the overhaul shops. Aircraft of this type visited Finland and France as part of an exchange project in 1978—without their IR sensor, however.

Specifications
Span (72° sweep), 7.779 m (25 ft 6.3 in); span (16° sweep), 13.965 m (45 ft 9.8 in); fuselage length (except probe), 15.65 m (51 ft 3.7 in); wheel track, 2.658 m (8 ft 8.7 in); wheel base, 5.772 m (18 ft 11.3 in); wing area (72° sweep), 34.16 m^2 (367.7 sq ft); wing area (16° sweep), 37.35 m^2 (402 sq ft); takeoff weight, 14,700 kg (32,400 lb); max takeoff weight, 17,800 kg (39,230 lb); wing loading (72° sweep), 430.4–521 kg/m^2 (88.2–106.8 lb/sq ft); wing loading (16° sweep), 393.6–476.6 kg/m^2 (80.7–97.7 lb sq ft); max operating limit load factor, 8.5 at ≤ Mach 0.85, 7.5 at > Mach 0.85.

Performance
Max speed in clean configuration (72° sweep), 2,500 km/h or Mach 2.35 (1,350 kt); max speed in clean configuration (16° sweep), 940 km/h or Mach 0.8 (508 kt); service ceiling, 18,500 m (60,680 ft); ferry range, 1,950 km (1,210 mi); with three 800-l (211-US gal) drop tanks, 2,820 km (1,750 mi).

MiG-23P / 23-14

The MiG-23P was a MiG-23ML specially modified for PVO intercept missions. The aircraft was guided to the interception point by ground stations. This guidance system, relayed by the AFCS navigation computer, set the best flight path and showed the pilot when to light up the afterburner and when to fire his cannons or missiles. Facts about the military situation were continuously transmitted by the aircraft to the ground stations. The MiG-23P's armament was identical to that of the MiG-23ML.

MiG-23MLD / 23-18

There were no MiG-23MLDs as variants with a separate existence: they were merely MiG-23MLs that, when returned to overhaul shops, had been fitted with a duplicated SOS-3-4 unit to replace the SOUA.

The MiG-23B differed from the MiG-23M only in its power plant and the shape of its nose. It retained the earlier aircraft's variable geometry air intakes.

This was an automatic device that, according to the aircraft's attitude, speed, and altitude, initiated the LE flaps' actuation for a 33-degree wing sweep angle. Extension of the LE flaps reached 20 degrees under 900 km/h (486 kt) for AOAs greater than 10 degrees. Beyond 900 km/h (486 kt) when the wing sweep angle was increased to 72 degrees, the LE flaps retracted and the device was neutralized.

The retrofit kit also included additional "spread wing" warnings beyond the 16-degree sweep, passive jamming system provided by chaff launchers, and radar warning receivers (for both ground and airborne radar). The MiG-23MLD could also be equipped with an air combat simulator that enabled the pilot to train himself to fire and guide missiles without actually having to fire any—the first device of its kind for a fighter. The economic advantages of this are obvious. Externally, this version could be identified by its wing-edge root extension and by small vortex generators on the Pitot probe.

MiG-23B / 32-24 MiG-23BN / 32-23
MiG-23BM / 32-25 MiG-23BK / 32-26

In 1969 the OKB prepared the preliminary design for a light attack aircraft (*shturmovik*) intended to destroy either isolated or multiple, fixed or mobile targets day and night. According to the design department, this aircraft would also be able to accomplish auxiliary missions such as attacking helicopters and transport aircraft at low and medium altitudes. At the time the Soviet tactical air command needed an attack

The low-pressure tires for the main gear increased the wheel track. The MiG-23B could carry up to three metric tons of bombs.

aircraft that could be mass-produced inexpensively and one that offered at least the same capabilities as the American Northrop F-5A, the Franco-British Jaguar, and the Italian Fiat G-91Y.

The initial plan called for a subsonic aircraft, but the concept was quickly modified because the aircraft had to be capable of supersonic speed dashes to get out of dangerous territory. It also had to be capable of attacking aircraft with its cannons and IR-guided missiles once it had dropped its bomb load. It was intended as a completely new type of aircraft; however, in the interest of production rationalization it was decided in the end that the MiG-23 airframe would be used. The new design—or *izdelye* 32-24—produced in 1970 the MiG-23B fighter-bomber. Externally, it differed from the MiG-23S only in the nose. Taking the aircraft's main role—attack of ground targets—into account, OKB engineers completely reshaped the aircraft's nose section after removing the radar in order to improve the pilot's sight forward and downward; hence that peculiar look that Soviet pilots have dubbed *out-konos* (duck bill).

The MiG-23B was unlike the MiG-23S in many respects:

—it was powered by the Lyulka AL-21F-300 turbojet rated at 7,840 daN (8,000 kg st) dry and 11,270 daN (11,500 kg st) with afterburner

—the Sapfir radar was replaced by the PrNK Sokol-23S nav-attack system that could find even the smallest ground targets; the PrNK was designed for level-flight, dive, or dive-recovery bombing and for level-flight cannon fire

—the front fuselage sides were covered with armor plates to protect the pilot against enemy fire

—the fuel tanks were filled by an inert gas as fast as the fuel was emptied to prevent explosions in case of a direct hit

—the aircraft was equipped with a complete array of active and passive radar-jamming devices for its own defense

Besides the twin-barrel GSh-23L cannon embedded in the fuselage, the aircraft could field a powerful cluster of weaponry at six store stations (four under the fuselage and two under the wing glove)—air-to-surface missiles, large-caliber rockets, automatic rocket pods, or multiple racks with four typical loads: eighteen 50-kg (110-pound) bombs, eighteen 100-kg (220-pound) bombs, eight 250-kg (550-pound) bombs, or six 500 kg (1,100-pound) bombs. For the first time on a Soviet fighter-bomber, the MiG-23B could carry beneath the wing glove two UPK-23-250 gun pods (the first figure gives the cannon's caliber, the second the number of rounds).

The MiG-23B made its first flight on 20 August 1970 with P. M. Ostapyenko at the controls. It passed the state trials and entered production in 1971.

The BN variant (32-23) used another engine—the R-29B-300 rated at 7,840 daN (8,000 kg st) dry and 11,270 daN (11,500 kg st) with afterburner—and the Sokol-23N nav-attack system, but externally the MiG-23B and MiG-23BN were identical. The MiG-23BM (32-25) differed from the BN in its computerized PrNK-23 nav-attack unit; the MiG-23BK (32-26) featured different equipment. Only twenty-four MiG-23Bs were built, but the aircraft's airframe, engine, and systems were upgraded a number of times by retrofit or other means.

Specifications

Span (72° sweep), 7.779 m (25 ft 6.3 in); span (16° sweep), 13.965 m (45 ft 9.8 in); fuselage length, 15.349 m (50 ft 4.3 in); wheel track, 2.728 m (8 ft 11.4 in); wheel base, 5.991 m (19 ft 7.9 in); wing area (72° sweep), 34.16 m² (367.7 sq ft); wing area (16° sweep), 37.35 m² (402 sq ft); takeoff weight in clean configuration, 15,600 kg (34,380 lb); takeoff weight with three 790-l (209-US gal) drop tanks and four UB-16-57 rocket pods, 18,600 kg (40,995 lb); max takeoff weight with six FAB-500 bombs, 18,900 kg (41,655 lb); internal fuel, 4,500 kg (9,920 lb); max landing weight, 15,200 kg (33,500 lb); wing loading (72° sweep), 456.7–553.3 kg/m² (93.6–113.4 lb/sq ft); wing loading (16° sweep),

417.7–506 kg/m^2 (85.6–103.7 lb/sq ft); max operating limit load factor, 7 at ≤ Mach 0.8; 6 at > Mach 0.8.

Performance

Max speed in clean configuration (72° sweep), 1,880 km/h or Mach 1.7 at 8,000 m (1,015 kt at 26,250 ft); max speed in clean configuration (45° sweep), 1,100 km/h or Mach 0.91 at sea level; max speed in clean configuration (16° sweep), 935 km/h or Mach 0.8 at 3,500 m (505 kt at 11,500 ft); radius of action, lo-lo-lo, 5 min on target with four 250-kg (550-lb) bombs, 600 km (370 mi).

MiG-27 Series

It does not appear that the MiG-23B, the first fighter-bomber descended from a family of genuine fighters, satisfied all of the hopes pinned on it. The main weaknesses were its power plant and its nav-attack system. The MiG-27 was therefore developed from a MiG-23B airframe—in this case, a MiG-23BM (32-25) powered by the new R-29B-300 turbojet rated at 7,840 daN (8,000 kg st) dry and 11,270 daN (11,500 kg st) with afterburner. But unlike the MiG-23, the MiG-27 had fixed air intakes and could be recognized by its very small splitter plates, set 80 millimeters (3.78 inches) away from the fuselage wall to act as boundary layer bleeds. The blow-in doors used at takeoff or in flight at low Mach numbers were retained.

Keeping in mind that the aircraft's new missions (bombing, deep tactical support) might require sudden changes in trim, the displacement speed of the type 3 wing between settings was modulated, so that during any aircraft acceleration or deceleration the wing could be given in due time the most favorable sweep angle, considering the aircraft's speed. The full-span trailing edge flaps were set to 25 degrees at takeoff and 50 degrees at landing (only when the wing was fully spread). The aircraft's standard AOA at landing was 15 degrees but could be reduced to 10 degrees for hard landings. It had a tail chute that covered 21 m^2 (226 square feet) and was housed in a canister at the base of the rudder.

The main equipment included the PrNK-23 nav-attack system, SAU automatic flight control system, SPS-141 thermal jammer, RI-65 vocal warning (for sixteen facts), SUA-1 AOA indicator, KN-23 navigation computer, SG-1 radar warning receiver, SO-69 transponder, SRZO/SRO-1P IFF interrogator/transponder, RV-5R/RV-10 radio-altimeters, and Fone range finder.

This MiG-27 carries three bombs and two rocket pods. Note the new shape of the engine air intakes.

The MiG-27's weaponry was quite impressive. It had one 30-mm six-barrel GSh-6-30 cannon with 260 rounds and could handle a total load of 4,000 kg (8,815 pounds) at seven store stations:

— two SPPU-22-01 gun pods carrying twin-barrel 23-mm guns that could be depressed to attack ground targets
— R-3S and R-13M air-to-air missiles
— Kh-23 or Kh-29 air-to-surface missiles
— 240-mm S-24/S-24B rockets
— UB-32A or UB-32-16 rocket pods
— twenty-two 50- and 100-kg (110- and 220-pound) bombs, eighteen 100-kg (220-pound) bombs, nine 250-kg (550-pound) bombs, eight 500-kg (1,100-pound) bombs, or tactical nuclear bombs of various sizes
— napalm containers

The internal fuel capacity was 5,400 l (1,426 US gallons), and the aircraft could also carry three drop tanks with 790 l (209 US gallons) apiece.

Specifications

Span (72° sweep); 7.779 m (25 ft 6.3 in); span (16° sweep), 13.965 m (45 ft 9.8 in); overall length, 17.076 m (56 ft 0.3 in); height, 5 m (16 ft 4.8 in); wing glove sweep at leading edge, 70°; wheel track, 2.728 m (8 ft 11.4 in); wheel base, 5.991 m (19 ft 7.9 in); wing area (72° sweep), 34.16 m² (367.7 sq ft); wing area (16° sweep), 37.35 m² (402 sq ft); empty weight, 11,908 kg (26,245 lb); takeoff weight, 18,100 kg (39,890 lb); max takeoff weight, 20,300 kg (44,740 lb); internal fuel, 4,560 kg (10,050 lb); wing loading (72° sweep), 529.9–594.2 kg/m² (108.6–121.8 lb/sq ft); wing loading (16° sweep), 485.7–544.7 kg/m² (99.6–111.7 lb/sq ft).

Performance

Max speed in clean configuration (72° sweep) at sea level, 1,350 km/h (729 kt); at 8,000 m (26,240 ft), 1,885 km/h or Mach 1.7 (1,018 kt); landing speed, 260–270 km/h (140–146 kt); radius of action, lo-lo-lo, with two Kh-29 missiles, 225 km (140 mi); with two Kh-29 missiles and three 790-l (209-US gal) drop tanks, 540 km (335 mi), plus 7% reserve fuel; takeoff roll, 950 m (3,115 ft); landing roll with tail chute, 900 m (2,950 ft); landing roll without tail chute, 1,300 m (4,265 ft).

MiG-27K

Just as the MiG-27 was developed from the MiG-23BM, the MiG-27K was developed from the MiG-23BK (32-26). Its new PrNK-23K nav-attack system could manage the aircraft's flight path and fire the cannon and missiles simultaneously. Compared with the MiG-23M's PrNK-23S, it offered new control possibilities: PMS mode (sighting from a maneuvering aircraft for bomb release as well as cannon and rocket fire) and PKS mode (time-tagged and corrected target tracking and bombing in blind flight according to navigation coordinates).

The twin-barrel 23-mm was also replaced by one GSh-6-30 six-barrel underside cannon. The SUV fire control system had many capabilities: programmed firing, missile and rocket firing (with emergency control), display of weapon availability, bomb release (cluster or individual), and cannon firing. The SUV also warned the pilot of the weapon racks' release. The aircraft was equipped with a flight management system (with automatic mode transfer), radar warning receiver, active radar jammer, and smoke-emitter. The MiG-27K could carry the same array of weapons as the MiG-23B plus laser-guided missiles. Production took place over several years.

Like the MiG-27, the MiG-23K could carry four metric tons of external military load.

Specifications

Span (72° sweep), 7.779 m (25 ft 6.3 in); span (16° sweep), 13.965 m (45 ft 9.8 in); fuselage length (except probe), 15.489 m (50 ft 9.8 in); wheel track, 2.728 m (8 ft 11.4 in); wheel base, 5.991 m (19 ft 7.9 in); wing area (72° sweep), 34.16 m² (367.7 sq ft); wing area (16° sweep), 37.35 m² (402 sq ft); max takeoff weight with eight FAB-500 bombs, 20,670 kg (45,555 lb); max takeoff weight on unprepared strip, 18,100 kg (39,890 lb); landing weight, 14,200 kg (31,295 lb); max landing weight, 17,000 kg (37,470 lb); on the load sheet, one 790-l (209-US gal) drop tank is worth 750 kg (1,655 lb), two are worth 1,530 kg (3,370 lb), and three are worth 2,280 kg (5,025 lb); wing loading (72° sweep), 605–529.9 kg/m² (124–108.6 lb/sq ft); wing loading (16° sweep), 553.4–484.6 kg/m² (113.4–99.3 lb/sq ft).

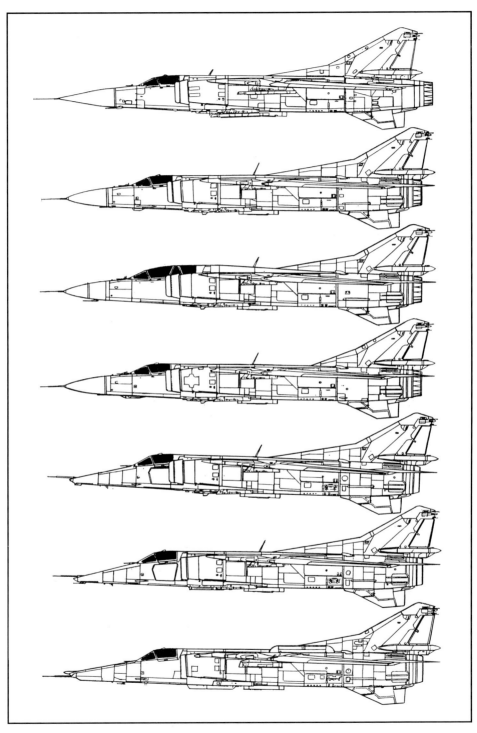

Top to bottom: 23-11, MiG-23M, MiG-23UB, MiG-23MLD, MiG-23BM, MiG-27, and MiG-27M (MiG OKB drawing)

The MiG-27M can be easily recognized by the dielectric lip located above the laser range finder window and by the leading edge root extension.

MiG-27D / 32-27   MiG-27M / 32-29   MiG-27L / 32-29L

These three versions were the most advanced of the MiG-27 family. They were all equipped with the upgraded PrNK-23M nav-attack system, which improved their operating range significantly. Their weaponry includes various containers such as the three-camera reconnaissance pod or SPPU-22 gun pods (for the 23-mm twin-barrel depressible cannon with 260 rounds).

The MiG-27D was equipped with the Klen ("maple") range finder (much more efficient than the MiG-23's Fone). The MiG-27L (32-29L) was the export version of the MiG-23M and is built under license by India's HAL as the Bahadur ("valiant"). Production of 165 machines was launched there in 1984; it seems very likely that this number will increase.

MiG-25 Series

The MiG-25 was a special case. Originating in the late 1950s as a response to the ambitious Lockheed A-11 project,* the aircraft that was to become the MiG-25—still referred to inside the OKB as the Ye-155—

---

*The Lockheed A-11 project would lead to the YF-12A interceptor and the SR-11A reconnaissance aircraft. The existence of the project was disclosed by President Johnson on 23 February 1964—but in fact it dated back to 1959, and the Soviets were aware of it as early as 1960. One prototype flew on 26 April 1962.

helped the Soviet aerospace industry to make great strides forward. And at the time technology was already progressing by leaps and bounds. Immediately after the first aircraft broke the sound barrier, everyone was already talking about level flight at Mach 3! And everyone knew that to reach that speed, another barrier had to be broken: the heat barrier.

On the MiG-19 at Mach 1.3 in 0° C (32° F) ambient air temperature, the airflow temperature at the nose reached 72° C (161.6° F). On the MiG-21 at Mach 2.05 that temperature increased to 107° C (224.6° F). At Mach 3 it would hit 300° C (572° F). The basic material used in aircraft manufacture, duralumin, could withstand temperatures of up to 130° C (266° F), but there were no semiconductors capable of surviving over 65° C (149° F). The new barrier seemed truly impassable. "The eyes are scared but the hands work," goes an old Russian saying— one the OKB engineers seemed to take to heart. Some started to make computations, others set out to visit suppliers, and in a short time the project started to take shape.

The engine was the first priority. A. A. Mikulin and S. K. Tumanskiy, his closest colleague, proposed an immediate answer: one derived from the 15K, an axial flow turbojet designed for a winged missile. The two engine manufacturers quickly developed the compressor, the combustion chamber, and the afterburner. They read the temperatures all along the gas channel and developed an adjustable-area nozzle. To obtain an exact fuel/air ratio for engine ratings subject to quick changes, the hydromechanical fuel metering valve was replaced by an electronic fuel control unit.

With the engine development seemingly well in hand, the time had come to deal with the airframe. The engineers' task was to create an aircraft whose flight envelope would be quite unusual—especially in terms of speed and ceiling—and one that would be equipped with many new systems. After testing several models in the TsAGI wind tunnels, one was selected. The next step was to choose the materials.

The forced abandonment of duralumin left only one option: titanium, which Lockheed used for the A-11 project. On the engineering drawings, the fuselage and the wing center section were to be used as built-in fuel tanks. Theoretically, those tanks could be made of duralumin because they were to be filled with a cold fluid; their walls would only warm up to dangerous levels once the tanks were empty. But to build such structures, rivets and sealer cement that could withstand high temperatures were vital—and they did not exist. Moreover, titanium was very difficult to machine, and cracks often formed after welding. Was steel a viable alternative?

At the same time, an unexpected obstacle cropped up: a shortage of qualified riveters. Few people wanted to do this unrewarding, unpleasant work. With welded steel, rivets would not be necessary. A number of steelworks cast high-quality, easy-to-weld steel that obviated the need for cement. Moreover, since World War II many welding schools had opened all over the country.*

After weighing the alternatives, Mikoyan made up his mind: the new aircraft would be made of steel. Everyone at the design office, the metallurgical industry's research institutes, and the specialized test laboratories went to work developing strong, corrosion- and heat-resistant, steels; new titanium-aluminum alloys for the less sensitive parts; and innovative machining, casting, stamping, and welding tools. Research was also conducted into microscopic metallurgy in a welding bath; the tendency of metal and welded assemblies to crack at different temperatures; the interaction of basic and added materials; crystallization laws; and crystallization process control for hard-to-weld materials. As fast as those problems were solved, all of the factory workshops were upgraded to use the new technologies: spot welding and seam welding, automatic or manual. All riveters were turned into welders.

A high-quality steel is three times more solid than duralumin but also three times heavier, so in order not to add weight to the aircraft's structure every structurally significant item had to be three times thinner. This forced the engineers to reconsider matters such as the strength of materials, aeroelastic stability, aerodynamic flutter, and so on. The whole process was as complicated as the shift from the antique wood airplane to the modern all-metal aircraft. Any move forward happened step by step, and workers constantly had to become acquainted with new methods for assembling panels and parts.

To start, only three wing structures were built. The first two were rejected because they did not withstand particularly severe static tests. The pessimists—and they were numerous—thought that the welded built-in fuel tanks would not hold out or that every landing would prompt disastrous cracks. The plexiglass of the canopy was so outdated that it melted. The hydraulic fluid decayed, and tires as well as rubber sealing rings lost their elasticity. Everything had to be questioned, adapted, or modified.

But eventually all the pieces of the jigsaw fell into place, and it became possible to build the first prototype. The technological results speak for themselves:

1. Material: structure made of tempered steel, 80 percent of the airframe weight; titanium alloys, 8 percent; structurally significant items made of D19 heat-resistant aluminum alloy, 11 percent
2. Assembly method: spot welding and seam welding, 50 percent (weld spot > 1,400,000); argon arc welding, 25 percent (4,000 m

---

*As early as the 1930s the Soviets had developed many forms of welding. During World War II the scholar Ye. P. Patone invented automatic welding methods that quintupled tank production.

[13,000 feet] of weld bead); fusion wolding and inert gas welding, 1.5 percent; assembly with bolts and rivets, 23.5 percent

The welded fuel tanks took up 70 percent of the fuselage volume. The seal was secured by welds whose reliability can be judged by the following anecdote: over one full year of welds—whose distance was equivalent to that between Moscow and Gorki (450 km [280 mi])—only one or two insignificant leaks were detected. The repair was no problem and, most important, could be made by field maintenance personnel.

The thermal problems were not completely settled for all of that. A full range of air-air and air-fuel exchangers, as well as turbine cooler units and other similar systems, had to be developed in order to lower the temperature of the air bled into the engine compressor from 700° C (1292° F) to the -20° C (-4° F) that had to be maintained near the electronic bay access door—and keep in mind that aircraft systems themselves emit a lot of heat. Even if the pilot's head was protected by fresh air sent by special nozzles, the canopy was far too hot to touch.

The engine bay was insulated by a heat shield made of silver-plated steel. Gosplan allocated 5 kg (11 pounds) of silver per aircraft—not a single ounce more. The silver was 30 microns thick, and its absorption factor was between 0.03 and 0.05. Other metals were tested such as gold and rhodium, but they were far too expensive even if their absorption factors were satisfactory. The 5 percent of heat absorbed by this silver-plated steel lining was held in fiberglass blankets to prevent it from escaping toward the fuel tanks. Even coatings made of basalt fibers were tested.

All of the big secrets of the MiG-25 are summed up above, and it takes just a few lines. On 16 March 1965 the world learned that Fedotov had topped the SR-71 records with a certain Ye-266; this was the somewhat spurious designation sent to the FAI authorities to have the MiG-25 records ratified. On twenty-one subsequent occasions, the FAI was notified of record attempts made by the Ye-266 or the Ye-266M. In 1993 nine of the records set by the MiG-25 in 1967, 1973, 1975, and 1977 still stand.

MiG-25P / Ye-155P / MiG-25PD / MiG-25PDS / 84

The Ye-155P high-altitude supersonic interceptor project was confirmed by a decree of the council of ministers in February 1962. But in fact the OKB had started the preliminary design two years earlier. Contrary to what was thought, the project was not intended to face the

Canard surfaces could be affixed to the auxiliary structures located near the top of the air intake duct on the Ye-155P, prototype of the MiG-25P. The wing pylons held R-40 air-to-air missiles.

threat represented by the American XB-70 Walkyrie Mach 3 bomber. Instead, it was a response to the Lockheed A-11.

By listing all of the advancements that the MiG-25 was about to inherit, its technical evolution can be better appreciated:

— a weapon system capable of intercepting any type of flying target, from cruise missiles at low altitudes to supersonic aircraft at very high altitudes
— a structure permitting the interceptor to break the heat barrier and fly long supersonic dashes
— high lift-to-drag ratio, good stability, and sharp maneuverability across a wide flight envelope where speed and ceiling were usually favored
— a new (for MiG) aerodynamic scheme with lateral air intakes, twin fins, and two ventral fins
— a structure in welded steel that featured high mass ratio, simple maintenance, good manufacturing regularity due to automation, high output factor for the materials, and lower production costs due to better productivity

One of the seven preproduction MiG-25s fitted with triangular winglets and antiflutter bodies at the wing tips. The wing had no anhedral.

—electronic fuel control (the first in the USSR) and a single refueling point
—a greater number of auto-flight control modes with (for the first time) a range of programming possibilities: for altitude, flights on preset paths, landing approaches, limitations in automatic or semiautomatic flight modes, and overspeed warning
—utilization of new materials and semifinished products in high-strength steel, titanium, and heat-resistant duralumin
—intensive employment of automatic control systems and flight data recorders
—new technological processes for the heat treatment of materials to alleviate strains and stresses, plus new control and maintenance practices
—a long lifetime and time between overhauls for a combat aircraft of this category

The Ye-155P-1 prototype was powered by two Mikulin-Tuman-skiy R-15B-300 turbojets originally rated at 7,350 daN (7,500 kg st) dry and 10,005 daN (10,210 kg st) with afterburner. Its main elements were developed on the R-15-300 that powered the Ye-150 experimental aircraft. Unfortunately, its service life was limited to 150 hours.

The MiG-25PD was equipped with the new RP-25 radar and four pylons under the wing for two R-40 and four R-60 air-to-air missiles.

The internal fuel capacity was considerable: 17,760 l (4,689 US gallons) distributed across built-in tanks in the fuselage and wing. The air inlet control was secured by a small rectangular flap in the lower lip of the air intake and by an internal door (both actuated by electronically controlled cylinders). There was a spill door in the upper panel of the air intake duct, and both turbojets were equipped with adjustable-area nozzles.

If one overlooks the materials and the production processes, the wing had a standard box structure with two main spars attached to fuselage bulkhead nos. 7 and 9, a front spar fixed to the no. 6B frame, and two rear spars fastened to bulkhead nos. 10 and 11. The hinge pin for the flaps was also fixed to bulkhead no. 11. The plain flaps occupied one-third of the trailing edge, as well as the two-section ailerons that measured 2.72 m² (29.28 square feet). There were two fences on the upper surface of each wing: one as long as the wing chord, roughly along the aileron/flap separation line; the other much smaller, along the aileron sections' separation line. The wing leading edge had a compound sweepback: 42 degrees, 30 minutes at the wing root on half of the LE span, then 41 degrees. On the P-1 prototype the wing had neither dihedral nor anhedral, and the wing tip was fitted with downward-canted winglets. The wing structure was made of welded steel, and its

The nose of the MiG-25PDS was lengthened by 250 millimeters (9.84 inches) to house the in-flight refueling probe. The added "slice" can be seen to the right of the "45."

skin was made partly of titanium (especially on the leading edge) and partly of D19 duralumin.

The structural backbone of the fuselage consisted of fourteen bulkheads (the first one level with the cockpit windshield, nos. 13 and 14 supporting the stabilator fulcrum pins on either side of the engine nozzles) and many frames and stringers. The air intake ducts were not added up but were built-in members. The nose, made of nonconductive material, housed the dish antenna for the Smerch-A ("whirlwind") radar, which could automatically lock on and track aerial targets within 50 km (31 miles). Behind the radar was the electronics compartment and the cockpit, whose canopy was hinged to open starboard. The inner walls of the air intake duct were separated from the fuselage to form a boundary layer bleed.

The airbrakes were located at the rear of the fuselage astride the engine nozzles, one atop the fuselage immediately ahead of the tail chute canister and one underneath it; they fit the curves of the nozzle closely. Both landing lights retracted into the lower wall of the air intake ducts. The tail unit comprised two fins canted outward (11 degrees) and a slab tailplane (sweep of 50 degrees at the leading edge and 9.81 m^2 [105.6 square feet] in area). Two large ventral fins—whose size was later reduced—were located under the engine nozzles. On the prototypes a canard surface could be installed on either side of the air intake duct to act as a destabilizing device on some flight regimes. This strategy was tested on the Ye-8. The tricycle gear consisted of a twin-wheel forward-retracting nose unit and a single wheel with high-pressure tires 1.3 m (51.2 inches) in diameter on each forward-retracting main unit. Those wheels were stowed vertically in the side walls of the air intake duct.

Though it was designed first, the Ye-155P-1 made its premier flight after the Ye-155R-1 reconnaissance variant. Engineers took advantage of the knowledge acquired during the R-1 flight tests to make a number of modifications:

—the canard surface was discarded as useless
—the area of the fins was increased significantly to 8 m² (86.1 square feet) apiece
—the chord of the ventral fins was reduced subsequently (they tended to touch at landing)
—the winglets were removed, but the wing tips were fitted with an antiflutter body
—the wing was given a 5-degree anhedral (before that modification seven preproduction machines had triangular end plates at the wing tips)
—after the displacement of its fulcrum pins, the slab tailplane had a taileron capability at high speeds, meaning that the two halves could operate in unison (for pitch) or differentially (for roll)

All of these modifications increased the maximum indicated airspeed to 1,300 km/h (702 kt). As was standard practice for new aircraft, the OKB tried hard to improve its operational availability, service life, and time between overhauls.

The Ye-155P-1 was first piloted by Ostapyenko on 9 September 1964 (six months after the Ye-155R-1) but was not certified until 1970; it entered service with the VVS only in 1973, though mass production had started four years earlier. The official decree by the council of ministers commissioning the aircraft for the Soviet air force was signed on 13 April 1972. Close scrutiny of those dates indicates that the MiG-25 suffered repeatedly from childhood diseases. This is not surprising in view of the project's many innovations. There were problems with the stabilator in its taileron mode. There were problems with the ailerons. There were problems with the dangerous asymmetry noticed at high speeds whenever a single missile was fired from a wing station. Automatic trim resolved the taileron shortcomings, and all others were settled by V. Gordyenko, the LII test pilot. There were also concerns about engine TBO that could not be solved—for want of money.

The Ye-155P-1 prototype was armed with two K-40 air-to-air mis siles, but the production MiG-25P could carry four of them plus two infrared-guided R-40Ts and two radar-guided R-40Rs. Its primary equipment included the Smerch-A radar and the K-10T associated weapon pointing device, the SOD-63 ATC transponder, the SRO-2M/SRZO-2 IFF (transponder/interrogator) whose antennae were flushed in the starboard fin, the Sirena-3 360-degree radar warning receiver whose antennae were set into the center of the antiflutter bodies at the wing tips and at the top of the starboard fin, the RV-UM or RV-4 low-altitude radio-altimeter for 0–600 m (0–1,970 feet), the ARK-10 automatic direction finder, the MRP-56P marker receiver, the SP-50 ILS, the RSBN-6S short-range navigation unit, the R-832M VHF-UHF transceiver, the Prizma HF transceiver, the Lazur command receiver, and the SAU-155 automatic flight control system. The ejection seat was the KM-1 (altitude, 0 m; speed, 130 km/h [70 kt]). The MiG-25P had two tail chutes with either a circular (60 m^2 [646 square feet]) or a cross-shaped canopy (50 m^2 [538 square feet]).

Taking into account the experience acquired in the air regiments as well as technological advances, a new version of the aircraft entered production in 1978: the MiG-25PD. It used a new power plant composed of two R-15BD-300s—each rated at 8,625 daN (8,800 kg st) dry and 10,975 daN (11,200 kg st) with afterburner—whose service life was extended in stages to 1,000 hours. The Smerch-A radar was replaced by the Sapfir-25 (RP-25), which offered better performance in the automatic tracking mode and true look-down/shoot-down capabilities. The MiG-25PD's armament was supplemented, comprising now two R-40 and four R-60 air-to-air missiles. Infrared sensors were placed under the forward section of the fuselage. The range was increased significantly by an auxiliary tank attached to the underbelly (developed for the Ye-155R) that could hold 5,300 l (1,400 US gallons).

As of 1979 all operational MiG-25Ps were upgraded to the MiG-25PD standard as quickly as they could be sent to overhaul workshops. These modified aircraft received the appellation MiG-25PDS. The MiG-25PD was mass-produced until 1982. Thanks to the weapon system modifications and improvements to the airframe and engine TBOs, the PVO command announced in 1990 that the MiG-25 would be still operational at the start of the next millennium. The MiG-25 was exported to Algeria, Iraq, Libya, and Syria.

The following details refer to the MiG-25P.

Specifications
Span, 14.015 m (45 ft 11.8 in); length (except probe), 19.75 m (64 ft 9.6 in); wheel track, 3.85 m (12 ft 7.6 in); wheel base, 5.139 m (16 ft 10.3

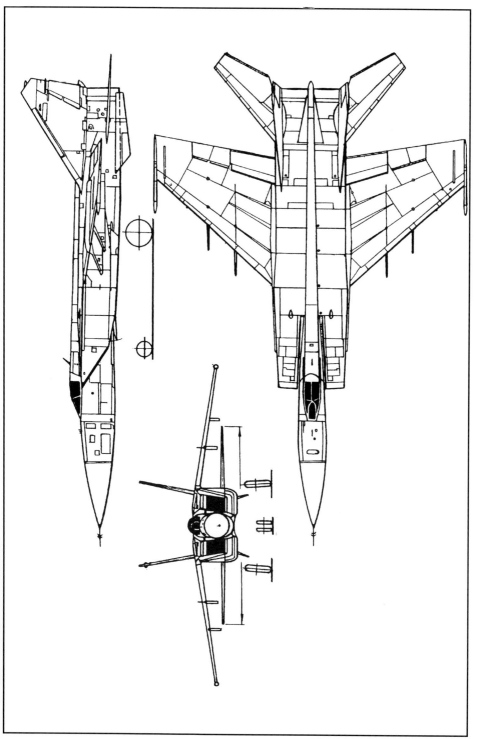

MiG-25P (MiG OKB three-view drawing)

in); wing area, 61.4 m^2 (660.9 sq ft); takeoff weight with four R-40 missiles and 100% internal fuel, 36,720 kg (80,930 lb); takeoff weight in clean configuration with 100% internal fuel, 34,920 kg (76,965 lb); fuel, 14,570 kg (32,110 lb); with 5,300-l (1,400-US gal) auxiliary tank, 18,940 kg (41,745 lb); wing loading, 598–568.2 kg/m^2 (122.6–116.5 lb/sq ft); max operating limit load factor at supersonic speed, 4.5.

Performance

Max speed, 3,000 km/h at 13,000 m (1,620 kt at 42,640 ft); max speed at sea level, 1,200 km/h (648 kt); max Mach number, 2.83; climb to 20,000 m (65,600 ft) in 8.9 min at Mach 2.35; service ceiling, 20,700 m (67,900 ft); landing speed, 290 km/h (157 kt); takeoff speed, 360 km/h (194 kt); range on internal fuel at supersonic speed, 1,250 km (775 mi); at subsonic speed, 1,730 km (1,075 mi); endurance on a coverage mission, 2 h 5 min; takeoff roll, 1,250 m (4,100 ft); landing roll with tail chute, 800 m (2,625 ft).

The Ye-266 Records

This label, invented solely for the purpose of submitting world-record corroboration documents to the FAI in Paris, applied not to one but rather to three aircraft: the Ye-155R-1, Ye-155R-3, and Ye-155P-1, all prototypes of the MiG-25 family. They were powered either by two R-15B-300 turbojets rated at 10,005 daN (10,210 kg st) with afterburner and renamed R-266 in the FAI file, where their thrust was listed at 9,800 daN (10,000 kg st), or by two R-15BD-300 turbojets rated at 10,975 daN (11,200 kg st) renamed RD and listed at 10,780 daN (11,000 kg st).

1. 16 March 1965. Speed over a closed circuit of 1,000 km (621 miles) with a 2,000-kg (4,400-pound) payload, 2,319.12 km/h (1,252.32 kt). Pilot, A. V. Fedotov
2. 16 March 1965. Speed over a closed circuit of 1,000 km (621 miles) with a 1,000-kg (2,200-pound) payload, 2,319.12 km/h (1,252.32 kt). Pilot, A. V. Fedotov
3. 16 March 1965. Speed over a closed circuit of 1,000 km (621 miles) without payload, 2,319.12 km/h (1,252.32 kt). Pilot, A. V. Fedotov
4.* 5 October 1967. Speed over a closed circuit of 500 km (310.5 miles), 2,981.5 km/h (1,610.01 kt). Pilot, M. M. Komarov. Absolute world record
5.* 27 October 1967. Speed over a closed circuit of 1,000 km (621 miles) with a 2,000-kg (4,400-pound) payload, 2,920.67 km/h

---

*These records were still standing as this book went to press.

(1,577.16 kt). Pilot, P. M. Ostapyenko

6. 27 October 1967. Speed over a closed circuit of 1,000 km (621 miles) with a 1,000-kg (2,200-pound) payload, 2,920.67 km/h (1,577.16 kt). Pilot, P. M. Ostapyenko

7. 27 October 1967. Speed over a closed circuit of 1,000 km (621 miles) without payload, 2,920.67 km/h (1,577.16 kt). Pilot, P. M. Ostapyenko

8.* 8 April 1973. Speed over a closed circuit of 100 km (62 miles), 2,605.1 km/h (1,406.75 kt). Pilot, A. V. Fedotov

9. 5 October 1967. Altitude with a 2,000-kg (4,400-pound) payload, 29,977 m (98,325 feet). Pilot, A. V. Fedotov

10. 5 October 1967. Altitude with a 1,000-kg (2,200-pound) payload, 29,977 m (98,325 feet). Pilot, A. V. Fedotov

11. 25 July 1973. Altitude with a 2,000-kg (4,400-pound) payload, 35,230 m (115,555 feet). Pilot, A. V. Fedotov

12. 25 July 1973. Altitude with a 1,000-kg (2,200-pound) payload, 35,230 m (115,555 feet). Pilot: A. V. Fedotov

13. 25 July 1973. Altitude without payload, 36,240 m (118,867 feet). Pilot, A. V. Fedotov. Absolute world record

14. 4 June 1973. Time to climb to 20,000 m (65,600 feet), 2 minutes, 49.8 seconds. Pilot, B. A. Orlov

15. 4 June 1973. Time to climb to 25,000 m (82,000 feet), 3 minutes, 12.6 seconds. Pilot, P. M. Ostapyenko

16. 4 June 1973. Time to climb to 30,000 m (98,400 feet), 4 minutes, 3.86 seconds. Pilot, P. M. Ostapyenko

MiG-25R / MiG-25RB /
MiG-25RBV / MiG-25RBT / Ye-155R / 02

The MiG-25R was a high-altitude supersonic reconnaissance aircraft cast in the same mold as the MiG-25P interceptor. It was designed and built in 1961 and 1962. Externally, the Ye-155R-1 was different from the Ye-155P-1 except for the forward fuselage (right up to bulkhead no. 1), which housed the reconnaissance systems, and the wing tip fuel tanks (capacity 1,200 l [317 US gallons]), which could not be removed because they held the winglets. The fin tips had a more square shape. The internal modifications were limited to the refurbishing of the cockpit and electronics compartment as well as the installation of additional antennae. To increase the operating range, the adjustable-area nozzles were fitted with larger flaps; the fuel capacity was increased by adding built-in tanks (350 l [92 US gallons] each) to both fins and by

---

*These records were still standing as this book went to press.

The Ye-155R-1, the first prototype of the MiG-25R. The wing—with no anhedral—had fuel tanks at the tips supporting a downward-canted winglet (the Soviets called it a flipper).

On the third prototype, the Ye-155R-3 numbered 3155, the wing tip fuel tanks were replaced by antiflutter bodies. The wing has a 5-degree anhedral.

A huge auxiliary fuel tank—capacity 5,300 l (1,400 US gallons)—could be mounted underneath the fuselage of the Ye-155R-3.

developing a huge auxiliary fuel tank (5,300 l [1,400 US gallons]) for the underbelly.

The MiG-25R-1 was first flown on 6 March 1964 by A. V. Fedotov, OKB chief pilot. It was powered by two R-15B-300s rated at 7,350 daN (7,500 kg st) dry and 10,005 daN (10,210 kg st) with afterburner. The first test flights led to a number of modifications that were introduced on the MiG-25P. These changes were made gradually. On the third prototype or Ye-155R-3 (the number 3155 tagged on its nose, it was one of the first four MiG-25s used for the Domodyedovo air display in July 1967) the wing tip fuel tanks and the winglets were replaced by antiflutter bodies, the wing chord was increased significantly, the leading edge compound sweepback was rubbed out (and replaced by a constant 41 degrees), the fin tips were given a bevel shape, and the canard surfaces were retained. Not until later was the fin area enlarged, the slab tailplane granted a taileron function, the ventral fin area reduced, and the aircraft powered by its definitive engines, two R-15BD-300s rated at 10,975 daN (11,200 kg st) with afterburner.

After passing the factory tests, the state acceptance trials, and the military acceptance inspections, the Ye-155 entered production in 1969 in the Gorki factory. It was also decided that year to give the aircraft a bombing capability; in 1970 the new version, the MiG-25RB, passed its tests and entered the production phase. Simultaneously, all MiG-25Rs already built were upgraded with retrofit kits to the standard of the MiG-25RB reconnaissance-bomber variant, which was the progenitor of many specialized subtypes such as the MiG-25RBK (k standing for Kub—"cube"—the nickname of its SLAR radar), MiG-25RBS (1972), MiG-25RBV (v standing for Virazh—"turn"—the nickname of its SLAR radar), and MiG-25RBT (1978). These later models differed only in their electronic intelligence or navigation systems.

When the basic MiG-25R (RB) was developed, the OKB had to face a number of difficult technical problems:

—the aircraft had to be capable of cruising for great distances at Mach 2.35 and flying at Mach 2.83 with its full external bomb load
—it had to be able to escape the interceptors and missiles of hostile air forces for the decade to come (1970–1980) by relying on its speed, ceiling, maneuverability, and electronic countermeasures (ECM) equipment
—a highly accurate, automatic homing bombing system had to be invented to attack ground targets at known coordinates from supersonic speeds and altitudes above 20,000 m (65,600 feet), around the clock and in any weather conditions
—a highly accurate inertial navigation unit had to be developed (the USSR's first) to tie together the DISS system (doplerovskiy izmeritel skorosti i snosa: a Doppler radar to compute ground speed and

This remarkable aerial photograph was taken near Cairo in 1971 by a MiG-25 flying at 22,000 m (72,160 feet) and 2,500 km/h (1,350 kt). The camera, with a 650-mm focal length, could cover a strip of ground equal to five times the aircraft's altitude—in this instance, 110 km (68 miles). Foreground, the pyramids.

The MiG-25RBs—both bombers and reconnaissance aircraft—were equipped with SRS-4A or 4B electronic intelligence systems.

drift) and other course correction devices; a bank of digital computers (another first) linked to the automatic flight control system initiated on a preset path the release of bombs or the activation of reconnaissance equipment

—three interchangeable bays had to be engineered to house various types of powerful high-resolution cameras capable of covering a strip 90 km (56 miles) wide

—electronic intelligence equipment had to be incorporated, such as the SRS-4A(4B) on the MiG-25R (RB) and the SRS-9 on the MiG-25RBV

—a network of ground stations had to be established to pick up the data transmitted by the aircraft

—the performance of the Peleng ("bearing") navigation system had to be improved

The MiG-25R had no armament (neither cannons nor missiles) and could rely only on its speed and ceiling attributes to escape any attacker.

For photo-reconnaissance missions, the MiG-25R might have two left-right rotating cameras in one of its three interchangeable bays. One camera could have a focal length of 650 mm and be capable of covering a strip of ground equal to five times the flight altitude, while the other might have a focal length of 1,300 mm to cover an area half that long. The two cameras shot obliquely through two port and two starboard ports. A vertical camera with a short focal length was located under the cockpit to make the linking shots.

The MiG-25RB could carry six 500-kg (1,100-pound) bombs, four under the wing and two under the fuselage. Structurally significant items were strengthened at the bomb-launcher attachment points. The MiG-25RB, RBK, and RBS were commissioned for the VVS in 1972 by the council of ministers. Those three versions as well as the MiG-25RBV were produced until 1972. The MiG-25 reconnaissance variant was exported to Bulgaria, Algeria, Syria, India, Iraq, and Egypt.* During the Iran-Iraq war, the Iraqi MiG-25Rs were upgraded to the RB standard by field service personnel.

The exceptional advantages of the MiG-25RB and RBV were greatly appreciated by their operators: extent of the ground area swept during a single flight by either the cameras or the elint equipment, high-speed long-distance flight, and near invulnerability to air defenses of the time. It is not widely known that MiG-21Rs were used by branches of the public authorities for tasks such as demarcating regions affected by forest fires, snow, or floods. They were so quick and economical that neither satellites nor aircraft built especially for aerophotogrammetry (such as the An-32) could ever compete.

Specifications
Span, 13.418 m (44 ft 0.3 in); length (except probe), 21.55 m (70 ft 8.4 in); height, 6.5 m (21 ft 3.9 in); wheel track, 3.85 m (12 ft 7.6 in); wheel base, 5.138 m (16 ft 10.3 in); wing area, 61.4 m² (660.9 sq ft); takeoff weight, 37,000 kg (81,550 lb); max takeoff weight, 41,200 kg (90,805 lb); fuel, 15,245 kg (33,600 lb); wing loading, 602.6–671 kg/m² (123.5–137.6 lb/sq ft).

Performance
Max speed, 3,000 km/h at 13,000 m (1,620 kt at 42,640 ft); max speed at sea level, 1,200 km/h (648 kt); max Mach, 2.83; climb to 19,000 m (62,320 ft) in clean configuration in 6.6 min; with a 2,000-kg (4,400-lb) bomb load in 8.2 min; service ceiling in clean configuration, 21,000 m (68,880 ft); range at supersonic speed, 1,635 km (1,015 mi); range at subsonic speed, 1,865 km (1,160 mi); range at supersonic speed with 5,300-l (1,400-US gal) auxiliary fuel tank, 2,130 km (1,320 mi); range at subsonic speed with 5,300-l (1,400-US gal) auxiliary fuel tank, 2,400 km (1,490 mi).

---

*A few months after entering service, the MiG-25R was seen in Egypt. Four aircraft of that type were delivered to Cairo by Antonov An-22 heavy cargo aircraft. The first reconnaissance missions over Israel set out from that airport in October 1971; the pilots and the field support crew were Soviets. The aircraft flew in pairs, their missions covering the Israeli coast and the Sinai Peninsula, at speeds above Mach 2.35 and altitudes above 20,000 m (65,000 feet). It was more a "loan" than a true export venture.

The first supersonic business jet could have belonged to the USSR. It was planned some twenty-five years ago, and it owed a great deal to the MiG-25.

MiG-25PD SL

Experience acquired with the MiG-25PD and PDS interception versions demonstrated that the aircraft could be operated at low and medium altitudes provided that they could be equipped with active jammers and IR countermeasures. One aircraft was modified and referred to as the MiG-23PD SL but did not go beyond the prototype stage.

A "Bizjet" Derivative of the MiG-25P

As development of the MiG-25 continued, various spinoff projects were considered. One of the most interesting—and unexpected—of these was a business jet (to be more precise and respect the Soviet terminology, "administrative" jet) designed between 1963 and 1965 to carry six passengers or an equivalent cargo load.

Even though Aeroflot might have evinced some interest in this aircraft, it was from the outset the OKB's project; but it was more than a stylistic effort, and the design was carried to an advanced stage. One of

On the MiG-25PU and MiG-25RU trainers, unlike the MiG-23UB, the flight instructor
sat in the front cockpit.

the main specifications required maximum commonality between the
bizjet and the MiG-25P interceptor. In fact, only the forward fuselage
had to be modified. This part of the aircraft was to be significantly
lengthened, its master cross-section increased, and the cockpit located
still farther forward than that in the front of the MiG-25PU trainer.

The cabin was planned to accommodate six passenger seats in a
straight line along a narrow aisle with the entrance door on the left side
of the fuselage, just behind the cockpit. This way it could quickly be
converted into a cargo hold. The master cross-section increase offered
another advantage: it increased the fuel capacity and consequently the
range to 3,000–3,500 km (1,860–2,175 miles) at cruising speeds of 2,500
km/h or Mach 2.35 (1,350 kt).

MiG-25PU / 39 / Ye-133   MiG-25RU / 22

During the flight tests of the family's basic aircraft, the MiG-25P and
MiG-25R, the pilots reported that—because of the peculiar flight envelope of those machines, especially its speed and altitude components—
it was imperative to build a separate two-seat trainer for each of the
two central missions, interception and reconnaissance. That is how
the MiG-25PU and RU project started. The first one was completed in
1968, the second in 1972.

Both versions were unarmed and had no combat capabilities. In
the preliminary design the OKB emphasized the greatest possible commonality not only between the two aircraft but also between the trainers and the basic versions. The MiG-25PU differed from the MiG-25RU
only by its cockpit firing simulator. Unlike in the MiG-23 two-seater,
the instructor was seated in the front cockpit. All control systems as
well as a number of other systems (static and dynamic pressure indica-

This MiG-25PU was in fact the Ye-133, which set several world records between 1975 and 1978.

tors, automatic flight control system, power plant and aircraft controls, fluid cooling units, air-conditioning ducts, and the like) were updated and modified to make possible all kinds of simulated failures. After improvements to the ejection devices, the instructor could himself eject the student pilot in an emergency. The performance data of the MiG-25PU and RU did not diverge greatly from those of the basic versions except for the never-exceed Mach number (MNE), lowered from 2.83 to 2.65 as a safety measure.

Ye-133 Records
Several female world records were beaten by a MiG-25PU renamed Ye-133 for this purpose and piloted by Svetlana Ye. Savitskaya.

1. 22 June 1975. Speed over a 15- to 25-km (9- to 15-mile) course at unrestricted altitude, 2,683.446 km/h (1,546.26 kt)
2. 31 August 1977. Altitude in horizontal flight, 21,909.9 m (71,864.47 feet)
3. 21 October 1977. Speed over a closed circuit of 500 km (310 miles), 2,466.31 km/h (1,331.81 kt)
4. 12 April 1978. Speed over a closed circuit of 1,000 km (621 miles), 2,333 km/h (1,259.8 kt)

MiG-25RBK / 02K
MiG-25RBS / MiG-25RBSh / 02S
MiG-25RBF / 02F

These aircraft were subtypes of the MiG-25R, equipped with the next generation of reconnaissance systems; but the MiG 25RBK project had received approval at the same time as the MiG-25R. The aircraft passed its combined state acceptance trials (reconnaissance and bomber missions) and was produced in the Gorki factory between 1971 and 1980. The RBS variant was launched a little later, in 1965; it passed its combined acceptance tests and was produced between 1971 and 1977. The MiG-25RB, RBK, and RBS were commissioned simultaneously.

More advanced systems were developed for the MiG-25RBS in the early 1980s. After this upgrade in 1981 the MiG-25RBS was renamed MiG-25RBSh, and all of the MiG-25RBSs were gradually brought up to the RBSh standard with retrofit kits as they came in for their major overhauls. The main difference between the MiG-25RB and the newer variants was the specialized electronic systems that replaced the cameras, and the modifications made to the cockpit, the power supply, and the air-conditioning unit. MiG-25RBFs (1981) were RBs that were updated to the RBK standard but still differed in some systems such as panoramic cameras, active and passive countermeasures, and the like. The specifications and performance of those subtypes were practically identical to those of the MiG-25RB.

Ye-155M / 99 / Ye-266M / Experimental Versions

While confirming the acceptance of the MiG-25RB, the decree signed by the council of ministers in 1972 outlined the path of future updates for the MiG-25 family. The VVS command was already asking for a range increase at medium and high altitudes, as well as more speed and a higher service ceiling. The Mikulin-Tumanskiy OKB proposed the R-15BF-2-300, an upgraded R-15B-300 rated at 13,230 daN (13,500 kg st) with afterburner—an increase of 3,225 daN (3,290 kg st)—that retained the size and connection points of the existing engine and reasonable specific fuel consumption.

Development of the new aircraft was to happen in two stages. First the range and rate of climb would be enhanced without structural modifications. The aircraft would be reengined after their operational life expired—a sure way to grow younger. Second, the aircraft structure would be modified, removing the little duralumin still used in the forward fuselage and the few non-heat-resisting wing elements so that the aircraft could fly at speeds above Mach 3. The MiG-25's never-exceed Mach number (MNE) of 2.83 was in fact somewhat theoretical: the lateral stability margin and the structural lifetime were supposed to diminish beyond that figure, but a number of pilots have (more or less intentionally) exceeded Mach 3 without causing damage to the aircraft or sending it to the overhaul shop to check for structural yielding.

The first stage was carried to a successful conclusion. The factory designation of the new product was Ye-155M, but the certification documents sent to the FAI after several record attempts in 1975 and 1977 called it the Ye-266M. Unfortunately, the excessive engine development time and the lack of factory availability delayed the second stage of the upgrade; as a result these modifications either remained experimental or did not go beyond the computational phase.

Nevertheless, the results obtained during the first step were very encouraging compared with the MiG-25P or R performance. The service ceiling increased to 24,200 m (79,375 feet), and the range at supersonic speed to 1,920 km (1,190 miles)—2,510 km (1,560 miles) if one adds the auxiliary tank's 5,300 l (1,400 US gallons). Another R & D channel consisted of powering the Ye-155M with two D-30F turbofan engines rated at 15,190 daN (15,500 kg st) with afterburner. It was developed by P. A. Solovyev out of the core engine of the D-30, the power plant capable of 6,665 daN (6,800 kg st) that had powered the Tupolev Tu-134 twin-jet airliner since 1963.

This engine change led to significant structural modifications that did not, however, change the aircraft's silhouette drastically; and the fuel capacity was raised to 19,700 l (5,200 US gallons). Two prototypes were constructed with two D-30Fs. They were used essentially as test beds for developing the engine that would later power the MiG-31. The takeoff weight of this variant reached 37,750 kg (83,200 pounds), the maximum takeoff weight 42,520 kg (93,715 pounds), and the internal fuel weight 16,270 kg (35,860 pounds). Due to the turbofan's better specific fuel consumption its range on internal fuel reached 2,135 km (1,325 miles) at supersonic speeds and 3,310 km (2,055 miles) at subsonic speeds. Its service ceiling topped out at 21,900 m (71,830 feet).

Ye-266M Records

The documents sent to the FAI showed that the Ye-266M was powered by two turbojets rated at 13,720 daN (14,000 kg st). In fact, the aircraft was powered by two R-15BF-2-300 turbojets at 13,230 daN (13,500 kg st). These six world records (including one absolute world record), established more than fifteen years ago, were still standing as this book went to press.

17 May 1975
Time to climb to 25,000 m (82,000 feet), 2 minutes, 34.2 seconds. Pilot, A. V. Fedotov
Time to climb to 30,000 m (98,400 feet), 3 minutes, 9.85 seconds. Pilot, P. M. Ostapyenko
Time to climb to 35,000 m (114,800 feet), 4 minutes, 11.7 seconds. Pilot, A. V. Fedotov

22 July 1977
Altitude with a 2,000-kg (4,400-pound) payload, 37,080 m (121,622 feet). Pilot, A. V. Fedotov
Altitude with a 1,000-kg (2,200-pound) payload, 37,080 m (121,622 feet). Pilot, A. V. Fedotov

31 August 1977
Altitude without payload, 37,650 m (123,492 feet). Pilot, A. V. Fedotov. Absolute world record

MiG-25BM / 02M

The goal of this project was to develop (from the MiG-25RB) an aircraft capable of destroying the enemy's air defenses, especially ground radars. Ordered by a decree of the council of ministers in 1972, the 02M product was equipped with powerful electronic countermeasures and Kh-58 antiradiation missiles. Those missiles took the place of the bombs under the wing pylons, and the elongated nose housed the ECM equipment. The cockpit instrumentation, the aircraft's power supply, and the air-conditioning system had to be modified because of the new missions.

The weights and performance of the MiG-25BM were practically identical to those of the MiG-25RB. After passing its certification tests, the aircraft was produced in the Gorki factory between 1982 and 1985.

MiG-31 Series

MiG-25MP / Ye-155MP / 83   MiG-31 / 01

The MiG-31 was intended to counter a very specific threat: that of American B-52 bombers carrying long-range cruise missiles, each bomber representing several potential dangers all by itself (and conse-

The Ye-155MP numbered 831—product 83, no. 1—was neither a MiG-25 nor a MiG-31. It has the latter's characteristic landing gear, but the airbrakes had to be moved and the wing strengthened.

The MiG-31 takes shape in its definitive silhouette. The airbrakes were moved under the air intake ducts. The staggered twin wheels of the main gear were distinctive.

The fairing for the 23-mm GSh-6-23 Gatling-type cannon is visible beside the air intake duct, under the wing leading edge root extension.

quently several targets). The future MiG-31 was to be capable of destroying multiple invaders at high or low altitudes in the forward and rear sectors and providing true look-down/shoot-down capability whatever the weather conditions, even if the invaders try to maneuver and use active countermeasures.

At the start the prototype was referred to as the Ye-155MP. On the aircraft's nose the number "831" was painted to indicate that it was the first example of the *izdeliye* 83, MiG's internal product number. The airframe of this prototype was closely related to that of the Ye-155M. This was a sturdy, time-tested structure, but the proportions of its metallic components were altered somewhat to 50 percent steel, 16 percent titanium, 33 percent duralumin, and a negligible 1 percent composite materials (including the radome). The new aircraft did not need to be faster than the MiG-25P/PD, but it did need to offer a longer range. This explains why the new aircraft—which was ten metric tons heavier than the MiG-25P/PD—was powered by two Solovyev D-30F6s rated at 9,310 daN (9,500 kg st) dry and 15,190 daN (15,500 kg st) with afterburner, the turbofan flight-tested on the Ye-155M.

In what ways did the MiG-25MP *izdeliye* 83 differ from the MiG-25 *izdeliye* 99?

1. It was a two-seater; a second crew member (the flight engineer) was needed to help manage its avionics
2. Both airbrakes were located under the air intake duct's outboard corners (in front of the main gear doors) and were obliquely hinged

3. The main gear retracted forward, but the single wheels were replaced by staggered twin wheels arranged so that the rear wheel never followed the furrow of the front wheel; because of the aircraft's weight it was important to distribute loads carefully, taking typical Russian weather conditions into account (winter snows and spring slushes)
4. Unlike the MiG-25, flaps and ailerons took up the whole trailing edge

The Ye-155MP was first piloted on 16 September 1975 by A. V. Fedotov, but four long years of tests were needed before starting production in the Gorki factory in 1979. Those tests led to several significant modifications before the MiG-25MP reached the *izdeliye* 01 MiG-31 production stage:

1. The whole of the wing's leading edge was fitted with slats in four sections
2. Small, sharply swept leading edge root extensions were added
3. The aircraft was equipped with a semiretractable refueling probe on the port side of the nose cone
4. The wing box was strengthened by a third main spar so that the MiG-31 could fly at high supersonic speeds near the ground
5. Both airbrakes were moved squarely under the air intake ducts and hinged in the vertical plane

All of the MiG-31's combat capabilities relied on its interception system, which consisted of the S-800 Zaslon ("flanker") phased array look-down/shoot-down radar in the nose, the infrared search-and-track device in a semiretractable pod under the nose, and the tactical situation display. The radar had effective ranges of 200 km (125 miles) in the forward clutter-free sector and 120 km (75 miles) in look-down mode. In the rear sector those figures are reduced to 90 km (56 miles) and 70 km (43 miles), respectively. Ten targets could be tracked simultaneously, and up to four could be simultaneously engaged. The simultaneous lock-on and firing sectors covered plus-or-minus 70 degrees in azimuth and –60 to +70 degrees in elevation.

As regards navigation, the position-finding accuracy was not influenced by the system's time in service. For long distances the Marshrut ("itinerary") system—similar to the West's Omega—was accurate to between 2 and 5 km (1.25 and 3.10 miles); this margin narrowed to between 1.8 and 3.6 km (1.1 and 2.2 miles) for flight distances between 2,000 and 10,000 km (1,240 and 6,210 miles). For medium-range navigation the Tropik—similar to the West's Loran—was accurate to between 250 and 1,300 m (820 and 4,265 feet); this margin narrowed to

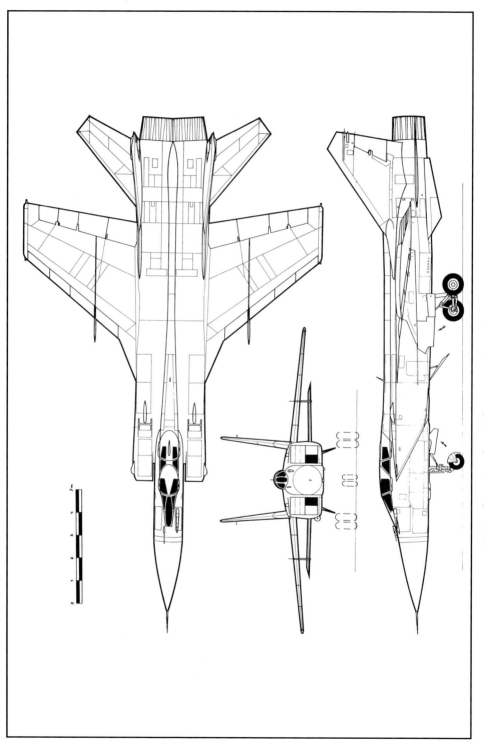

MiG-31 (three-view drawing by Jean Molveau from MiG OKB documents)

The four long-range R-33 air-to-air missiles hang in pairs under the fuselage of the MiG-31.

The complex structure of the MiG-31's main gear. *Foreground*, the two wheel axles.

between 130 and 1,300 m (425 and 4,265 feet) for flights under 2,000 km (1,240 miles).

The combat capability of the MiG-31 also relied upon information transmitted by a ground network of automatic guidance stations (ASU) operating in the following modes: remote guidance, semiautonomous guidance (coordinated support), isolated operation, and group operation (a flight of four interceptors exchanging information automatically). This latter mode requires some explanation. The leader is the only aircraft linked to the ground automatic guidance network designated AK-RLDN, but it can exchange information with the other three aircraft. Each aircraft is kept 200 km (124 miles) apart. The four are therefore lined up to cover a space 600 km (372 miles) wide; but because of their radar scanning angles (140 degrees) and the overlap of the scanned sectors, the overall zone swept by the four aircraft is 800–900 km (495–560 miles) wide.

The APD-518, a powerful digital data signaler, gives the leader and the wingmen a continuous flow of exchangeable information. All problems of guidance, target identification, and coordination between interceptor flights were then managed by the MiG-31's avionics via the automatic data exchange between aircraft. The target allotment—just before attacking—is carried out by the flight leader according to the information provided by the tactical situation display.

The MiG-31's armament was quite impressive:

1. Air-to-air missiles. Four long-distance (110-km [68-mile]) radar-guided R-33s under the fuselage, or four R-33s and two medium-range infrared-guided R-40Ts, or four R-33s and four short-range infrared-guided R-60Ts, or two R-60s and two 2,500-l (660-US gallon) drop tanks under four wing pylons. A new method of carrying the four bigger missiles—placing them in pairs beneath the fuselage, one behind the other—reduced drag considerably. Before firing, the missiles split off from the aircraft by means of special AKU ejecting pylons. The missile's engine is ignited and its homing device enabled once it is a safe distance out of the aircraft's flight path.

2. Fixed armament. One Gatling-type 23-mm GSh-6-23 cannon with 260 rounds, fed by a linkless ammunition belt. The rate of fire is presently 6,000 plus-or-minus 500 rounds per minute (but should later be raised to 8,000 rounds per minute), and the initial speed is 700 meters per second (2,300 feet per second). This weapon is located on the right side of the fuselage behind the main landing gear.

The aircraft entered production in 1979 at the Gorki factory, and the first MiG-31–equipped regiments were operational by 1982. The

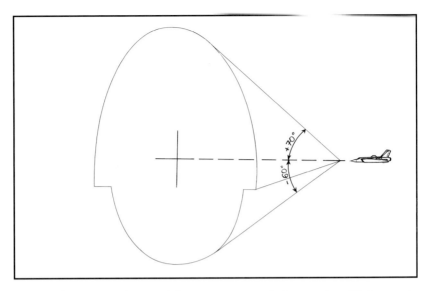

The simultaneous tracking and fire sectors covered plus-or-minus 70 degrees in azimuth and –60 to +70 degrees in elevation.

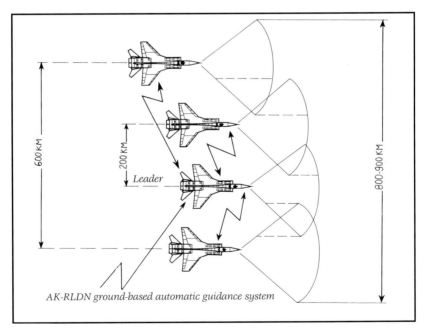

AK-RLDN ground-based automatic guidance system

Only four MiG-31s are needed to sweep a zone 800–900 km (495–560 miles) wide. The leader is the only aircraft linked to the ground stations, but all information can be exchanged between the four aircraft via the APD-518 automatic data signaler.

MiG-31 project was managed by R. A. Belyakov, assisted by bright team
members such as G. Ye. Lozino-Lozinskiy, V. A. Arkhipov, K. K.
Vasilchenko, and A. A. Belosvyet.

Specifications
Span, 13.464 m (44 ft 2.1 in); overall length, 22.688 m (74 ft 5.2 in);
height, 6.15 m (20 ft 2.1 in); wing area, 61.6 m² (663 sq ft); empty
weight, 21,820 kg (48,105 lb); takeoff weight with 100% internal fuel,
41,000 kg (90,365 lb); internal fuel, 16,350 kg (36,035 lb); takeoff
weight with 100% internal fuel and two 2,500-l (660-US gal) drop
tanks, 46,200 kg (101,825 lb); wing loading, 665.6–750 kg/m²
(136.5–153.8 lb/sq ft); max operating limit load factor at supersonic
speed, 5.

Performance
Max speed, 3,000 km/h at 17,500 m (1,863 kt at 57,400 ft); max speed
at sea level, 1,500 km/h (810 kt); max cruising speed, Mach 2.35; economical cruising speed, Mach 0.85; service ceiling, 20,600 m (67,570
ft); climb to 10,000 m (32,800 ft) in 7.9 min; landing speed, 280 km/h
(151 kt); ferry range with 100% internal fuel and two 2,500-l (660-US
gal) drop tanks, 3,300 km (2,050 mi); max endurance with two 2,500-l
(660-US gal) drop tanks, 3.6 h; endurance with one in-flight refueling,
6–7 h; radius of action at Mach 2.35, 100% internal fuel, and four R-33
missiles, 720 km (448 mi); at Mach 0.85, 100% internal fuel, and four
R-33 missiles, 1,200 km (745 mi); at Mach 0.85, 100% internal fuel,
four R-33 missiles, and two 2,500-l (660-US gal) drop tanks, 1,400 km
(870 mi); at Mach 0.85, same conditions with one in-flight refueling,
2,200 km (1,365 mi); takeoff roll at 46,200 kg (101,825 lb), 1,200 m
(3,935 ft); landing roll, 800 m (2,625 ft).

MiG-31M / 05

Still listed as highly classified when the French edition of this book was
published in 1991, the MiG-31M was first shown publicly in February
1992. It looks very much like the MiG-31, but after looking closely one
notices many differences. After all, the basic MiG-31 was developed
with early-1970s technology; it is no wonder that serious updates were
needed.

First known within the OKB as *izdeliye* 05, the new aircraft has kept
most of the MiG-31 structure. The most striking change involves the
complete reshaping of the cockpit and the dorsal spine running aft

The 057 identification number on this MiG-31M means that it is the seventh prototype of "product 05." The one-piece windshield greatly improves the pilot's view. (Photo RR)

Seen from this angle, the most noteworthy features of the MiG-31M are the deeper and broader dorsal spine, the greater area of the rudders, and the ECM/ECCM wing tip fairings. (Photo RR)

from it, which is now broader and deeper—no doubt saddle tanks were added to increase the internal fuel capacity. The three-window windshield has been replaced by a single-piece of rounded glass that greatly improves the pilot's view. But surprisingly, the systems operator has only two small view ports. Other modifications include a greater leading edge root extension, a reduction of the wing fence's depth on the upper surface, and the addition of a "Christmas tree" of dielectric panels and bodies to the wing tips (possibly to house a number of ECM and ECCM antennae). Those wing tip units are practically identical to those tested some twenty-nine years before on the first seven preproduction MiG-25Ps, but at that time they were used more as winglets and anti-flutter bodies.

The height of the fins was increased slightly. Because the rudders were enlarged as well, the base of these surfaces is not inserted into the fin any longer. All systems have been upgraded, and the aircraft is now fitted with digital flight controls, cathode-ray-tube (CRT) multifunction displays in the cockpit, and a more advanced phased-array radar (developed by Fazatron) that has a bigger diameter, altering both the shape and volume of the radome. The semiretractable refueling probe is now located on the right side of the fuselage.

Armament of the MiG-31M includes six long-range air-to-air missiles beneath the fuselage and four new medium-range RVV-AE air-to-air missiles on the four wing pylons. Apparently the MiG-31M has no additional cannons.

105-11

The MiG OKB was handed the task of developing the EPOS experimental manned orbit vehicle, a delta-body aircraft, designed within the context of the Spiral space project managed by G. Ye. Lozino-Lozinskiy since 26 June 1966 to observe the handling characteristics and to study the abandon-orbit and landing procedures of the future Soviet space shuttle. The wing, the fin, and the flaps were all set at the rear of the lifting body. The retractable gear had four skid-equipped legs, and the pilot was seated in a pressurized capsule.

Seen on a planform view, the lifting body/fuselage revealed a sweepback of 78 degrees at the leading edge. Seen on a sectional view, its upper part was distinctly rounded—but the base was practically flat. The front part of the rounded body, as the site of airflow impact, was rather bulky. The shapes of the lifting body, the wing, and the fin were designed for optimum performance whatever the flight regimes and permissible skin temperatures generated by frictional heating. The

The 105-11 was an experimental prototype built to clear the way for the future Soviet space shuttle. Its role was to assess handling, abandon-orbit, and landing procedures.

fuselage had sufficient internal space to house all of the systems as well as the necessary test meters.

The fuselage structure was composed of three main parts:

1. The external sheet and extruded sections reinforced the articulated heat shield; the latter could lose its shape to any direction, could not generate any thermal stress in uneven heat, and was lined with insulating material
2. The structure itself, composed of tubes and extrusions, withstood all stresses; the shield was attached to this structure, and so were the capsule, power plant, wing, fin, and equipment racks
3. The removable panels included the escape hatch (the inspection holes providing access to the equipment), the access door to the turbojet air intake duct, and the lateral fuselage panels

According to flight regime, the wing panels (55-degree sweepback at the leading edge) could be rotated and set at an angle between 90 and 60 degrees off the vertical. The fin and rudder—area, 1.7 m² (18.3 square feet); leading edge sweepback, 60 degrees—were attached on the top of the turbojet bay. The airbrakes were hinged on the upper surface of the rear fuselage.

The flying controls (elevons and rudder) were manually operated. The control column and the rudder pedals were of the standard variety. The turbojet was controlled by a throttle lever; the jet reaction control nozzles were electrically controlled by a special lever. The manual controls worked with the automatic controls of the SNAU (automatic navigation and control system). The wings were rotated by an electric engine through a ball-screw actuator. The turbojet's air intake shutter was controlled by a two-position pneumatic cylinder.

The aircraft's angle of attack was quite high at landing. The rear skids touched the ground first, before the aircraft tipped forward onto the front skids. The four struts of the gear were fitted with shock absorbers. The front legs retracted into the lateral fuselage panels, above the heat shield, while the rear legs retracted into the rear part of the fuselage.

The compressed air necessary to extend or retract the gear and the flaps was stored in the front oleo struts. The cockpit consisted of a pressurized metallic capsule lined with insulating material. The rear part of the pilot's capsule was protected by a heat shield close to the "emergency exit" in the atmosphere. The cockpit's glass panels offered sufficient outward vision in orbit, on approach, and at landing. The capsule was mounted on two rails anchored in the fuselage structure and had a pyrotechnic ejection device. The pilot could decide to eject at any time during the flight, from takeoff to landing.

The jet reaction control nozzles (GDU) could be used in orbit or in the earth's atmosphere at supersonic and hypersonic speeds. The nozzles were located at the rear of the fuselage on the sides of the power units and were protected by fairings. Each power unit fed three large nozzles rated for 15.68 daN (16 kg st) and five smaller nozzles for 0.98 daN (1 kg st); there were therefore six large and ten small nozzles in all. Four of the large nozzles, set vertically, controlled the pitch and roll axis just like elevons; the other two, set horizontally, controlled the yaw axis. To provide perfect orbit stabilization, the smaller nozzles were distributed according to the control channels: four in pitch, four in roll, and two in yaw. The nozzles were controlled by electric valves that received signals from the SNAU unit and the lever.

The rocket engine was intended for performing maneuvers in orbit and for braking to abandon orbit. With its group of turbopumps, it was rated at 1,470 daN (1,500 kg st). It was located in the rear fuselage, and its thrust force was directed to the aircraft's center of gravity. It had two auxiliary combustors capable of 39.2 daN (40 kg st) that could be used to brake and abandon orbit in case of main engine failure. The propellant tanks were located in the fuselage's center section near the aircraft's center of gravity.

The Kolyesov RD-36-35K turbojet, rated at 1,960 daN (2,000 kg st), was used at takeoff up to Mach 0.8 as well as at landing. A fairing located between the upper part of the fuselage and the base of the fin housed this turbojet. The air intake duct was plugged by a cylinder-operated flap that was opened just before the engine started. The fuel tanks were located in the center section ahead of the center of gravity. All systems were housed in two containers flanking the rear fuselage; for access, one simply removed the containers. Inside there, normal operating conditions were maintained (pressure, 760 mm Hg; temperature, 10–50° C [50–122° F]).

For the first test phase, planned for 1974, the OKB built a full-scale experimental machine. Referred to as the *izdeliye* 105-11, it was unlike the orbital plane in several respects:

- neither the rocket engine controls nor the jet reaction control nozzles were installed
- the electrical governors of the control nozzles were replaced by the ARS-40 electrohydraulic actuation unit
- for the ground-roll and leapfrog tests, the front skids were replaced by wheels
- the instrument panel was equipped with standard instruments (gyrocompass, altimeter, and the like)

In 1975 the turbojet was revved up on one of TsIAM's test benches (ignition tests: V = 300 km/h [162 kt], α [angle of attack] = −5°/+20°, ß [sideslip angle] = ±5°). Once all of these experiments were carried out successfully, it was time to start the flight-test phase. Between 1976 and 1978 the 105-11 made a series of ground rolls and actually went a few feet into the air, enabling the pilot and the engineers to assess its stability and handling qualities as well as the ground's effect on its maneuverability.

On 11 October 1976 the MiG test pilot A. G. Festovets took off in the 105-11 (with half-skid/half-wheel gear), climbed to 560 m (1,835 feet), and landed a few minutes later at another airfield 19 km (12 miles) away. After the data was analyzed it was decided to proceed with the basic tests in order to sharpen the approach path for final and landing sequences. On 27 November 1977 the 105-11 with Festovets at the controls was released from a Tupolev Tu-95K bomber at 5,000 m (16,400 feet) and landed on a specially prepared unpaved strip.

The aircraft flew only eight times between November 1977 and September 1978, but this was enough to assess its subsonic flight envelope. After full analysis of the test results, it was decided to proceed with the project. The 105-11 can be seen today in the VVS museum on the Monino airfield near Moscow.

Specifications
Wing area, 6.6 m² (71.04 sq ft); lifting body/fuselage area, 24 m² (258.3 sq ft); empty weight, 3,500 kg (7,715 lb); takeoff weight, 4,220 kg

(9,300 lb); landing weight, 3,700 kg (8,155 lb); fuel + oil, 500 kg (1,100 lb); wing loading, 640 kg/m² (131.2 lb/sq ft).

Performance
Landing speed, 250–270 km/h (135–146 kt).

MiG-29 Series

MiG-29 / 9-01 to 9-12

After the MiG-23 multipurpose fighter and the MiG-25 interceptor, the MiG engineers became aware of new trends in fighter development and studied the lessons of two decades of local military conflicts. They focused their thoughts on the next-generation aircraft: a highly maneuverable frontline fighter that would remain true to the long-standing MiG tradition of the MiG-15 and MiG-21. This aircraft had to be the end

The 9-01, the first prototype of the MiG-29, without the fin extensions forming ventral fins (à la Su-27) that normally characterize it.

A large mudguard was necessary on the 9-01 because of the forward location of the front gear leg, which greatly increased the risk of foreign objects being sucked into the engine air intakes.

At first the MiG-29 was armed with a twin-barrel cannon, hence the two ports visible on the port side of the fuselage's blended area.

product of the latest advances in technology, supplemented by the OKB's customary savoir-faire.

Some very peculiar external considerations influenced their decisions. The new fighter was intended to counter a trio of American fighters developed during the 1960s and 1970s: the F-15, F-16, and F-18. Needless to say, the engineers had set for themselves the goal of developing an aircraft that would not only meet the requirements of the Soviet air force but also become a tough competitor on the international market. Their aim would be achieved.

The LFI project (*Legkiy Frontovoy Istrebityel*: frontline light fighter) was launched in the early 1970s. The specification asked for a fighter capable of the following tasks:

1. Destroying hostile fighters in air combat, thus demonstrating air superiority
2. Destroying hostile aircraft, attacking troops and targets on the front line
3. Destroying enemy reconnaissance, AWACS, and ECM aircraft
4. Protecting aircraft bound for other missions
5. Opposing all of the enemy's aerial observation assets

The aircraft should also be available for reconnaissance missions and for ground support missions against small targets, under visual flight rules and with such weapons as bombs, rockets, and cannons.

The LFI gave rise to a number of quite different proposals for the aircraft's overall architecture. One called for lateral air intakes to feed two turbojets housed in the fuselage, à la MiG-25. As is now widely recognized, the layout finally selected was particularly original, and no fewer than nineteen prototypes (numbered 9-01 to 9-19) were needed to develop the engine, all of the systems, and some variants.

The MiG-29's architecture was called "integral aerodynamic design" by the Soviets. It had no fuselage, or at least nothing recognizable as one by contemporary standards; but it could carry a lot of weight (twice what the previous generation of fighters could), had a wide flight envelope, and was capable of sustained maneuvers up to 9 g. The high thrust-to-weight ratio (1.1) allows excellent takeoff performance and vertical climbs in building up speed. Its rational architecture, thrust-to-weight ratio, and safe automatic flight control system give the MiG-29 outstanding maneuverability. Piloting the aircraft is simple and "comfortable"—an achievement that can be credited to several of R. A. Belyakov's close assistants, namely, A. A. Chumachenko, V. A. Lavrov, and M. R. Valdenberg.

The wing shows a leading edge sweep of 42 degrees on outer panels and has a 3.5 aspect ratio. Its outstanding lift capability is the result of various lift devices: wing twist, computer-controlled maneuvering flaps (programmed for different flight regimes) across the leading edge, and plain trailing edge flaps. The aerodynamic design helps to unload the wing significantly, because 40 percent of the effective lift is provided by the lift-generating center part of the fuselage.

The wing contains two integral tanks, each with a capacity of 350 l (92 US gallons). The center body—it should be considered more a lifting body than a fuselage—consists of (from nose to tail) the radar and its radome, the forward electronics compartment, the cockpit and the lower electronics compartment, the rear electronics compartment, fuel tank no. 1 (705 l [186 US gallons]), fuel tank no. 2 (875 l [231 US gallons]), fuel tank no. 3 (1,800 l [476 US gallons]), the engine bay, and fuel tank no. 3A (285 l [75 US gallons]). The center body's leading edge features a sweep angle of 73 degrees, 30 minutes. In the center of the boat-tail and the engine nozzle throats, the tail chute is placed in the middle of the jaws formed by the two-cylinder actuated, forward-hinged airbrakes (one opens upward, the other downward). Tanks nos. 1 and 3 form a major carry-through double stress box of the body structure that is made of aluminum-lithium, measures 3,105 x 3,000 x 830 mm (122.2 x 118.1 x 32.7 inches), and weighs 220 kg (485 pounds).

The cockpit is fitted with the 10-degree-inclined K-36DM zero/zero ejection seat, and the pilot has a forward angle of vision limited to 14 degrees. Three internal mirrors provide rearward view. The vertical tail surfaces are carried on slim booms alongside engine nacelles. The tail fins are canted 6 degrees outward. Their leading edge (featuring a sweep angle of 47 degrees, 50 minutes) can be extended forward to form overwing fences that contain BVP-30-26M flare launchers.

Compared to that of the prototype and first-series aircraft, the rudder was made larger by extending its chord. Some of the prototypes, including the 9-01 and a few production aircraft, had two ventral fins; they were quickly removed. The all-moving horizontal tail surfaces on either side of the booms that carry the fins travel between +15 and –35 degrees (either symmetrically or differentially). The total span of the horizontal tail surfaces is 7.78 m (25 feet, 6.3 inches), and their sweep angle at the leading edge is 50 degrees.

The gear is of the retractable tricycle type. The nose gear has two driven wheels (tires 570 x 140 mm) and retracts rearward between the engine air intakes. On prototype 9-01 the longer front leg was hinged farther forward—practically under the pilot's seat—and held gear door elements. There was a big mudguard to the rear of the prototype's nosewheels that was later reduced in size to avoid any cramming effect. The nosewheels are steerable: plus or minus 8 degrees for taxiing, takeoffs, and landings, plus or minus 30 degrees for slow ground maneuvers. The main gear retracts forward into the wing roots, the

This interesting photograph shows the unexpected comeback on prototype 9-17 of ventral tail fins, which never seemed to disappear once and for all.

This first-series MiG-29 carries six air-to-air missiles—two R-27s and four R-60s. Note the peculiar shape of the mudguard; it will later be modified.

wheels (tires 840 x 290 mm) turning 90 degrees to lie flat above the legs. The gear is hydraulically powered and can be extended mechanically in case of emergency.

The power plant consists of two RD-33 two-spool turbofans developed by the Klimov OKB. Each is rated at 4,940 daN (5,040 kg st) dry and 8,135 daN (8,300 kg st) with afterburner and fed by two ducts that reveal a very distinctive feature. Because this fighter was designed to operate from rough strips near the front line, the engine ducts are canted slightly and have wedge intakes that stand away from the lower part of the center body to form a boundary layer bleed. They also house a multisegment ramp that allows the pilot to modify the ducts' size to suit the aircraft's speed and flight conditions up to the maximum indicated airspeed of 1,500 km/h (810 kt) or Mach 2.3. The overall internal fuel capacity—4,365 l (1,153 US gallons)—can be complemented by a 1,500-l (396-US gallon) auxiliary tank located under the fuselage, between the engine ducts.

This ramp device includes a top-hinged, perforated forward door that closes the duct while the aircraft is taxiing, taking off, or landing. At that point the engines are fed mainly via louvers on top of the outer parts of the center body and via the perforations of the door. When the aircraft's speed reaches 200 km/h (108 kt) at takeoff, the nose gear's shock strut expands and opens the door; compression of the same shock strut at touchdown closes the door. Both engines are thus protected against foreign object damage (FOD). The louvers also have an air inlet control function, sometimes disymmetrical, and behind them are three lattice ports that are in fact spill doors.

Both engines drive the accessory gearbox. The GTDE-117 auxiliary power unit is a small turbine engine weighing 40 kg (88 pounds) and delivering 98 ch equivalent shaft horsepower to start the turbofans or 70 ch for other duties. The air scoop for this APU can be seen above the rear fuselage on port side. Exhaust passes through the underbelly fuel tank when in place. There is a single-point pressure refueling through receptacle in the port wheel well but there are also overwing receptacles for manual gravity fueling.

The mechanical flying controls, hydraulically powered and outstandingly efficient, ensure steadiness throughout the flight envelope. The pilot thus can reach the appropriate angle of attack and load factor quickly—a point of the utmost importance for a fighter. Moreover, they can be serviced by field support crews. This flight control system includes an AOA limiter set at 26 degrees to prevent spins and roll-offs and to maintain control of the roll and pitch attitude. In symmetrical maneuvers that do not involve banking, an AOA of 30 degrees can be safely reached.

During flight demonstrations of the MiG-29 at Farnborough in 1988, specialists noticed with interest its 360-degree sustained level

This photograph of prototype 9-10 emphasizes the cleanness of the MiG-29's "integral aerodynamic design."

The 9-17 was used for testing larger rudders. Here, it carries overwing fin extensions containing IRCM flare launchers.

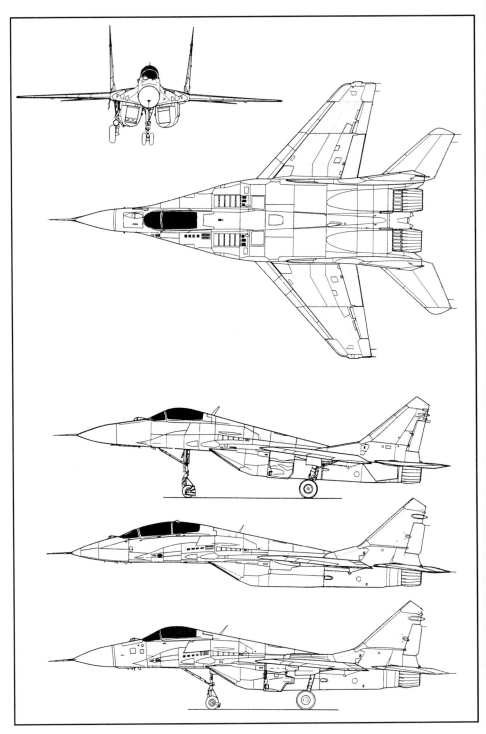

Top to bottom: side views of the MiG-29 (9-01), MiG-29UB, and production MiG-29 (RR drawing from MiG OKB documents)

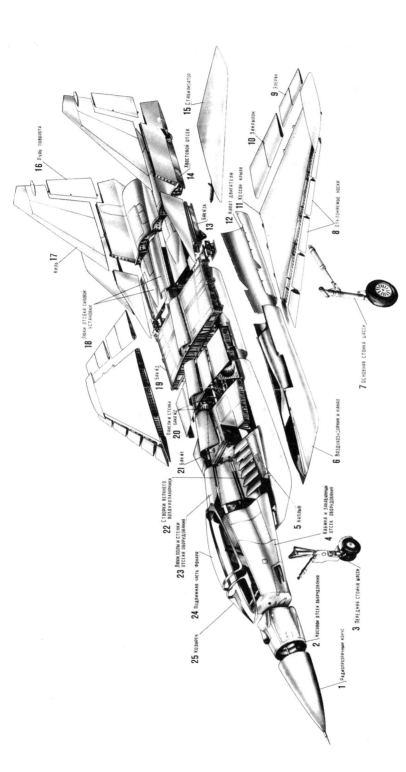

Exploded view of the MiG-29. (1) Radome. (2) Forward electronics compartment. (3) Nose gear strut. (4) Cockpit and lower electronics compartment. (5) Apex. (6) Air intake duct. (7) Main gear leg. (8) Leading edge flaps. (9) Aileron. (10) Flaps. (11) Wing box. (12) Port engine cowling. (13) Fuel tank no. 3A. (14) Rear bay. (15) Slab tailplane. (16) Rudder. (17) Tail fin. (18) Engine access panels. (19) Fuel tank no. 3. (20) Panels and walls of fuel tank no. 2. (21) Fuel tank no. 1. (22) Air intake louvers. (23) Access panels and walls of the central equipment bay. (24) Cockpit canopy. (25) Windshield (MiG OKB document). This view depicts one of the early production aircraft, as indicated by the gear door element on the nose gear strut, smaller wheels (530 x 100 for the nosewheels, 770 x 200 for the main gear), smaller rudders, five-section leading edge flaps, and two-section ailerons and flaps.

On recent production aircraft the base of the nose probe is fitted with small vortex generators.

turns as well as its 350-m (1,150-foot) radius of turn at 800 km/h (432 kt) and its 225-m (740-foot) radius of turn at over 400 km/h (216 kt)—with, in both cases, a 3.8 load factor. In a sustained level turn at 10,000 m (32,800 feet) and Mach 0.9, both the pilot and the airframe had to withstand between 4.6 and 5 g.

If turn rate is essential to a fighter, linear acceleration is even more so. That of the MiG-29 at Mach 0.85 at sea level is 11 m/sec² (36 ft/sec²), which means that the aircraft needs 13 seconds to accelerate from 500 to 1,000 km/h (270 to 540 kt). Its linear acceleration is still 6.5 m/sec² (21.3 ft/sec²) at Mach 0.85 at 6,000 m (19,680 feet).

On production aircraft the front gear strut was moved back to between the engine air intakes. The mudguard was replaced by a simple scraper.

The SUV multiform fire control unit is one of the most interesting features of the MiG-29. For the first time anywhere, a fighter was equipped with a fire control unit that employed three different channels for target acquisition: pulse Doppler radar linked to a laser range finder, infrared search and tracking system (IRST), and helmet-mounted target designator. All of these systems work together with the help of on-board computers. The fire control system is thus entirely automatic, ensuring efficiency and discretion at the time of attack and increasing the combat capabilities of the aircraft as it engages hostile targets in a countermeasure environment. The IRST system measures

The whole MiG-29 optoelectronics suite (KOLS) is contained in this small ball, located
in front of the windshield: infrared search-and-track system plus laser range finder.

the target coordinates with the highest accuracy; the first salvo of cannon shells seldom if ever fails to hit its objective.

The N 019/RP-29 (Sapfir 29) pulse Doppler radar can acquire a target with the radar cross-section of another fighter at a distance of 100
km (62 miles) and track it to within 70 km (44 miles). It can also track
ten hostile aircraft simultaneously but has no mapping mode. The helmet-mounted target designator is used for off-axis direction of air-to-air
missiles.

Standard weaponry includes either six R-60T or R-60MK IR-guided
close-range air-to-air missiles or four R-60s and two R-27R-1 radar-guided
medium-range (50–70 km [31–44 miles]) air-to-air missiles at six wing
store stations. But it can also carry other weapons such as R-73A/R-73E
close-range air-to-air missiles, which can be fired as long as the load factor is under 8. The missiles' homing heads are hardened against enemy
ECM.

The cannon armament is limited to a single 30-mm GSh-301 with
150 rounds. It is located in the forward part of the port glove formed by
the center body ahead of the wing (the 9-01 prototype had a twin-barrel
cannon at the same place). The maximum war load of 3,000 kg (6,600
pounds) may also include four FAB-500 or eight FAB-250 bombs, four
B-8M-1 rocket pods (20 x 80 mm), four S-24B 240-mm rockets, four 3B-

Takeoff of a MiG-29. The gear is not yet fully retracted, and the leading edge flaps are still extended; but the trailing edge flaps have already retracted, and the slab tailplane is deflected upward. This aircraft has only four store stations instead of the usual six.

This MiG-29 (9-10) still has the older type of rudders and carries six air-to-air missiles (two R-27s and four R-60s).

The canister for the MiG-29's tail chute sits between two airbrakes, one above the fuselage and one under it. *Bottom,* the tail chute is inserted into the canister.

500 napalm bombs, four KMGU-2 submunitions dispensers, or a combination of these weapons.

The MiG-29 was first piloted on 6 October 1977 by A. V. Fedotov. The aircraft's official acceptance certificate was signed in 1984, but mass production had started as early as 1982. The MiG-29 was photographed by a U.S. satellite in November 1977 at the Ramenskoye flight test center and given the provisional Western designation Ram-L. The second prototype was flown in early June 1978 and lost due to engine fire on 15 June with Menitskiy at the controls; the pilot was saved by its KM-1 ejection seat. The fourth prototype was lost with Fedotov at the controls on 31 October 1980 due to engine problems also, and once more the pilot was saved by the ejection seat.

In June 1983, when the first MiG-29s were delivered to the air regiments, 2,000 test flights had taken place. The aircraft was continuously updated, and if not for the recent events in the ex-USSR it could have had a bright future. The production version did correspond to the *izdeliye* 9-12 standard.

The MiG-29 was exported to eleven countries: Cuba, Czechoslovakia, East Germany, India, Iran, Iraq, North Korea, Poland, Rumania, Syria, and Yugoslavia. It is interesting to note that eighteen single-seaters and five two-seaters delivered to the former East Germany are now flown by pilots of the 5th Luftwaffe division in the FRG.

Specifications
Span, 11.36 m (37 ft 3.2 in); overall length, 17.32 m (56 ft 9.9 in); fuselage length, 14.875 m (48 ft 9.6 in); height, 4.73 m (15 ft 6.2 in); wheel track, 3.09 m (10 ft 1.7 in); wheel base, 3.645 m (11 ft 11.5 in); wing area, 38 m^2 (409 sq ft); takeoff weight, 15,240 kg (33,590 lb); max takeoff weight, 18,500 kg (40,775 lb); wing loading, 401–486.8 kg/m^2 (82.2–99.7 lb/sq ft); max operating limit load factor, 9 at $\leq$ Mach 0.85; 7 at > Mach 0.85.

Performance
Max speed at sea level, 1,500 km/h (810 kt); max permissible operating speed, 2,450 km/h or Mach 2.3; takeoff speed, 220 km/h (119 kt); approach speed, 260 km/h (141 kt); landing speed, 235 km/h (127 kt); climb rate at sea level, 330 m/sec (64,945 ft/min); service ceiling, 17,000 m (55,760 ft); range in clean configuration, 1,500 km (930 mi); with 1,500-l (396-US gal) auxiliary tank, 2,100 km (1,300 mi); takeoff roll, 260 m (855 ft) with afterburner, 600 m (1,970 ft) without afterburner; landing roll with tail chute, 600 m (1,970 ft).

The one-piece canopy of the MiG-29UB is hinged at the rear and opens upward.

Though not equipped with fire control radar, the MiG-29UB has retained limited combat capabilities.

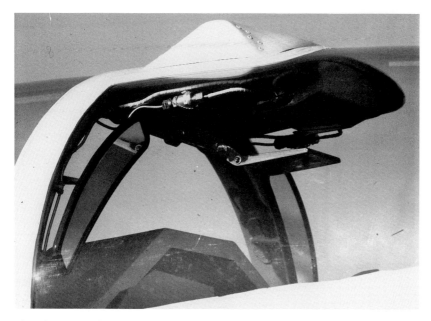

This photograph depicts the wide-screen periscope of the MiG-29UB in its folded disposition. Notice the two rearview mirrors on the canopy post.

MiG-29UB / 9-51

In order to have the proper tool to train pilots and field support crews for the MiG-29, the OKB developed the MiG-29UB two-seat variant concurrently. The *u* stands for *uchebniy* (training) and the *b* for *boyevoy* (combat)—meaning that this trainer retained at least limited combat capabilities. The radar was removed, but the cannon, IRST complex, laser range finder, and wing store stations of the single-seater remained. At 17.42 m (57 feet, 1.8 inches) the MiG-29UB is 100 millimeters (3.94 inches) longer than the single-seater. This is the only structural difference necessitated by the second cockpit. The one-piece canopy is hinged at the rear and opens upward. There a periscope that provides the occupant of the rear cockpit with a wide field of vision. Both ejection seats are of the K-36DM type.

The MiG-29UB was first piloted on 29 April 1981 by A. G. Fastovets with a mannequin strapped to the second seat. The test schedule moved briskly, Mach 1.4 being reached on the fourth flight and Mach 1.9 on the ninth flight. Production got under way in 1982 for the VVS and for export.

The MiG-29S has something of a fatback silhouette because its bigger electronics bay and slightly enlarged no. 1 fuel tank occupy the extended dorsal spine.

MiG-29S / MiG-29SE / 9-13

The first modified version of the basic model was produced concurrently and became operational in the same units that received the standard fighter. Known at first as the *izdeliye* 9-13 (the OKB's internal name), the MiG-29S was first flown on 23 December 1980 with V. M. Gorbunov at the controls.

The only external differences from the MiG-29 are a slightly humpbacked dorsal spine behind the cockpit to hold additional avionics and a bigger no. 1 tank with a capacity of 780 l (206 US gallons). The computer-controlled leading edge flaps are divided in five segments, instead of the four on the MiG-29. But the MiG-29 differs from its predecessors in many other aspects as well:

—the conventional flying controls were optimized to increase the AOA operating range (up to 28 degrees), to augment the aircraft's steadiness in flight and controllability at high AOAs, and to move back the trigger level of unintentional stalls and spins
—this new version could carry under the wings two jettisonable extra fuel tanks having a capacity of 1,150 l (304 US gallons)

The first variant of the MiG-29, the MiG-29S differs from the basic aircraft by its somewhat humpbacked fuselage and the wing that is piped for two 1,150-l (304-US gallon) drop tanks.

The louvers that partly feed the engines while the aircraft is taking off or landing are covered in this photograph by protective equipment. Compare the wing thickness at the leading edge with that of the MiG-29M.

The weaponry fitted to the wing of this sixth prototype of the MiG-29M comprises four Kh-31P air-to-surface missiles under the inner panels, and two RVV-AE and two R-73A air-to-air missiles under the outer panels.

Unlike the MiG-29 and MiG-29S, the MiG-29M has no louvers above the leading edge root extensions. The sharp leading edge of the wing should be compared with that of the MiG-29S.

This photograph of the MiG-29M shows some of the distinctive features of this version: the "fat back," the deeper cockpit canopy, and the notched tailplane of greater area.

Located in front of the windshield, the ball that houses the OLS-M system (protected by a removable fairing on nonoperational flights) differed in shape from that of the MiG-29S.

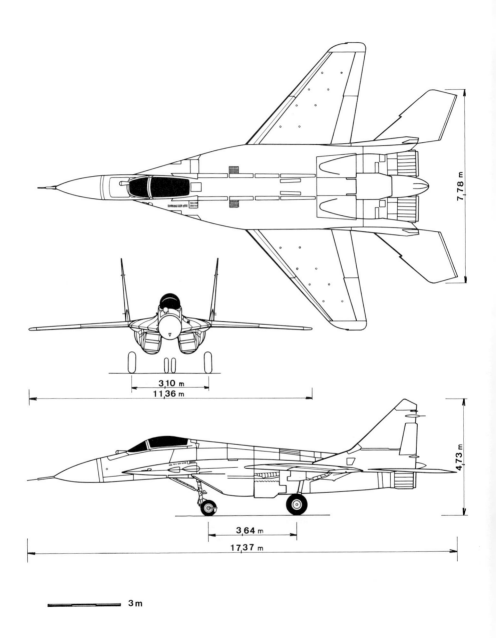

7,78 m

3,10 m
11,36 m

4,73 m

3,64 m
17,37 m

3 m

MiG-29M (MiG OKB three-view drawing)

apiece, bringing the aircraft's overall fuel capacity to 8,240 l (2,177 US gallons) and its maximum range to 2,900 km (1,800 miles)

The weapons load was increased to 4,000 kg (8,800 pounds). The MiG-29S can carry most of the MiG-29's weaponry but is wired for new armament such as the new RVV-AE medium-range active air-to-air missiles or other R-27E semiactive and IR homing air-to-air missiles offering improved range. It can also carry four B-13 rocket pods that can each fire five 122-mm S-13T or S-13OF munitions. The aircraft is capable of engaging two targets simultaneously with its active and IR homing missiles and of firing its GSh-301 cannon when fitted with the underbelly tank. The N 019M radar is of the coherent pulse-Doppler type, an improved version of the RP-29 that has a built-in test set but no mapping mode.

The MiG-29SE is the export version of the MiG-29S. Its radar unit is the N 019ME, a somewhat downgraded version of the N 019M.

The dimensions and wing areas of the MiG-29 and the MiG-29S are identical. The only difference is the weight: takeoff weight of the MiG-29S is 15,300 kg (33,730 pounds), and its maximum takeoff weight is 19,700 kg (43,430 pounds). The overall performance of both aircraft is identical as well except for the maximum range, which on the MiG-29S reaches 2,900 km (1,800 miles) when the aircraft is fully fueled with the internal tanks (4,440 l [1,173 US gallons]), the underbelly tank (1,500 l [396 US gallons]), and the wing drop tanks (2,300 l [608 US gallons]). All operational MiG-29s can be updated to the 9-13 (MiG-29S) standard.

MiG-29M / 9-15

Western technicians thought it odd that the MiG-29—a new aircraft to those who saw it for the first time at Farnborough—was still equipped with conventional hydraulically powered flying controls. They did not realize that the blueprints for the aircraft dated back some twenty years, and that in the early 1970s fly-by-wire (FBW) controls were far from being fully developed. But they are developed now, and it should be no surprise that the second-generation MiG-29 is FBW-equipped. This new variant, initially called the *izdeliye* 9-15, is now referred to as the MiG-29M. But the modifications are not limited to quadruplex analog-computed FBW on the pitch channel (triplex on the roll and yaw channels). This machine may still look like a MiG-29, but in fact it is an entirely new aircraft.

To increase the aircraft's range—markedly too short—more room was needed inside the aircraft to increase the fuel capacity. Because the fuselage was already chock-full, the whole structure had to be completely rethought and rebuilt with new materials. Major structural changes included a welded aluminum-lithium section in front of the main landing gear, a welded steel section behind, and more elements made of composite materials. Moreover, the louvered upper surface auxiliary air intakes were deleted, and the number of rounds for the GSh-301 cannon was reduced to 100 from 150. Those last two measures alone permitted an increase in the capacity of tank no. 1 from 705 l (186 US gallons) in the MiG-29 to 1,710 l (452 US gallons) in the MiG-29M.

The wing, which was given a new airfoil section with a sharp leading edge and new ailerons extended out to the wing tips (to help improve handling characteristics at high AOAs), was also structurally modified to increase to 400 l (105 US gallons) the capacity of each of the two tanks it houses. The aircraft's internal fuel capacity now totals 5,810 l (1,535 US gallons), distributed as follows: tank no. 1, 1,710 l (452 US gallons); tank no. 2, 840 l (222 US gallons); tank no. 3, 1,800 l (475 US gallons); tank no. 3A, 530 l (140 US gallons); tank no. 3B (an additional tank), 130 l (34 US gallons); wing tanks, 800 l (210 US gallons). That represents a 33 percent increase in internal fuel capacity over the MiG-29. And that is not all. The wing is piped (like that of the MiG-29S) to receive two 1,150-l (304-US gallon) drop tanks under pylons. If one adds the 1,500-l (396-US gallon) underbelly tank, the overall fuel capacity totals 9,610 l (2,539 US gallons).

Externally the MiG-29M departs from the basic model in its slightly lengthened nose; its broader, deeper, and longer dorsal spine, terminating in a spade-shaped structure that extends beyond the jet nozzles; and its single paddle-type airbrake, hinged on the top of the rear fuselage and hydraulically actuated. The all-moving (collectively and differentially) horizontal tail surfaces have a greater area and a notched leading edge.

The MiG-29M is powered by two Klimov/Sarkisov RD-33K engines rated at 5,390 daN (5,500 kg st) dry and 8,625 daN (8,800 kg st) with afterburner. They are equipped with a full-authority digital engine control. The air intakes have a greater section and a hydraulically actuated lip at their forward bottom to modify the mass flow. The FOD exclusion doors of the MiG-29 have been replaced by lighter deflector grilles.

The aircraft's avionics suite was entirely updated. The new radar, the Fazatron N 010 Zhuk ("beetle"), is a multimode system that, with its 680-mm (26.77-inch) dish antenna, can provide:

—uniform-scale ground mapping with specified resolution, scale enlargement, and freeze capabilities

—measurements of the aircraft's velocity and the coordinates of ground marks for navigation updating

—measurements of selected ground or sea mark coordinates and target designation information for air-to-surface missiles, rockets, bombs, and guns

—air-to-air modes (look-up and look-down) and control of the launches of missiles equipped with active, passive, or semiactive radar homing heads, as well as rocket launches and gunfire

—a close air combat mode against visual targets

—target detection (up to ten), track-while-scan capabilities on multiple targets (up to four), and simultaneous multimissile attack

—automatic terrain-following and terrain-avoidance modes for low-level operation

The radar is also compatible with the aircraft's automatic guidance systems. Its detection range is greater than 100 km (62 miles)—the exact number is still kept secret.

The two other elements of the SUV fire control system housed in the ball fairing in front of the windshield are: the OLS-M (*optiko-lazernaya sistem*), an optoelectronic detection and sighting system consisting of an IRST that includes a TV capability collimated with a laser rangefinder/target illuminator; and the helmet-mounted target designator (NSTs) whose computational capability was increased fourfold over that of the MiG-29. The three autonomous elements of the SUV can be fully interconnected. It was designed for HOTAS (hands on throttle and stick) use.

The cockpit canopy is slightly more humped, and the pilot's seat was raised to give a better forward view (angle of vision, 15 degrees). The whole cockpit instrument panel was rethought and is now equipped with two multifunction CRT displays (on which no primary instrument information is displayed) and a HUD. Despite the FBW, the pilot has a center stick with a stick force that is reduced by half.

The weapon load was increased to 4,500 kg (9,920 pounds), and there are now eight store stations under the wing (instead of three) to carry a wide variety of loads: air-to-air missiles (up to eight), air-to-surface missiles (up to four), rocket pods, bombs, and the like. On the MiG-29M no. 155 exhibited in 1992 at Machulishche, there were four Kh-29T air-to-surface missiles under the inner panels of the wing and four new RVV-AE medium-range air-to-air missiles under the outer panels. The first of six prototypes (numbered 151 to 156) was first piloted on 25 April 1986 by V. Ye. Menitskiy, and the test flights showed that the type's legendary maneuverability had improved still further.

Besides, as the new variant has inherited all the good features of the basic MiG-29 supplemented by the many improvements detailed above, the operating limits of the MiG-29M have leapt forward tremendously, as indicated by the following statistics: preflight check, 30 minutes; turnaround time (depending on armament), 15–25 minutes; ground staff, 7; operational availability, 90 percent; mean time between failures (MTBF), 8 hours; man-hours per flying hour, 15; routine maintenance cycle, 200 hours; mean troubleshooting time, 1.2 hours; engine change time including depreservation, installation, and ground test, 2.2 hours; man-hours for engine change, 5.3; accident rate after six to eight years of operation, one every 150,000 hours; airframe design life, 2,500 hours or 30 years (possible prolongation up to 4,000 hours); time controlled overhaul, 1,000 hours; engine TBO, 700 hours; engine service life, 1,400 hours.

Specifications
Span, 11.36 m (37 ft 3.2 in); overall length, 17.37 m (56 ft 11.8 in); height, 4.73 m (15 ft 6.2 in); wheel track, 3.09 m (10 ft 1.7 in); wheel base, 3.645 m (11 ft 11.5 in); wing area, 38 m² (409 sq ft); takeoff weight, 15,000 kg (33,000 lb); max takeoff weight, 22,000 kg (48,500 lb); wing loading, 394.7–578.9 kg/m² (80.83–118.57 lb/sq ft).

Performance
Same as MiG-29, except for: range in clean configuration, 2,000 km (1,245 mi); range with one 1,500-l (396-US gal) and two 1,150 l (304-US gal) auxiliary tanks, 3,200 km (1,990 mi).

MiG-29KVP   MiG-29K / 9-31

In the early 1980s the Soviet Navy announced its desire to equip its future cruisers–aircraft carriers* with highly maneuverable, well-armed supersonic fighters. This aircraft type was to have the operational radius to intercept and destroy airborne invaders as far away from its floating base as possible and to be capable of short takeoffs at full load. The OKB engineers and pilots under the leadership of M. R. Valdenberg thought that their best choice for this project was to "navalize" the MiG-29 airframe, and they transformed it into a proof-of-concept aircraft.

The essential modifications concerned the gear, which was strengthened to withstand higher sink rates at touchdown, and the tail section of the fuselage, which was reinforced to receive an arrester hook. The new aircraft was named MiG-29KVP (*Korotkii Vzlet*

---

*This shrewd label authorizes the carrier to proceed through the Dardanelles. As "carrier" alone, she could not pass in accordance with international agreements and would therefore be trapped in the Black Sea.

MiG-29K no. 311 lands on the deck of a carrier. It is just about to catch the first arresting cable with its hook.

The wing's folding axis runs between the flaps and the ailerons. Structural modifications were reduced to the bare essentials.

A MiG-29K takes off from the aircraft carrier's deck in fewer than 100 m (300 feet) thanks to the ski-jump technique and the special booster regime of its RD-33K turbofans.

The MiG-29KVP was a proof-of-concept aircraft used to prepare the design and development of the MiG-29K.

The MiG-29K is fitted with a retractable refueling probe.

i Posadka: short takeoff and landing, or STOL) and was first piloted by Fastovets on 21 August 1982. A long test campaign then began, leading to the conclusion that to fulfill all its requirements the aircraft needed a greater wing area, more efficient lift devices, more powerful engines, and a far greater fuel capacity. Since the OKB was concurrently developing the 9-15 project (the future MiG-29M), which met most of the requirements, it was decided to develop the naval aircraft from that airframe.

Like the MiG-29KVP, the modifications for this new project concerned the strengthening of the long-stroke gear (shortening links on the main gear legs) and the reinforcement of the rear section of the fuselage to receive the arrester hook (entailing the removal of the brake chute). In addition, engine thrust was boosted at takeoff thanks to the ChR system, and the wing had to be thoroughly redesigned. The wing area of the MiG-29M still fell short of the navy's needs, so the wing chord was extended ahead of the leading edge by adding a front

The second prototype of the MiG-29K with its full external armament: four R-73A short-range air-to-air missiles and four Kh-31P long-range air-to-surface missiles.

spar to support the two-segment LE flaps; also, the wing flaps were extended beyond the trailing edge, and the ailerons were shortened. That is how the wing area was increased by 3.6 m² (36.6 square feet). The wing was also made foldable (the fold axis lies between the aileron and the flap).

The wing tips were bulged to house electronic support measures (ESM) equipment. Because the danger of ingesting foreign objects on takeoff and landing is practically nonexistent on a carrier, the overbody louvers were deleted and the multisegment ramp system was readjusted; but to prevent seabirds from being sucked in, the air intakes are still equipped with lightweight deflector grids that can be retracted in flight. Due to the special nature of the environment in which the aircraft would operate, engineers had to take exceptional anticorrosion measures, make sure all bays and access doors sealed tightly, and include the stowage devices usually found on carrier-based aircraft.

The 9-31, renamed MiG-29K (*Korabelniy:* ship-based), is, like the MiG-29M, powered by two RD-33K turbofans rated at 8,625 daN (8,800 kg st). When taking off in a fully loaded aircraft, the pilot can use a special afterburning rating similar to the ChR (*Chrezvichainiy Rezhim:* exceptional rating) used on the MiG-21SMT and MiG-21bis that adds 588 daN (600 kg st) to each engine. The overall fuel capacity is 9,610 l

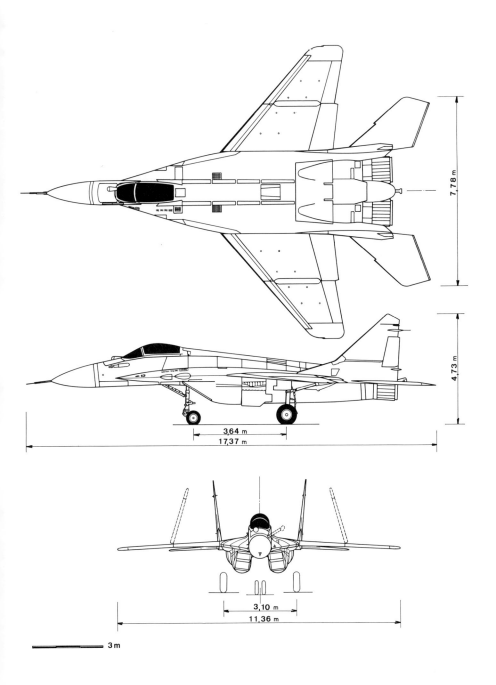

7,78 m

4,73 m

3,64 m

17,37 m

3,10 m

11,36 m

3 m

MiG-29K (MiG OKB three-view drawing)

451

(2,540 US gallons), and the aircraft is fitted with a retractable refueling probe on the left side of the nose. Like the MiG-29M, the MiG-29K is fitted with the N 010 multimode pulse-Doppler radar.

Prototype no. 311 was first flown on 23 June 1988 by test pilot T. Aubakirov at Saki naval airfield on the Crimean Peninsula, where one of the runways was equipped with a ramp very similar to the ski jump of the 65,000-tonne (67,000-ton) aircraft carrier *Tbilisi.* In November 1989 the aircraft made its first deck takeoffs and landings from that ship, which in the meantime had been renamed *Admiral of the Fleet Kuznetsov.* It was a great premiere for both the Soviet Navy (VMF) and its aeronautical branch (MA) and was much to the credit of test pilot Aubakirov, who had to use the ski jump of the deck's forward end instead of the usual carrier catapult to take off.

In 1990 the focus of the tests was the development of an automated landing procedure and the necessary electromagnetic compatibility between the aircraft and the carrier's radioelectric systems. The second prototype, no. 312, exhibited at Machulishche near Minsk in February 1992, carried four R-73E short-range air-to-air missiles under the wing's folding panels and four Kh-31P air-to-surface missiles fitted to AKU-58M ejector pylons under the fixed panels. But it was only one of the many combinations of weapons that the MiG-29K could carry because on the sole export version no fewer than twelve different types of missiles are offered: air-to-air missiles, R-73E, R-27R1, R-27T1, R-27E(R), R-27E(T), RVV-AE; air-to-surface missiles, Kh-29T, Kh-29L, Kh-25P, Kh-25ML, Kh-31P, Kh-31A. The aircraft could also carry various types of bombs, including the KAB-500KR "smart" bomb, or B-8 and B-13 rocket pods. Its armament contains the GSh-301 (9A4071K) cannon with 100 rounds, located above the port body leading edge.

Specifications
Span, 11.36 m (37 ft 3.2 in); overall length, 17.37 m (56 ft 11.8 in); height, 4.73 m (15 ft 6.2 in); wheel track, 3.09 m (10 ft 1.7 in); wheel base, 3.645 m (11 ft 11.5 in); wing area, 41.6 m^2 (447.78 sq ft); takeoff weight in clean configuration, 18,480 kg (40,705 lb); max takeoff weight, 22,400 kg (49,340 lb); wing loading, 444.2–538.4 kg/m^2 (90.57–110.27 lb/sq ft).

Performance
Same as MiG-29M, except for: range in clean configuration, 1,600 km (1,000 mi); range with one 1,500-l (396-US gal) and two 1,150 l (304-US gal) auxiliary tanks, 2,900 km (1,800 mi).

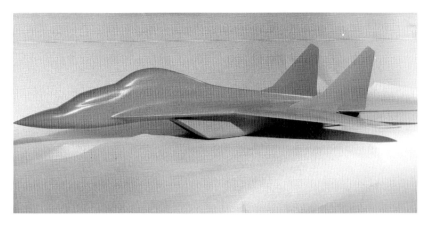

This wind tunnel model of the MiG-29KU ship-based trainer shows the especially high location of the rear seat.

MiG-29KU

Training ship-based aircraft pilots is a rather delicate task, especially if they are to fly sophisticated machines with high wing loadings. While landing or taking off using a carrier's deck, mistakes are usually fatal. Thus, a specialized trainer aircraft was needed. The MiG OKB engineers proposed a two-seat variant of the MiG-29.

Why not simply "navalize" the MiG-29UB? Add stronger gear legs and the arrester hook, some suggested. This seemed like a logical approach—and yet it would have proved shortsighted in every sense of the word. Even though the angle of attack that the aircraft has to take at landing because of its short gear legs is relatively small, the forward view from the rear cockpit would still be totally inadequate to make a precise landing on a carrier deck. So on the KU project the rear seat was raised noticeably, giving the aircraft a distinctly hunchbacked silhouette. Like the K, the MiG-29UK was to be powered by two RD-33K turbofans rated at 8,625 daN (8,800 kg st). But this project has since been abandoned.

Mikoyan and the *Konversiya*

It took until the end of the MiG OKB's fifth decade and the start of the *konversiya* for the first real civil project to find its way to the design bureau's drawing boards. This project was not initiated either by

Aeroflot or by a foreign airline. It was instead a purely homemade product searching out its own customers—or even partners. And it is not the only such project in the OKB's files.

101M Multirole Twin-Engine Aircraft

This lightweight twin-engine was designed to carry passengers or cargo to and from any unpaved strip 400 m (1,300 feet) long and having a minimum strength of 5 kg/cm² (71.1 pounds per square inch). The aircraft was intended for around-the-clock, all-weather use. Its APU supplies the necessary power for all loading and unloading operations.

Its power unit—two TV7-117 turboprops rated individually at 1,840 kW (2,500 ch-e)—and fuel system were specially designed to allow a limited use of diesel oil. The engines drive reversible-pitch propellers. The aircraft can fly and land with one engine inoperative, and it can be equipped with floats or skis. The twin-boom architecture with a high-set tailplane was used for ease of entry to the rear fuselage; the rear end opens upward, clearing the way for direct access to the cargo hold: length, 4 m (13 feet, 5.4 inches); width, 1.48 m (4 feet, 10.3 inches); height, 1.6 m (5 feet, 3 inches); volume, 6 m³ (211.89 cubic feet). At 1.5 m (4 feet, 11.1 inches), the sill height of this hold permits direct transfers to and from truck beds. All other loading problems are handled by the integral ceiling hoist.

The 101M was created to handle five basic missions:

—transport of field hospitals that can be set up quickly in case of emergency (disasters, accidents, epidemics)
—evacuation of casualties and the critically ill
—transport of supplies, medicines, and relief workers in the affected areas
—transport of geological expeditions and the like to remote or inaccessible locales
—forest fire extinguishment

To fulfill its purpose, the aircraft could carry a variety of loads:

—everything required for a complete airmobile field hospital in eight containers attached to the underwing store stations, plus the necessary medical staff (ten to twelve persons); total weight, 2,000 kg (4,400 pounds)
—eight to twelve sick or wounded persons on stretchers, plus the medical assistant; medical personnel, survivors, badly burned persons, and the like; total weight, 1,000 kg (2,200 pounds)
—various other loads, solid or liquid

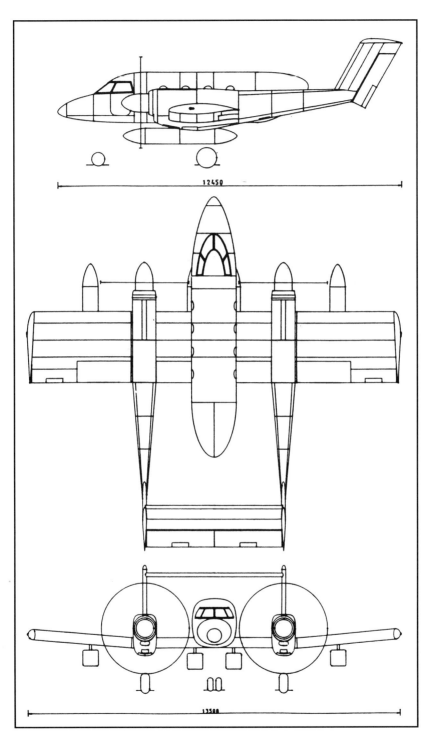

101M (MiG OKB three-view drawing)

455

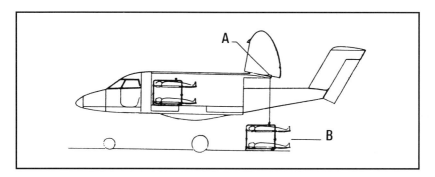

Loading a stretcher holder with a ceiling hoist. (*A*) Electrical hoist on rail. (*B*) Stretcher holder (two or three persons).

For the first layout, the following setup times were planned: 30 minutes to install the eight containers; 15 minutes for aircraft turn-around; 10 minutes for a quick change of the cabin layout to evacuate wounded persons; 10 minutes for a quick change of the cabin layout to transport loads; and 15 minutes to load eight wounded persons on stretchers.

The airmobile field hospital created for this aircraft includes:

—four inflatable-frame tents at 50 m^2 (538.2 square feet) apiece
—four electronic monitors, surgical instruments, stretchers, oxygen tanks, and other medical equipment
—the emergency power unit that burns kerosene out of the aircraft's supply to provide the necessary overpressure, lights, and climate controls in the tents
—eight to twelve stretchers, monitors with the appropriate connections for the stretchers, anesthetics, various life-support devices, and other evacuation materiel

The tents, medical equipment, and emergency power unit (but not the monitors or the stretchers) are carried in eight standardized containers set in pairs under four wing store stations. Those containers can be either lifted or transported on wheels. The field hospital and all of its equipment weighs 1,200 kg (2,645 pounds) and takes up 200 m^2 (2,150 square feet). The first tent can be erected in fifteen minutes; and it takes one and one-half hours to set up the entire hospital, which can be heated or cooled to a constant 22° C (plus or minus 5° C). The hospital is self-sufficient between five and six days with six to eight medical attendants and four technicians.

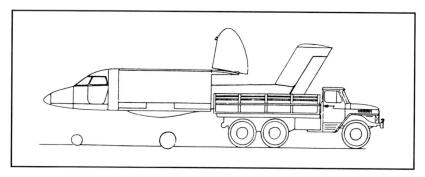

Loading directly out of a truck bed.

Specifications
Span, 13.5 m (44 ft 3.5 in); overall length, 12.45 m (40 ft 10.2 in); height, 4.4 m (14 ft 5.2 in); wing area, 33.53 m² (360.92 sq ft); takeoff weight with 2,000-kg (4,400-lb) payload, 9,000 kg (19,835 lb); max payload, 4,000 kg (8,800 lb); max fuel, 2,000 kg (4,400 lb).

Design Performance
Economical cruising speed for range of 2,700 km (1,680 mi), 530 km/h at 11,800 m (286 kt at 38,800 ft); economical cruising speed for range of 1,300 km (810 mi), 530 km/h at 200 m (286 kt at 650 ft); max cruising speed for range of 1,800 km (1,120 mi), 670 km/h at 7,000 m (362 kt at 22,960 ft); takeoff/landing roll, 150–200 m (490–655 ft).

101N Multirole Twin-Engine Aircraft

Another plane intended to transport passengers, the 101N project is also another twin-boom twin-engine, but the short rectangular wing of the 101M is replaced by a tapered wing of higher aspect ratio; the power unit consists of two TVD-1500 turboprops rated at 957 kW (1,300 ch-e).

The 101N is a commuter aircraft designed to hold nineteen passengers, but different cabin layouts can accommodate mixed cargo, all cargo up to 1,700 kg (3,750 pounds), training for civil aviation schools, ambulance runs, parachute drops, and other transport roles. The aircraft can be equipped with special systems such as 360-degree scanning radar, rotating infrared and ultraviolet radiometers, thermal sights, and cameras for various missions: around-the-clock, all-weather maritime surveillance (including over the 200-nautical-miles zone); ecological sur-

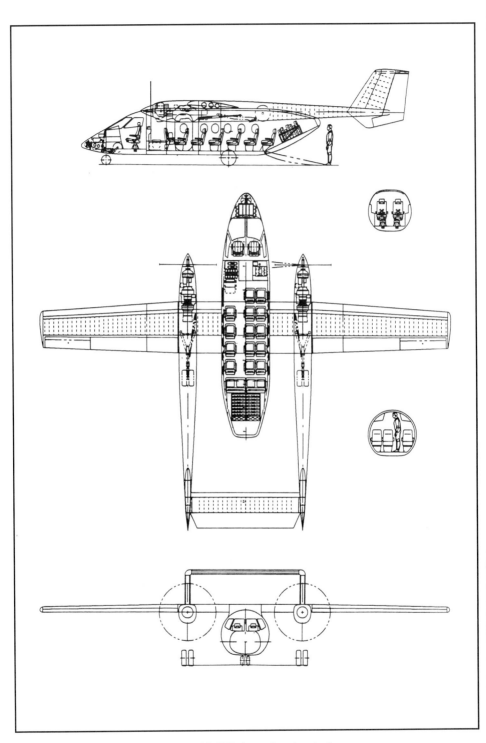

101N (MiG OKB three-view drawing)

veillance; maritime search-and-rescue (plus survival kits can be dropped by parachute); shoal mapping, or forest fire surveillance.

The aircraft can be operated in adverse weather conditions, in cold climates, and on unpaved strip whose strength is at least 6 kg/cm² (85.3 pounds per square inch). The aircraft's APU starts the engines and provides the necessary power for some on-board systems, the loading and unloading devices, and the cockpit's heat when the plane is on the ground. Like the 101M, the 101N can be equipped with floats or skis.

Specifications

Span, 17.6 m (57 ft 8.9 in); length, 14.3 m (46 ft 11 in); height, 4.64 m (15 ft 2.7 in); wing area, 34.82 m² (374.8 sq ft); takeoff weight, 8,000 kg (17,630 lb); payload, 1,700 kg (3,750 lb); fuel, 550 kg (1,210 lb); max takeoff weight, 9,000 kg (19,835 lb); max fuel, 2,000 kg (4,400 lb); wing loading, 229.8–268.5 kg/m² (47.11–55.04 lb/sq ft).

Design Performance

Cruising speed, 550 km/h at 8,000 m (297 kt at 26,240 ft); range, 500 km (310 mi) with 1,700-kg (3,750-lb) payload and 45 min of fuel reserves; max range, 2,000 km (1,240 mi); service ceiling, 9,000 m (29,500 ft); takeoff roll on paved runway, 320 m (1,050 ft); on rough strip, 380 m (1,250 ft).

SVB, a Mountaineer

This commuter was engineered to carry fifty passengers or cargo in hot mountainous regions. It can be operated in hot climates up to 40° C (104° F) and out of high-altitude airfields up to 4,000 m (13,120 feet) above sea level. The SVB is powered by two TV7-117 turboprops rated at 1,840 kW (2,500 ch-e); they drive low-noise SV-34 six-blade airscrews. The cabin, which maintains a constant width from the cockpit rear bulkhead, is pressurized and dimensioned to accommodate ten rows of five seats—separated into three and two by the aisle, 400 mm (15.75 inches) wide—at a pitch of 780 mm (30.71 inches). The cabin is 2.96 m (9 feet, 8.5 inches) wide, with headroom of 2 m (6 feet, 6.7 inches). The aircraft has a flight crew of two plus two cargo handlers or cabin attendants, as appropriate.

In its all-freight setup the SVB can carry a payload of 5,000 kg (11,000 pounds). At the rear of the fuselage is a loading ramp for various types of vehicles. An integral ceiling hoist helps to manipulate the freight inside the cargo hold. The aircraft is equipped with a digital

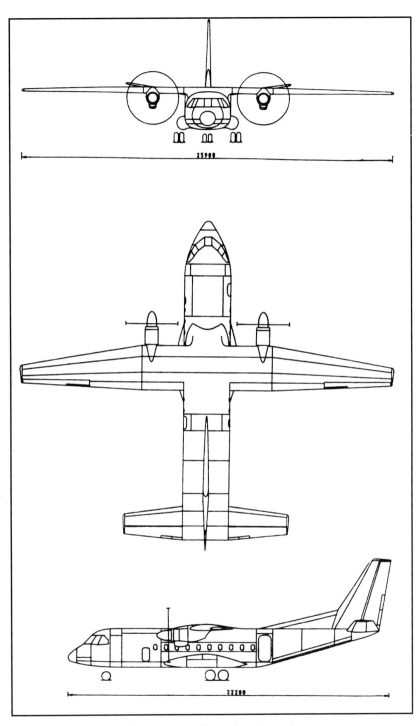

SVB (MiG OKB three-view drawing)

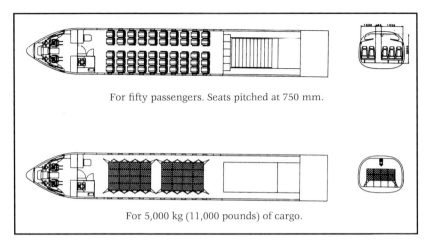

For fifty passengers. Seats pitched at 750 mm.

For 5,000 kg (11,000 pounds) of cargo.

Possible arrangements of the SVB's interior.

flight management system and integrated communications capabilities, allowing all-weather and around-the-clock operations.

Specifications
Span, 25.9 m (84 ft 11.7 in); length, 22.2 m (72 ft 10 in); height, 8.07 m (26 ft 5.7 in); wing area, 62 m² (667.37 sq ft); takeoff weight, 19,400 kg (42,760 lb); payload, 5,000 kg (11,000 lb); fuel, 2,100 kg (4,630 lb); wing loading, 312.9 kg/m² (64.15 lb/sq ft).

Design Performance
Cruising speed, 550 km/h at 6,000 m (297 kt at 19,680 ft); range, 1,500 km (930 mi) with 5,000-kg (11,000-lb) payload and 45 min of fuel reserves; required field length, 1,800 m (5,900 ft) on rough strip capable of 5–6 kg/cm² (71.1–85.3 lb/sq in) at 2,100 m (6,890 ft) and 30° C (86° F); max field altitude, 4,000 m (13,120 ft); energetic efficiency, 23.6 g/pax-km.

MiG Engine Developers

Vladimir Yakovlyevich Klimov (1892–1962), member of the USSR Academy of Science, Hero of Socialist Labor, winner of the USSR State Prize, engineer major-general of aeronautical engineering.

Klimov was educated in the automobile engine laboratory managed by the academician Ye. A. Chukadov. When the TsIAM was set up, he managed the gasoline engine department and was appointed professor. He lectured simultaneously at the Zhukovskiy military air academy. Then he went to France to negotiate the manufacturing license for the Hispano-Suiza twelve-cylinder V-type engine, which became in the USSR the M-100 rated at 750 ch. Derivatives of this engine—the VK-103, VK-105PF, and VK-107A—powered all of the Yakovlev fighters of World War II and the Petlyakov Pe-2 bomber. At the end of the war Klimov was developing the VK-108, but this engine never made it to production.

During one of the first postwar Paris air shows, Mikoyan and Klimov were able to examine the first Rolls-Royce centrifugal compressor jet engines. To pursue this technology, they went to Great Britain to order a small number of Nene turbojets. With these few engines, Klimov developed in the USSR the RD-45 for the MiG-15 fighter, the VK-1 for the MiG-17, and the reheated VK-1F for the MiG-17F. Later he designed the VK-5 and VK-7 experimental engines and developed the

VK-3, one of the very first turbofans, in close collaboration with one of his best assistants, Sergei Piotrovich Izotov.

Management of the Klimov OKB was entrusted to Izotov after its founder died in 1962. After graduating from the Leningrad polytechnic institute Izotov joined the OKB in 1941, where he became known over two decades for the turboshaft engines and reduction gears that he developed for the Mil and Kamov helicopters. But his first achievements included the GTD-350 turboshaft, the VR-2 reduction gear for the Mi-2, and the TV2-117 and VR-8 reduction gear for the Mi-8. Izotov worked hard at the same time on powering tanks with gas turbine engines, and he succeeded in developing the RD-33, a very reliable turbofan used for the MiG-29. Izotov died in 1983 and was succeeded by V. Stepanov and A. Sarkisov, who is today in charge of the design bureau.

Aleksandr Aleksandrovich Mikulin (1895–1985), member of the USSR Academy of Science, Hero of Socialist Labor, winner of the USSR State Prize, engineer major-general.

The first 100-percent Soviet engine, the AM-34 was developed under Mikulin's supervision in the 1930s. The first MiG fighter was powered by a Mikulin engine, the AM-35. He went on the develop experimental high-altitude turbocharged engines such as the AM-39 and AM-42B. During World War II his AM-38 and AM-42 engines powered tens of thousands of Il-2 and Il-10 *shturmovik* aircraft. After the war he developed the AM-3 turbojet for the first Soviet jet airliner, the Tu-104. His military turbojets powered mass-produced aircraft such as the MiG-19, MiG-21, and MiG-25, and he knew how to surround himself with talented assistants, namely, B. Stechkin, G. Livshits, S. Tumanskiy, N. Metskvarshivili, V. Gavrilov, and K. Kachaturov.

Sergei Konstantinovich Tumanskiy (1901–73), member of the USSR Academy of Science, Hero of Socialist Labor, winner of the Lenin Prize and the USSR State Prize.

After graduating and soldiering at the Leningrad technical military school, the Zhukovskiy military air academy, and the TsIAM (central institute for airplane engine construction), Tumanskiy developed his first engine in 1938—the M-88, which powered the Il-4 bombers. In the early 1940s he joined the Mikulin OKB. In 1956 he was handed responsibility for developing the R-11F-300 twin-spool turbojet for the MiG-21. He kept the twin-spool layout for the R-11F2-300, rated at 6,000 daN (6,120 kg st), and the R-13-300, rated at 6,360 daN (6,490 kg st). Then he developed the R-27, rated at 7,645 daN (7,800 kg st), for the MiG-23 and the R-15B-300, rated at 10,000 daN (10,210 kg st), for the MiG-25.

Pavel Aleksandrovich Solovyev (1917–), chief constructor, corresponding member of the USSR Academy of Science, doctor of technical science, professor.

After graduating from the Rybinsk aeronautical institute, Solovyev joined the Shvetsov design bureau in 1939. Two years later he helped to launch the production of the ASh-82 radial engine and, later still, that of the ASh-82FN for the La-5 and La-7 fighters and that of the ASh-82T for the Il-12 and Il-14 transport aircraft. The ASh-82 was selected in 1942 to power the DIS-200 (IT), and Solovyev personally supervised the installation of the engines on the MiG prototype. Then he worked with Shvetsov on the development of the most powerful piston engine of the time, the ASh-2K rated at 3,460 kW (4,700 ch).

Solovyev succeeded Shvetsov in 1953. The next year he began developing turbofan engines for airliners, the first of which was the D-

20 for the Tu-124. He worked simultaneously on the D-25V turboshaft engine and its R-7 gearbox for the Mi-6 and Mi-10 heavy helicopters. Pursuing his turbofan line of products, between 1964 and 1966 he designed the D-30, rated at 6,665 daN (6,800 kg st), for the Tu-134 airliner. It was the first Soviet engine to be certified worldwide. Surprising as it may seem, the D-30 is still the core of the D-30F6 reheated turbofans, rated at 15,190 daN (15,500 kg st), that power the MiG-31 interceptor. Today the design bureau, located in Perm, works under the leadership of Yuri E. Reshetnikov on various versions of the PS-90 turbofan to power several types of airliners.

APPENDIX 1.
The Primary Piston Engines That Powered MiG Fighters

Engine designation	No. of cylinders	Cylinder arrangement	Pressure ratio	Takeoff power ch (kW)	Specific fuel consumption g/ch.h	Dry weight kg (lb)	Takeoff rating rpm	Length mm (in)	Width mm (in)	Height mm (in)	Diameter mm (in)
AM-35A	12	V 60°	7.0	1,350 (993)	330–360	830 (1,829)	—	2,402 (94.57)	866 (34.09)	1,089 (42.87)	—
AM-38	12	V 60°	6.8	1,600 (1,177)	305–335	860 (1,895)	2,150	2,289 (90.12)	875 (34.45)	1,082 (42.60)	—
AM-38F	12	V 60°	6.0	1,700 (1,250)	305–325	880 (1,940)	2,350	2,287 (90.04)	878 (34.57)	1,084 (42.68)	—
AM-39B	12	V 60°	6.0	1,750 (1,287)	330–335	1,040 (2,292)	2,350	2,415 (95.08)	890 (35.04)	1,142 (44.96)	—
AM-42	12	V 60°	5.5	2,000 (1,471)	330–345	1,012 (2,230)	2,500	2,290 (90.16)	875 (34.45)	1,153 (45.39)	—
VK-107A	12	V 60°	6.75	1,650 (1,214)	280	769 (1,695)	3,200	2,166 (85.28)	867 (34.13)	962 (37.87)	—
M-11FM	5	single row	—	110 (81)	255–280	172 (379)	1,800	1,010 (39.76)	—	—	1,075 (42.32)
M-82	14	double row	7.0	1,700 (1,250)	305	850 (1,873)	2,400	2,005 (78.94)	—	—	1,260 (49.61)
ASh-82F	14	double row	7.0	1,850 (1,361)	325	938 (2,067)	2,500	1,986 (78.19)	—	—	1,260 (49.61)

Source: MiG OKB and TsIAM.

APPENDIX 2.
Some of the Jet Engines That Power MiG Fighters

	Type[a]	Mass flow kg/sec	Pressure ratio	Bypass ratio	Thrust with afterburner kg st (daN)	Specific fuel consumption kg/kg st.h	Max dry thrust kg st (daN)
D-30F6	TF, AB, TS	—	21.5	0.50	15,500 (15,190)	1.9	9,500 (9,310)
RD-33	TF, AB, TS	76.3	21.9	0.47	8,300 (8,134)	2.0	5,040 (4,940)
R-35-300	TJ, AB, TS	110	13	—	13,000 (12,740)	1.96	8,550 (8,380)
R-29B-300	TJ, AB, TS	104	12.4	—	11,500 (11,270)	1.8	8,000 (7,840)
R-29-300	TJ, AB, TS	110	13	—	12,500 (12,250)	2.0	8,300 (8,135)
R-27F2M-300	TJ, AB, TS	95	10.5	—	10,000 (9,800)	1.9	6,900 (6,760)
AL-21F-3	TJ, AB, SS	104.5	14.75	—	11,500 (11,270)	1.82	8,000 (7,840)
R-25-300	TJ, AB, TS	67.9	9.55	—	7,100 (6,960)	2.25	4,100 (4,020)
R-13-300 (R-13F-300)	TJ, AB, TS	65.6	8.9	—	6,490 (6,360)	2.093	4,070 (3,990)
R-11F2S-300	TJ, AB, TS	65.2	8.9	—	6,175 (6,050)	—	3,900 (3,820)[d]
R-11F2-300	TJ, AB, TS	65	8.72	—	6,120 (6,000)	2.19	3,950 (3,870)
R-11F-300	TJ, AB, TS	63.7	—	—	5,740 (5,625)	2.18	3,880 (3,800)
R-21F-300	TJ, AB, TS	74	8.7	—	7,200 (7,055)	2.35	4,700 (4,605)
R-15B-300	TJ, AB, SS	144	4.75	—	10,210 (10,000)	2.7	7,500 (7,350)
RD-36-35	LJ, SS	40.4	4.4	—	—	—	2,350 (2,300)
RD-45F	TJ, SS	40.5	4	—	—	—	2,270 (2,225)
AM-9B (RD-9B)	TJ, AB, SS	43.3	—	—	3,250 (3,185)	1.6	2,600 (2,550)
AM-5F	TJ, SS	37	6.1	—	2,700 (2,645)	1.8	2,150 (2,105)
AM-5	TJ, SS	37.5	5.8	—	—	—	2,000 (1,960)
TR-1	TJ, SS	31.5	3.16	—	—	—	1,350 (1,325)
AL-5	TJ, SS	95	4.5	—	—	—	5,030 (4,930)
VK-7	TJ, SS	—	6.3	—	6,270 (6,145)	—	4,200 (4,115)
VK-5F	TJ, AB, SS	—	—	—	3,850 (3,775)	1.9	3,000 (2,940)
VK-3	TF, AB, SS	98.4	12.7	0.12	8,440 (8,270)	1.9	5,730 (5,615)
VK-1F	TJ, AB, SS	—	—	—	3,380 (3,310)	2.0	2,650 (2,595)
VK-1	TJ, SS	48	4.2	—	—	—	2,700 (2,645)

Notes:
[a]*TJ = turbojet, LJ = lift jet, TF = turbofan, AB = afterburner, SS = single spool, TS = twin spool.*
[b]*EHM = electro-hydromechanical, HE = hydroelectronic, HM = hydromechanical.*
[c]*First figure is low-pressure stage number, second is high-pressure stage number.*
[d]*When the SPS (flap blowing) system is in use, 3,330 daN (3,400 kg st).*

Specific fuel consumption kg/kg st.h	Turbine inlet temperature K	Compressor type	Structure Compressor	Turbine	Weight kg (lb)	Flow control[b]
0.73	1,660	axial	5 + 10	2 + 2	2,416 (5,324)	HE
0.77	1,540	axial	4 + 9[c]	1 + 1[c]	1,050 (2,314)	HE
0.96	1,493	axial	5 + 6	1 + 1	1,765 (3,890)	EHM
0.94	—	axial	5 + 6	1 + 1	—	EHM
0.95	1,423	axial	5 + 6	1 + 1	—	EHM
0.97	1,373	axial	5 + 6	1 + 1	1,650 (3,637)	EHM
0.9	1,400	axial	14	2	1,700 (3,747)	EHM
0.96	1,330	axial	3 + 5	1 + 1	1,210 (2,667)	EHM
0.931	1,233	axial	3 + 5	1 + 1	—	EHM
—	—	axial	3 + 5	1 + 1	—	EHM
0.94	1,211	axial	3 + 5	1 + 1	1,117 (2,462)	EHM
0.94	1,173	axial	3 + 5	1 + 1	1,182 (2,605)	EHM
1.01	1,233	axial	3 + 5	1 + 1	1,220 (2,689)	EHM
1.25	1,215	axial	5	1	2,625 (5,785)	HE
1.33	1,230	axial	6	1	176 (388)	HM
1.07	1,140	centrifugal	1	1	814 (1,794)	HM
0.93	—	axial	9	2	695 (1,532)	HM
0.99	—	axial	8	2	624 (1,375)	HM
0.93	—	axial	8	2	445 (981)	HM
1.315	1,050	axial	6	1	815 (1,796)	HM
0.95	1,050	axial	7	1	1,848 (4,073)	HM
0.95	—	centrifugal	—	—	1,135 (2,502)	HM
1.1	—	centrifugal	1	1	980 (2,160)	HM
0.82	1,190	axial	2 + 8	3	1,850 (4,077)	HM
1.15	—	centrifugal	1	1	978 (2,155)	HM
1.07	1,170	centrifugal	1	1	870 (1,917)	HM

APPENDIX 3.
Machine Guns and Cannons on MiG Fighters

	Caliber mm	Weight kg (lb)	Rate of fire rpm	Initial speed m/sec	Weight of ammo. g
ShKAS	7.62	10 (22)	1,800	825	9.6
BS (UBS)[a]	12.7	24.6 (54.2)	800	860	48
BK (UBK)[b]	12.7	21.5 (47.3)	1,000	860	48
N-57	57	135 (297.3)	230	600	2,000
N-37	37	103 (226.9)	400	690	735
NS-23	23	37 (81.5)	550	690	200
NR-23	23	39 (85.9)	850	690	200
NR-30	30	66 (145.4)	900	780	410
GSh-23L[c]	23	51 (112.3)	3,200	700	200
GSh-6-30[d]	30	145 (319.4)	5,400	850	380
GSh-6-23[d]	23	—	8,000	700	200
GSh-301 (9A4071K)	30	46 (101.3)	1,800	860	380

Source: MiG OKB.
Notes:
[a]Synchronized machine gun.
[b]Wing machine gun (nonsynchronized).
[c]Twin-barrel cannon.
[d]Six-barrel cannon.

APPENDIX 4.
Some of the Radars Installed on MiG Fighters

Designation	Code	Manufacturer	Aircraft	Azimuth	Scanning Elevation	RCS 16 m²
Izumrud	RP-1	Tikhomirov	MiG-17P, MiG-19P	target coordinates		12 km (7.45 mi)
TsD-30	RP-21	Nyenartovich	MiG-21P	±30°	—	20 km (12.43 mi)
Sapfir-21, S-21	RP-22	Volkov	MiG-21P	±30°	20°	30 km (18.64 mi)
Sapfir-23, S-23	—	Kunyavskiy	MiG-23P	±30°	6°	70 km (43.50 mi)
Smerch-A		Volkov	MiG-25P	±60°	6°	100 km (62.15 mi)
Sapfir-25	RP-25	Kirpichev	MiG-25PD	±56°	6°	100 km (62.15 mi)
Zaslon	S-800	—	MiG-31	±70°	+70°/−60°	200 km (124.30 mi)

Source: MiG OKB.

Aircraft	No. of weapons x rounds	Year certified
MiG-1, MiG-3	2 x 600	1932
MiG-1, MiG-3	1 x 300	1939
MiG-3	2 x 145	1940
MiG-9	1 x 40	1946
MiG-9	1 x 40	1947
MiG-9 MiG-15 (until 1950)	2 x 80 2 x 60	1944
MiG-15 (from 1950), MiG-15 bis, MiG-17	2 x 60	1949
MiG-19 MiG-21	3 or 2 x 65 2 or 1 x 60	1955
MiG-21 MiG-23	1 x 200 1 x 200/250	1965
MiG-27	1 x 265	1974
MiG-31	1 x 260	—
MiG-29	1 x 150	1980

Tracking	Systems	Fire sequencing display	Remarks
2 km (1.24 mi)	two antennae: scanning and tracking	no	—
10 km (6.21 mi)	single dish antenna	yes	TsD-30 experimental, RP-21 production
15 km (9.32 mi)	single dish antenna	yes	S-21 experimental, RP-22 production
55 km (34.18 mi)	single dish antenna	yes	Doppler, detection despite ground clutter
50 km (31.07 mi)	single dish antenna	yes	detection at low altitudes
75 km (46.61 mi)	single dish antenna	yes	Doppler, detection despite ground clutter
120 km (74.58 mi)	phased-array antenna	yes	look-down/shoot-down capabilities

Index

READERS OF THIS BOOK may be somewhat disconcerted by the great variety of designations—seldom simple, more often than not unintelligible—given to airplanes. To begin with, the service designation of a Russian military airplane is an acronym of two or three letters abbreviating the manufacturer's name, followed by an odd number for the fighters—MiG-15, for instance—or an even number for all other types of airplanes, even though there are a few exceptions to the rule. One, two, or three letters added to that acronym will help to explain the airplane's specialization.

When a new design takes form on the manufacturer's drawing boards, a preliminary designation is allotted to the new project. The manufacturer refers to the future airplane by either one or two letters, one or two numbers, or a mixture of both. The MiG-15 was first named S, or more precisely *izdeliye* S (product S). The MiG-9 was first named *izdeliye* F, and the MiG-29 *izdeliye* 9. The MiG-25R was first called *izdeliye* 02 but the MiG-31 started as *izdeliye* 01. This apparent disorder cleverly conceals a certain form of misinformation. There was also an I register—I standing for *istrebityel* (fighter)—that gave a sequential ordering to all MiG fighters and interceptors until 1955. It started in 1939 with the I-200 or MiG-1 and closed with the I-500 or MiG-21.

Lastly, remember that there was a cold war and that, because of the Russian predilection for secrecy, Western intelligence agencies knew little of all these details. To facilitate an unambiguous reporting of Soviet airplanes, the NATO Air Standards Coordinating Committee (ASCC) had to devise a designation system of its own for each type of airplane. Consequently fighters have names beginning with F, bombers and reconnaissance airplanes with B, transport airplanes with C, helicopters with H, and miscellaneous types with M. The idea was to choose reporting names that could not be confused, even over a poor radio link. The system was extended to the main airplane systems and to the missiles. All NATO reporting names assigned to MiG airplanes are listed in this index, but for quicker identification they have also been added to the corresponding Russian designation (for instance, MiG-21SMT Fishbed-K).

The U.S./NATO reporting names of the air-to-air (AA) and air-to-surface (AS) missiles mentioned in this book appear following the index. A few missiles were never assigned such names, either because they were not known to the West or because they never reached production status.

Air-to-Air Missiles

Precertification designation	Service designation	U.S./NATO designation
K-5	RS-2U	AA-1a Alkali
K-5M	RS-2US	AA-1b Alkali
K-55	R-55	
K-8	R-8	AA-3 Anab
K-9		AA-4 Awl
K-13T	R-3S	AA-2a Atoll
K-13R	R-3R	AA-2c Atoll
K-13	R-3A	
K-13M	R-13M	AA-2d Atoll
K-23R	R-23R	AA-7a Apex
K-23T	R-23T	AA-7b Apex
K-60	R-60	AA-8 Aphid
K-60M	R-60M/MK	AA-8 Aphid
	R-60T	AA-8 Aphid
K-80	R-4	AA-5 Ash
	R-27R1	AA-10a Alamo
	R-27RE	AA-10c Alamo
	R-27T1	AA-10b Alamo
	R-27TE	AA-10d Alamo
	R-33	AA-9 Amos
	R-40R	AA-6 Acrid
	R-40T	AA-6 Acrid
	R-73A	AA-11 Archer
	R-73E	AA-11 Archer
	RVV-AE	AA-12

Air-to-Surface Missiles

Service designation	U.S./NATO designation
KS-1	AS-1 Kennel
Kh-23	AS-7 Kerry
Kh-66	AS-7 Kerry
Kh-25ML	AS-10 Karen
Kh-25P	AS-12 Kegler
Kh-29	AS-14 Kedge
Kh-29L	AS-14 Kedge
Kh-29T	AS-14 Kedge
Kh-31A	AS-17 Krypton
Kh-31P	AS-17 Krypton
Kh-58	AS-11 Kilter

About the Authors

Rostislav Apollosovich Belyakov, born in 1919, joined the MiG OKB in 1941. He became the general designer after the death of A. I. Mikoyan in 1970 and remains at the helm today, as the Russian aerospace industry struggles to adapt to an entirely new economic environment.

Jacques Marmain, born in 1924, devoted most of his professional life to the aviation press. He cofounded *Aviation Magazine International,* one of the world's best-known aviation publications, and retired after forty years of service. A recognized expert on Soviet aviation for nearly half a century, he died in December 1993.